Thomas Nemeth
Russian Neo-Kantianism

New Studies in the History and Historiography of Philosophy

Edited by
Gerald Hartung and Sebastian Luft

Editorial Board
Karl P. Ameriks (Notre Dame University, West Bend, IN, USA), Margaret Atherton (University of Wisconsin, Milwaukee, WI, USA), Frederick Beiser (Syracuse University, Syracuse, NY, USA), Fabien Capeillères (University of Caen Normandy, Caen, France), Faustino Fabbianelli (University of Parma, Parma, Italy), Daniel Garber (Princeton University, Princeton, NJ, USA), Rudolf A. Makkreel (Emory University, Atlanta, GA, USA), Steven Nadler (University of Wisconsin, Madison, WI, USA), Alan Nelson (University of North Carolina, Chapel Hill, NC, USA), Christof Rapp (LMU Munich, Munich, Germany), Ursula Renz (University of Klagenfurt, Klagenfurt, Austria), Wilhelm Schmidt-Biggemann (FU Berlin, Berlin, Germany), Denis Thouard (HU Berlin, Berlin, Germany), Paul Ziche (University of Utrecht, Utrecht, Netherlands), Günter Zöller (LMU Munich, Munich, Germany)

Volume 10

Thomas Nemeth

Russian Neo-Kantianism

Emergence, Dissemination, and Dissolution

DE GRUYTER

ISBN 978-3-11-135829-1
e-ISBN (PDF) 978-3-11-075540-4
e-ISBN (EPUB) 978-3-11-075553-4
ISSN 2364-3161

Library of Congress Control Number: 2022930308

Bibliographic information published by the Deutsche Nationalbibliothek
The Deutsche Nationalbibliothek lists this publication in the Deutsche Nationalbibliografie;
detailed bibliographic data are available on the Internet at http://dnb.dnb.de.

© 2023 Walter de Gruyter GmbH, Berlin/Boston
This volume is text- and page-identical with the hardback published in 2022.
Printing and binding: CPI books GmbH, Leck

www.degruyter.com

Man muss den Begriff ‚Neukantianismus' nicht substanziell,
sondern funktionell bestimmen. Es handelt sich nicht um
eine Art der Philosophie als dogmatisches Lehrsystem,
sondern um eine Richtung der Fragestellung.

Ernst Cassirer

Contents

Introduction —— 1

Chapter 1
A Neo-Kantianian Look on Physics: Aleksandr Vvedenskij —— 12
1.1   An Early Neo-Kantian Philosophy of Science —— 12
1.2   A Short Argument for Kantian Criticism —— 25
1.3   The Evolution of Vvedenskij's Neo-Kantianism —— 30

Chapter 2
The Problem of Other Minds: Lapshin, Khvostov, Lappo-Danilevskij —— 38
2.1   Some Biographical Data on Lapshin —— 38
2.2   Space and Time as *a priori* Categories —— 41
2.3   On Other Minds —— 48
2.4   A Historian Weighs in on Other Minds —— 52
2.5   An Eclectic Neo-Kantianism —— 57

Chapter 3
The Psychologist as a Transcendental Realist —— 64
3.1   The Preparatory Years in Kiev —— 64
3.2   The *a priori* Understood Psychologically —— 72
3.3   A Defense of Transcendental Realism —— 79
3.4   A Neo-Kantian *malgré lui* —— 84

Chapter 4
Neo-Kantian Marxism: A Curious and Unstable Blend —— 92
4.1   From Kant to Jesus —— 92
4.2   A Flirtation with a Realist Neo-Kantianism —— 100
4.3   From Ethical Marxism to Russian Religious Idealism —— 109
4.4   A Note on S. L. Frank —— 117
4.5   An Obscure Neo-Kantian Marxist —— 122

Chapter 5
Baden versus Marburg on Russian Soil —— 128
5.1   Natural Law in Russian Jurisprudence —— 129
5.2   Novgorodcev on Natural Law —— 134
5.3   A Marburgian Philosophy of Law —— 140
5.4   A Russian Dispute between Marburg and Baden —— 147

Chapter 6
Baden Makes Inroads —— 155
- 6.1 A Ukrainian Activist Finds the Baden School —— 155
- 6.2 Russian-Language Works —— 159
- 6.3 A Historian's Use of Baden Tenets —— 170
- 6.4 A Forgotten Neo-Kantian —— 179

Chapter 7
Baden School Philosophers Who Scattered —— 188
- 7.1 The Critical-Realist Objection to Baden —— 188
- 7.2 From Neo-Kantianism to a Realist Phenomenology —— 190
- 7.3 The Journal *Logos* —— 198
- 7.4 A Disciple of Rickert's —— 202
- 7.5 A Neo-Solov'ëvian Critique —— 208
- 7.6 An Amorphous Neo-Kantianism —— 216

Chapter 8
The Marburg School's Influence in Imperial Russia —— 220
- 8.1 Cohen in Russia —— 221
- 8.2 Two Ethnic Germans —— 229
- 8.3 A Cohen-Inspired Reading of Kant —— 230
- 8.4 The "Last of the Mohicans" in Soviet Russia —— 237
- 8.5 *Dii Minores* —— 242
- 8.6 From Neo-Kantianism to the Early Husserl —— 248

Chapter 9
One-Time Neo-Kantians Who Stayed: Sakketti, Two Rubinshtejns, Kagan —— 255
- 9.1 The Marburgian with Italian Lineage —— 255
- 9.2 From Marburg to Soviet Psychology —— 258
- 9.3 In the Shadow of Baden —— 265
- 9.4 From Baden to Lebensphilosophie —— 269
- 9.5 Excursus on the Nevel School —— 275

Chapter 10
One-Time Neo-Kantians Who Strayed: Vejdeman, Jakovenko, Sezeman —— 285
- 10.1 From Pan-Methodism to Pan-Logicism —— 285
- 10.2 Transcendental Idealism or Skepticism? —— 295
- 10.3 From Neo-Kantianism to Phenomenology —— 306

Chapter 11
Concluding Remarks —— 316

Bibliography —— 323

Index —— 340

Introduction

In a previous study, this author focused on the reaction to Kant's philosophy – its interpretation and criticism – by a host of figures during the late Imperial Russian era. My concern here is with the philosophical ideas of the generally young philosophers who broadly and explicitly accepted what they took to be the meaning of Kant's thought, but found it needed updating, modification, or completion in some respect without departing from what each took to be the "spirit" of that thought. This, as a sweeping generalization, was and is the essence of neo-Kantianism regardless of the particular direction each "school" took. Some pivoted their constructions on a realization that Kant had based his philosophy on the natural sciences of the day, primarily Newtonian mechanics. They saw their task in the wake of the first *Critique* to be a reexamination and reassessment of Kant's work based on the current state of natural science, while employing what they took to be his essential methodology. Of course, a divergence in opinion as to just what that essential methodology was immediately came to the fore. Others believed that Kant had not addressed all possible fields of scientific study, notably the rise and development of the social sciences. And again, of course, differences arose not simply with regard to the approach to be taken toward the social sciences but also over which social science should be taken as paradigmatic. These philosophers largely had no quarrel with Kant's treatment of the physics of his day, but for them, the task awaiting contemporary philosophy was to extend Kant's proclaimed transcendental methodology, in their understanding of it, to the newly emerging fields. They saw that the key to their endeavor could be found in Kant's second and third *Critique*s rather than the *Critique of Pure Reason*, though they certainly appreciated the latter too.

As with the study of any intellectual movement in history, let alone a philosophical one, a more precise definition of neo-Kantianism, both in terms of its central ideas and of its proponents or sympathizers, though a desideratum, presents a daunting challenge if it is even possible. And as with many historical trends, attitudes, eras, and, yes, movements, the more we attempt to confine them within a definition the more they appear to slip from our grasp. Taken as a historical phenomenon, it is virtually inevitable that controversies will eventually erupt between scholars concerning what the movement upheld, who its protagonists and antagonists were, and what those individuals saw as the movement's proper tasks. In a study such as this, some, undoubtedly, would argue that the ground to be explored must first be delimited lest the resulting investigation embrace so many figures and ideas that the central ones, those most instrumental in shaping the movement, be eclipsed or smothered by those on the periphery, those who contributed little or even nothing to the movement's dynamic nature. It is possible, on the other hand, that a precise delimitation at the start will unduly narrow our understanding of the movement in such a way that we risk losing sight of its fluidity and of the vitality of the ideas promoted within it.

As a case in point, German neo-Kantianism consisted of more than two branches, even though only two exhibited some cohesion across more than a single generation. Thus, they understandably have drawn the lion's share of attention from intellectual historians, who gravitate toward the transmission of ideas rather than their veracity. The first of these branches or directions within neo-Kantianism was promulgated predominantly by two philosophy professors, Hermann Cohen and Paul Natorp, in the university town of Marburg and is generally referred to as the Marburg School. There were others who intellectually were part of this direction, foremost among them being Ernst Cassirer, even though he never taught in Marburg. However, Cassirer rose to prominence within the neo-Kantian movement only when the Russian neo-Kantian movement was well underway. For many of the Russians, he was a diligent and committed fellow traveler along the neo-Kantian road rather than a teacher, but there is no evidence that they looked upon him with awe or toward him for inspiration. Cassirer's early writings on the history of modern philosophy, with their emphasis on the theory of cognition, amazingly received at most scant recognition, itself a telling sign regarding the concerns of those within Russia. We should point out that most young philosophy students from Imperial Russia who went for further instruction to Marburg went there specifically to hear Cohen, though not necessarily to learn about his interpretation of Kant. Cohen had a certain renown in Russia for his treatment of Plato. Although Natorp's reading of Plato is better known today than Cohen's, it attracted less attention in Russia. The philosophical audience there valued Natorp as a teacher and a promoter of the Marburg School agenda, but none from Russia went to Marburg specifically to study with him. We find this reflected in the Russian secondary literature, in which Cohen's ideas are discussed with hardly a mention of Natorp's. We should also not discount the role of the respective personalities in the transmission of ideas. Even from afar, Cohen's enthusiasm and proselytizing on behalf of his reading of Kant seeped into Moscow and St. Petersburg, becoming a small but steady stream. Nor should we forget that his unabashed and unapologetic Judaism drew attention and amazement from young Jews, whether observant or not, in Russia familiar with career restrictions, if not outright discrimination, in their home country.

The other major branch of German neo-Kantianism is largely associated with what then was the state of Baden in south-west Germany. Whereas the Marburg philosophers demonstrated a marked concern for reflection on the natural sciences of the day, particularly physics, the Baden philosophers, primarily Wilhelm Windelband and Heinrich Rickert, focused attention more directly on the methodology of the social sciences and on the role of values within the theory of cognition. These concerns, rather than those associated with Cohen and Natorp, were more closely attuned to the traditional interests of Russian thought, and as a result, many promising students from Imperial Russia flocked to the Baden universities and initially adhered more closely to the overall framework of Baden neo-Kantianism than did those who went to Marburg. But by the same token, their commitment to it was less firm. With fewer canonical and systematic texts to look to for inspiration and guidance, those Russians who studied with the Baden philosophers often drifted away sooner from philosophy than those who studied in Marburg.

If we were to restrict ourselves to the philosophers associated with just these two German schools of neo-Kantianism, thereby omitting such figures as, for example, Leonard Nelson and Alois Riehl, we would have a greatly diminished and far less varied conception of the German neo-Kantian movement than was the case and one that all agree dominated German academic philosophy during the Wilhelmian era. But neo-Kantianism was the dominant philosophical movement in German academia for some fifty years and not just at two universities. Although Riehl was certainly well-known within the country's higher educational institutions at the time and his works were read both domestically as well as internationally, neither he nor Nelson created a separate neo-Kantian school. Yet both Riehl and Nelson held distinctive positions within the movement that make them stand apart from the others and, of course, from one another. No history of German neo-Kantianism could afford to do without some discussion of their ideas and of the active role each played in the events, squabbles, and disputes of the time. Nevertheless, Nelson will not appear in the pages that follow. His work, apart from a single long essay that appeared in 1913 in translation, was largely unknown to the Russian philosophical audience, which in any case would have reacted in horror to what they surely would have understood as support for a psychologized neo-Kantianism. The young Russian neo-Kantians were unanimous in opposition to psychologism.

Riehl plays only a small role in our study. He had no significant following in Imperial Russia despite the appearance of one of his major writings in translation at a comparatively early date. Why this is so is by no means obvious. Riehl's international recognition in the early twentieth century largely did not significantly extend into Imperial Russia, although Lenin – known to some by his real name, Vladimir Ulyanov – found it fit to remark that Riehl had "purified" Kant by way of Hume. Lenin, we can be sure, did not mean that as a compliment.

Riehl's decidedly and distinctively realist viewpoint held the attention for a time of Lenin's one-time rival Petr Struve, an economist and political activist, alone – largely owing to that very realism – but Struve's mind was too dynamic, too multi-dimensional, too interested in practical activity to be confined for long to any focused theory of science. Riehl himself recognized that his early work was overly associated with theoretical issues at the expense of moral and cultural issues and tried to make amends. However, even though his writings on Nietzsche and Giordano Bruno were translated during the years under study here, they failed to make any special impression in Russian philosophical circles. Riehl was too indebted to a hard-nosed perspective for the major portion of the idealistic and intellectually restless Russian audience.

No discussion, however brief, of German neo-Kantianism should omit the contributions and activities of Hans Vaihinger and Benno Erdmann, both of whom contributed greatly to Kant-scholarship in the narrow sense and more in the years covered in this study. In the case of Vaihinger, his enormously patient study from 1881 *Commentar zu Kants Kritik der reinen Vernunft*, which, in two thick volumes totaling more than one-thousand pages, never progressed further than the first seventy-five pages of Kant's work, had no a parallel in Russian philosophy – nor for that matter anywhere else in that era. His later *Philosophie des Als Ob* (1911) as such went unnoticed in Russia, but its central idea can be dimly seen in the Russian symbolist movement prominent at that time, a fact that stemmed from their mutual appreciation of Nietzsche. Erdmann's general philosophical views received some quite limited recognition in the Russian philosophical and theological journals, but no one was willing to engage with him on the basis of his narrowly focused texts dealing with Kant exegesis. No graduate students from Russia ventured to Germany to learn the craft of punctilious Kant-interpretation, although some did write purely expository tracts without themselves actively contributing to an updating or an alleged correction to Kantian philosophy. The generation coming of age in late Imperial Russia was, with few exceptions, much too eager to witness and, hopefully, participate in the furthering of a revolt against the positivism that immediately preceded them. If we sought to measure one's allegiance to an amorphous Russian neo-Kantianism in terms of doctrinal allegiance to the original teachings of Kant, such figures as Vladimir Solov'ëv and Boris Chicherin would have to be included owing to their fundamental commitment to many aspects of Kant's moral theory. Yet, their inclusion within the neo-Kantian movement would only make a mockery of the entire proceeding. The general metaphysical thrust of the systems each attempted to construct was fundamentally at odds with that of Kant and of his transcendental idealism. Solov'ëv's intent was to reinvigorate Christianity. Kantian practical philosophy was correct insofar as it was that of Christianity, not vice versa. Chicherin's thought had its roots in a vigorous, though not technically rigorous, neo-Hegelianism that predominated during his student years. In contrast, whereas Riehl in Germany sought to retain the "thing in itself" in his neo-Kantianism – much to the horror of Baden and Marburg – he attempted to justify his position in terms of his understanding of Kant's texts, something fundamentally absent from the metaphysical systems of Solov'ëv and Chicherin.

We should also mention, even though it has been pointed out by others, that German neo-Kantianism was an exclusively academic phenomenon. Thus, Russian students were conveniently able to hear directly from their doctrinal source about the latest ideas in German philosophy, not just read about them in books. All of the proponents of German neo-Kantianism held professorships, and as professors at governmental institutions, they were state employees with a stake in not just the survival of the government but also in its furtherance. This may well have contributed to an overt nationalism and to their widespread support of government policies, particularly in the instance that mattered most, namely, the outbreak of war in 1914, when even the most prominent un-converted Jew (Cohen) in German academia lined up fulsomely in support of military action and remarkably in some cases (Windelband, Riehl) even in the face of such atrocities as the sack of Louvain in August 1914. The consequence of this was that some Russians, in effect, not just symbolically but even militarily confronted their German counterparts in the early years of the War. It also meant that many young Russian philosophers had to defend their appreciation of German thought and culture in the face of an overt assault on all things German in the popular media.

Despite the multifariousness of German neo-Kantianism, which has only fairly recently captured any attention at all in English-language scholarship, both historical and philosophical, there is currently no study that seeks to put the neo-Kantian movement into an international context. Such an undertaking, naturally, would be enormous. However, we should bear in mind that the German movement, contrary to most literature was not an isolated phenomenon. We can meaningfully speak of French neo-Kantianism, Italian neo-Kantianism, and, yes, Russian neo-Kantianism, as the present study hopes to show. The features of each of these neo-Kantianisms were as different as were the respective concerns, but it is instructive to bear in mind, for example, that many of today's social sciences had roots deep within neo-Kantianism. Max Weber, for example, was closely aligned with the Baden philosopher Rickert and simply adopted his friend's methodology as his own in practice.

In France, the co-founder of sociology as a discipline unto itself, Émile Durkheim, also had intellectual roots deep in neo-Kantianism, albeit of the French variety, a fact often ignored by those with little interest in theoretical matters.

Durkheim's transformation of the Kantian categories into sociological ones, spearheaded by the French movement in which he was reared, has only recently gained recognition in English-language philosophical literature. Certainly, one may raise serious concerns and even doubts as to what degree such individuals as Charles Renouvier and Émile Boutroux, the latter of whom was Durkheim's dissertation advisor, can legitimately be classified as neo-Kantians, but based on the same criteria we could raise similar objections against such pillars of German neo-Kantianism as Hermann Cohen, the author of three systematic works in the early twentieth century. Lest we throw out the baby with the bathwater and dismiss neo-Kantianism entirely, Renouvier and Boutroux were not alone in reacting to the dismissal of metaphysics by the positivism of Comte and in confronting the rapid advance of natural science and mathematics in their time. Cassirer already in 1912 recognized that Renouvier held first place outside Germany in grasping the task set by Kant. Admittedly, neither of the two French philosophers established a unified "school" comparable to the German ones, but we can in retrospect see in their writings and from then on through Durkheim the foundations of a sociological neo-Kantianism.

In Italy, the neo-Hegelian and later Fascist philosopher Giovanni Gentile, writing also at this time, rejected in its entirety the originality of his country's somewhat earlier neo-Kantian movement, seeing it as offering no greater insight into reality than that offered by the positivists. Of course, Gentile had his reasons for this assessment and his outlook, but he could not deny the presence in Italy of neo-Kantians in the second half of the nineteenth century. Italian neo-Kantianism had, to be sure, its own unique national characteristics in comparison with German and French neo-Kantianism. Unlike, for example, the Marburg philosophers, Carlo Cantoni found Kant to be insufficiently psychological and reproached him for not providing a psychological examination of the issues he had raised. In doing so, Cantoni and other Italian neo-Kantians, such as Francesco Fiorentino, thought the connection of "Critical Philosophy" to realism had to be tightened, rather than loosened, if not severed, as Cohen hoped to accomplish in the early twentieth century through his re-interpretation of the *Ding an sich*. Fiorentino went further than Cantoni in thinking that the Kantian *a priori* could be explained in terms of the psychological notion of association and the biological ideas of heredity and evolution. He saw the *a priori* of experience as a human construct that arises in response to our gradual adaptation to our objective environment. With this picture, he opened himself up to all the possible objections the Germans and, for that matter, Russians, might pronounce against reductionism. Even if we should dismiss much of both French and Italian neo-Kantianism as philosophically untenable, it does show us today that neo-Kantianism was by no means a monolithic product of a certain era in German thought.

My intent here is not to write the history of international neo-Kantianism, only the essentials of Russian neo-Kantianism and thereby correct any mistaken and even absurd view that the "return to Kant" in the half century between 1875–1925 was an exclusively German phenomenon. But if the figures discussed in the following pages were indeed representatives of a Russian neo-Kantianism, were they even Russian? Just as the definition of neo-Kantianism must remain rather amorphous, so too must we be granted some latitude in our conception of "Russianness." Whereas certainly many figures who participated in the movement were undoubtedly of ethnic Russian stock, a number were far less so. One was proudly Ukrainian, another was half-Russian, others were of Baltic German ancestry, and still another was an American, whose engineer-father happened to work in Russia. This study seeks to be broadly inclusive on the issue of "Russianness," taking into account those individuals who contributed to a philosophical neo-Kantianism and have been recognized as contributors in one or another secondary study in the history of Russian philosophy. Some figures not discussed were omitted for the reason that they wrote very, very little or because, in this author's opinion, the connection of their ideas to neo-Kantianism is exceedingly tenuous. The reader will find in the following pages a veritable plethora of names that is already more than ample. Some readers will undoubtedly disagree with the choices here. Some individuals discussed herein may not be viewed by all as neo-Kantians but were included based on another prominent figure within Russian philosophy, such as Losskij or Jakovenko, doing so.

A standard means of portraying the emergence of German neo-Kantianism is to see it as a reaction and in opposition to the metaphysical excesses of German Romanticism and its speculative – to put it kindly – philosophy of nature. In a similar manner, we can see French neo-Kantianism arising against the backdrop of Comtian positivism, and Russian neo-Kantianism, or at least its representatives stemming from Moscow, opposing the metaphysical speculations of such figures as Vladimir Solov'ëv and Lev Lopatin. At least in their younger days, most young neo-Kantians sought to distance themselves from the overtly religious outlook then predominant in the numerically small philosophical community within Imperial Russia and welcomed what they took to be the sheer philosophical professionalism of the German professors. Whatever their innermost reason for going westward, particularly to the universities in Marburg and Baden, the young Russian converts and adherents of neo-Kantianism made the virtually obligatory study-tour of German universities to learn the latest developments in their field. An additional reason, from the government's viewpoint, for going abroad was to avoid intellectual incest. Inasmuch as neo-Kantianism reigned supreme at the German institutions, it can hardly come as a surprise that the Russian students heard and became intellectually attached to the latest philosophical trend. The German scene appeared to them to exude originality and a sheer dynamism, features that were noticeably absent at their home institutions in Russia.

There were exceptions, of course, but perhaps the most remarkable feature was that many of the Russian graduate students studying philosophy were not products of Russian university philosophy departments, but of the respective law faculties. Viewed simply as a statistical fact, this should not come as a surprise. The society of late Imperial Russia, which had only recently entered the Industrial Age and was still feeling its growing pains, needed more lawyers than philosophy or physics professors. From the individual's standpoint, we can easily see why one would choose law as a profession over academic teaching: lawyers received greater compensation than university professors, all of whom, apart from those with independent means, had to supplement their university salaries by teaching additionally at other higher educational institutions. We should not be surprised, then, that many of the Russian neo-Kantians had as undergraduates studied law, albeit with a special interest in legal theory, and as a result were more concerned with the social sciences than with the natural sciences. This, in itself, would have a telling effect on their form of neo-Kantianism. Few, very few, would have been able to converse with their German counterparts, such as Cassirer, on the latest developments in symbolic logic, mathematics, and physics or would have had any interest in doing so.

Whatever impression one may get of the philosophical worth of the neo-Kantian movement in Imperial Russia, the pages that follow will show that there was such a philosophical movement. True, many, indeed most, of its participants were inspired to take up the mantle even if only for a relatively short time as a result of their study with the German neo-Kantians. But by no means were all Russian neo-Kantians converted through some indoctrination while studying in Germany. As we shall see, a number of them participated in the *Zeitgeist* more or less independently of the German scene, although they took reassurance in their own efforts when they glanced at developments there. The world which the Russian neo-Kantians inhabited is now long gone as a result of political events well beyond what they could have imagined. Most recognized from early on that the Revolution in late 1917 entailed a decisive break with the intellectual and cultural world they had known and enjoyed up to that time. A few, admittedly, thought they could adapt; their wish to survive left them little choice. Some, at least initially, surely clung to the slowly but surely fading hope that they would not need to adapt, that the post-Revolutionary world they physically inhabited would change and revert to what they had known and loved. It was not to be. In the minds of a number of its initial proponents, Russian neo-Kantianism did not die so much as fade into a blend with the ruling ideology until it was completely eviscerated.

Unlike in my earlier study on Russian Kant-interpretation, which considered all of the nineteenth century and part of the twentieth, this study could not proceed in either some more or less strict chronological order or by educational institution. The Russian neo-Kantian movement, for one thing, lasted but some four decades at most. Nevertheless, we can start in Chapter 1 at the beginning with the unique neo-Kantianism of the St. Petersburg philosopher Aleksandr Vvedenskij. He almost alone in Imperial Russia came to the study of Kant with an interest in physics. However, his concern was focused not so much on the mathematization of nature as on an understanding of physical terms and their ontological import. His protégé and successor at the University, Ivan Lapshin, is the central figure of Chapter 2, but Lapshin had no special interest in further developing a Kantian-inspired philosophy of science. Instead, he was intrigued by the problem of other minds, to which he would return several times though seemingly oblivious to the work on it being done elsewhere, particularly in Germany, at the time. Apart from that, however, his broad cultural interests quickly diverged from neo-Kantianism and even from traditional philosophical concerns. In Chapter 3, we turn to the transcendental realism of the Kiev and then Moscow professor Chelpanov. His inclusion among the neo-Kantians is certainly not unanimous. With an interest as much if not more so in psychology than in philosophy in some narrow sense, he was well-positioned to inaugurate a school of thought. That he did not do and showed no interest in doing so owed much to his interest in being above all a good educator, rather than a system-builder. His philosophical writings stand out in the literature as a model of a simplistic style, approachable, unlike Cohen's, to even the beginning undergraduate student of philosophy. Although a realist, he took his inspiration from Wilhelm Wundt rather than Riehl.

With Chapter 4, we encounter the attempt to meld the staid academic philosophy of neo-Kantianism with the revolutionary doctrine of Marxism. The brief existence of this neo-Kantian Marxism should not distract us from recognizing its moral mission. Had it not been born in the over-heated political environment of late Imperial Russia, it, through its most capable representative, might have nurtured the country along a liberal-democratic and humane path. The emergence of tensions between Russians educated in and attracted to the Baden School with those educated in Marburg forms the focus of Chapter 5. The former held the upper hand institutionally, making an advocacy of the Marburg version of neo-Kantianism in social science a perilous endeavor at the cost of one's career prospects. Chapter 6 looks at three figures who employed Baden School tenets in the fields of history and law, and Chapter 7 deals with several, nearly forgotten youthful figures within the neo-Kantian movement, in particular the emergence of the new international journal *Logos*, which hoped to promote a fruitful exchange between German and Russian philosophy. Its very conception would prove to be its undoing as the war clouds became increasingly ominous leading to its ultimate demise.

We shift our attention in Chapter 8 to the influence of the Marburg School within Imperial Russia. Boris Vogt, or Fokht, as his name is rendered in transliteration from Russian and as we will refer to him hereafter, seemingly remained true to Cohen's overall view of Kant to the end. Remaining in the Soviet Union, he amazingly survived through the Stalinist era, largely remaining silent until the end. His close friend Gavriil Gordon threw his lot in with the Bolsheviks and would not prove so lucky. Both studied for a time in Marburg, and both returned to Russia brimming with enthusiasm over what they had heard there. In Chapter 9, our attention is centered especially on two Rubinsteins, or Rubinshtejns as we transliterate the name here. There was no discernable family relation between the two men, and indeed they came to neo-Kantianism from two different directions, one associated with Marburg philosophy, the other with Baden. Each maneuvered his intellectual path away from narrow philosophical issues, one toward psychology, the other toward pedagogy with a leaning toward a new philosophical fashion, *Lebensphilosophie*. Chapter 9 also looks at two names familiar to those with backgrounds in literary criticism Mikhail Bakhtin and his early associate Matvej Kagan. Both paid homage to the Marburg School, although Bakhtin's idiosyncratic usage of neo-Kantianism within a framework far from that of the neo-Kantians themselves forces us to question his commitment. Kagan's Marburg neo-Kantianism was more genuine and honest, but he abandoned philosophy entirely and abruptly. Finally, in Chapter 10 we take a look at three one-time neo-Kantians who emigrated of their own volition and remained deeply interested in philosophy and the development of their own ideas to the end. In two of the cases, their dedication to intellectual integrity remained even in the face of extreme monetary difficulties and in the third case of the almost unimaginable hardships of a Siberian labor camp.

As I have remarked in previous studies, the rendering of Russian names into English presents a measure of perplexity in that various systems have been employed over the years. Adding to the difficulty is the fact that a number of the philosophers discussed in the following pages published not only in Russian but also in the German language – one even in Lithuanian – and rendered their names as they thought fit for that language. Since many of the names that will appear here also appeared in my *Kant in Imperial Russia*, to render them differently now could lead to confusion and bewilderment. Some readers, understandably, will prefer a different rendering of the Russian names, but hopefully, they will recognize the spelling found here and will not misidentify the individual intended.

No work such as this could be accomplished without either the resources of numerous libraries or the previous work of outstanding scholars. A deep expression of thanks certainly must be extended to the New York Public Library, whose outstanding collection I have noted on many occasions, to New York University and, in particular, to the Jordan Center for the Advanced Study of Russia, at which I was a visiting scholar for two years. Many of the details in the pages that follow, especially those concerning the young Russians who studied in Marburg, were gleaned from the many highly informative writings of Nina Dmitieva and Vladimir Belov. A number of the chapters that follow underwent a critical reading by Frederic Tremblay, who suggested innumerable corrections and offered a wealth of comments. To him, I must extend a special note of gratitude. And to my wife Anne, who read over the entire manuscript correcting my blatant grammatical mistakes and supported this endeavor throughout, I owe more than a few words of thanks could possibly convey. Finally, I would like to thank all those at the publishing house of Walter De Gruyter who helped in the publication of this work. In particular, thanks go to Christoph Schirmer, Mara Weber, and Antonia Mittelbach.

Chapter 1
A Neo-Kantianian Look on Physics: Aleksandr Vvedenskij

Unlike in Germany and to some extent elsewhere on the European continent, Russian neo-Kantianism seemingly appeared out of nowhere. It had no precedents. Technical philosophy in Russia drew little attention in educated circles. Thus, it may be all the more surprising that neo-Kantianism in Imperial Russia turned with its first breath toward reflecting, as it would in Germany, on philosophy of natural science. Yet away from the swirling maelstrom of the political extremes pitting reactionaries against revolutionaries, neo-Kantianism gained a foothold, albeit at a price. Competing for an audience against entrenched forces, Russian interest in philosophy of science, as witnessed by the case of Vvedenskij, soon dissipated.

1.1 An Early Neo-Kantian Philosophy of Science

Throughout much of the first half of the nineteenth century, the Russian Imperial government and the officially sanctioned Orthodox Church viewed Kant's philosophy with suspicion and even outright hostility. The government saw in Kantianism a defense of Enlightenment values and concomitantly of critical thinking, a toleration of opposing viewpoints, and universal human values. Owing to these features, it feared that Kantianism posed a threat to the established order. The Russian Orthodox Church, in turn, perceived it as challenging the role of faith, Biblical revelation, and the need for mediation between the individual and the Deity. The Church, thereby, saw Kantianism as contesting the need for sanctioned ecclesiastic institutions. From today's perspective, it must appear ironic that Aleksandr I. Vvedenskij (1856–1925), an avowed disciple of Kant, should be appointed professor of philosophy at the university located in the very heart of Imperial Russia, St. Petersburg, well before the end of the century and while Tsar Alexander III, an arch-conservative, if not reactionary, sat on the throne.

Vvedenskij was born in Tambov, a small city southeast of Moscow, and attended the local *gymnasium* (secondary school). He enrolled initially in the department of natural sciences at Moscow University in 1876 but also attended lectures given in the liberal arts department. Whatever be the reason, he transferred after just one year to St. Petersburg University. His future academic career might have been compromised in 1879 when he was implicated in political activities, for which he was arrested, spending a number of months in the infamous Peter and Paul Fortress. However, through the intercession of Mikhail Vladislavlev, the professor of philosophy, and Konstantin Bestuzhev-Ryumin, who taught history at the University, Vvedenskij was released rather than exiled to Siberia. Wisened by this experience, he never again participated in such suspect actions. Rather, Vvedenskij continued his studies at the University, graduating in 1881, and began teaching at a secondary school in the capital, while preparing for the *magister*'s degree examination in philosophy. He spent two years in the mid-1880s in Germany (Leipzig, Berlin, and Heidelberg), and while there his first published article appeared in January 1886. He debuted as a *privat-docent* at St. Petersburg University in the academic year 1887/1888, and defended a *magister*'s thesis in 1888.[1]

Taken alone, Vvedenskij's article, "Leibniz's Theory of Matter in Connection with His Monadology," from 1886 hardly reveals his later distinctive form of neo-Kantianism. However, we can observe in retrospect that Vvedenskij was paving the way for the philosophy of science that he would expound a mere two years later. Much of Vvedenskij's concern with Leibnizian physics is of little interest to us today, but clearly, he strongly objected to the Cartesian view of extension and impenetrability as the fundamental attributes of matter. Neither Vvedenskij nor Kant objected to conceiving matter in terms of extension. Indeed, if matter is simply defined as an object of the outer sense, i.e., as occupying space, it, therefore, must have extension. But Vvedenskij – again like Kant – objected to ascribing absolute impenetrability to mass. Vvedenskij questioned the basis for viewing impenetrability as a passive property. He argued that a physical body in preventing another from penetrating it (i.e., from occupying the space it does) testifies to that body's activity. Whereas the inertia of a body at rest is a passive resistance to motion, in the case of a body in motion its inertia is an active resistance to change. Yet to Vvedenskij, both cases must be seen as a manifestation of an activity, and consequently as a force within matter. Vvedenskij's intention was to show the defective character of Leibniz's position in comparison to Kant's, but without ever so much as mentioning Kant by name or by invoking his explicit position.

---

1 Tikhonov (2013), 201–204. Tikhonov writes that Vvedenskij's stay in Germany extended from January 1885 to May 1887. Pustarnakov writes that it extended from 1884 to 1886. Pustarnakov (2003), 334. Dmitrieva concurs with Tikhonov, but none of them provides the basis for their respective assertions that would help us adjudicate the matter. See Dmitrieva (2013), 41.

Vvedenskij found a number of weaknesses in Leibniz's depiction of mass and its essential properties. These deficiencies, he reasoned, stemmed from the latter's relationship to mechanics. Leibniz considered everything that cannot be explained by means of it to be a miracle.[2] Nonetheless, in detailing Leibniz's explanation of the hardness of a body, Vvedenskij found no "miracle." Rather, in Vvedenskij's telling Leibniz contradicted his own fundamental mechanism by appealing to the imperceptible movements of the particles comprising a body. Vvedenskij thought this explanation to be inadequate, for "pushing liquid particles together will not produce solidity."[3] The "miracle" he detected in Leibniz's explanation, then, was his own construal of how he understood Leibniz's alleged inconsistency.

Vvedenskij ended his piece, however, on a positive note listing five indubitable merits of Leibniz's theory: (1) its critique of Cartesianism, which would have materiality defined by extension alone; (2) its critique of materialistic atomism; (3) its powerful arguments in favor of the superiority of the spiritual over the material; (4) its correct formulation of the problem of interaction, although the problem remained unresolved; and (5) its new view of space and time. What he left unstated here was that all of these points would be further developed after Leibniz by Kant in his 1786 *Metaphysical Foundations of Natural Science*. Vvedenskij's implicit argument, in other words, was to demonstrate that Kant's work was an important advance over his predecessor.

The best evidence for Vvedenskij's initial philosophical interest in philosophy of science is his *magister*'s thesis, which dealt with two quite disproportionate themes. Its initial chapters were devoted to a justification of epistemology as first philosophy, whereas the bulk of the work continued Vvedenskij's elaboration of the concept of matter. He held that at the time there was "a chaos of opinion" in our understanding of it.[4] Therefore, our conceptualization of matter requires a philosophical elaboration on the basis of the data contemporary physics offered. We have at the start, then, a promising approach in that Vvedenskij saw philosophy not as legislating to physics, but as working cooperatively with it to achieve coherence and clarity, which will, in turn, provide a basis for further advancements.

It is most disconcerting, then, in light of his initial pronouncement to see Vvedenskij in the body of his text state his conception of philosophy along lines reminiscent more of Fichte than of Kant. He wrote, "There exists something

---

2 Vvedenskij (1886), 48.
3 Vvedenskij (1886), 49.
4 Vvedenskij (1888), v.

that is indubitable. It is precisely the I or consciousness and its acts or states. The indubitable I must serve as the starting point of philosophy."[5] To the I is given sense data, which collectively form the object of experience. But these data, the manifold contained in sensations, correspond in some ill-specified manner to something existing independently of us and of our subjective states. The manifold originates from things in themselves. Vvedenskij readily admitted that we determine the existence of things in themselves by means of the law of causality. Where philosophical "Criticism" – a common expression at the time for the essential sense of Kantian philosophy – enters is that the dogmatic approach assumes the universal applicability of certain general principles, such as causality, to both appearances, i.e., subjective states, as well as to things in themselves on faith, i.e., dogmatically. "Critical Philosophy," on the other hand, "investigates such principles with respect to their validity, the scope of their applicability, and also of our understanding or meaning of them."[6] Thus, it is not its ontological conclusions that distinguishes Critical Philosophy from, say, naïve realism, but simply its methodological approach to those conclusions.

We, today, recognize that the role of the *a priori* intuitions of space and time was of the greatest importance for Kant in establishing his "transcendental idealism." Indeed, one may argue that his argument for transcendental idealism largely rested on his view of the nonspatiality and nontemporality of things in themselves and that, therefore, the objects of our ordinary cognition must be mere appearances. Instead of beginning his treatise with arguments for the apriority of space and time as in the *Critique of Pure Reason*, Vvedenskij started with causality. It, for him, makes all experience possible and lies at the basis of the very distinction between the I and everything else, i.e., the non-I. Without causality, cognition would be impossible. Therefore, it is *a priori*, i.e., a logically *a priori* condition of cognition, and not a feature of the physio-psychological organization of the mind. Vvedenskij wrote, "An explanation of the origin of innate ideas by means of the psycho-physical organization [of the mind] suffers from the same defects as earlier, theological rationalism. That is to say, it leads to skepticism or to the substitution of knowledge by faith."[7]

---

5 Vvedenskij (1888), 2–3. For more on Vvedenskij's Fichtean conception of philosophy, at least as outlined in these opening pages of his thesis, see Nemeth (2017), 224–227.
6 Vvedenskij (1888), 5.
7 Vvedenskij (1888), 41. Vvedenskij made this statement in the course of his discussion of the position of F.A. Lange, the founder of Marburg neo-Kantianism, charging him with promoting a psychologized version of pre-Kantian rationalism. It is an odd charge given Vvedenskij's own reliance, as we shall see, on faith, which will become even more prominent in his later writings, and arguably his own psychologism. A Russian translation of the third German edition

Vvedenskij, unlike Kant, neither provided a complete accounting of the *a priori* conditions of experience nor did he aim to do so. But regardless of how many conditions there may be, they are intrinsically involved in cognition, and therefore the objects of consciousness are its products. This allowed Vvedenskij to concur with Schopenhauer that the world of appearances is my representation.[8] This empirical world, the world of appearances is set against consciousness, made objective by consciousness itself through its inherent processes. Nevertheless, the objective world does exist in itself, since it is the cause of my representations of it, the cause of the very fact that I am conscious of it. Vvedenskij made no attempt to escape the vicious circle he had set down by making objectivity the cause of itself, except by saying that philosophical reasoning is incapable of establishing any existence apart from consciousness. Existence must be accepted on faith alone.[9]

Both Vvedenskij and Kant held that space and time are *a priori*. However, for the latter, they are also intuitions, whereas for the former they, like sensations, are produced (*proizvodjatsja*) by consciousness, though by a conceptually different consciousness than the empirical one. Vvedenskij wrote that we "objectify" our inner states, i.e., situate them in time, and thereby understand them as independent of consciousness.[10] This, in his mind, was equivalent to the law of contradiction: the I is not the not-I. Hence, our earlier characterization of his conception of philosophy is more Fichtean than Kantian. Still, Vvedenskij was unwilling to demarcate sharply things in themselves from the cognizable. We cannot determine *a priori* what we can cognize and to what extent it can be cognized. The limits of possible cognition must be established *a posteriori*.

Vvedenskij was not entirely clear why he claimed the limits of cognition cannot be established *a priori*. For he held that all *a priori* concepts are logically derivable from an objectification of the contents of consciousness. Nevertheless, he saw the first task of Critical Philosophy to be the derivation of all *a priori* ideas or principles that make experience possible.[11] The second task is the construction

---

Lange's Geschichte *des Materialismus* appeared already in 1881–1883. See Lange (1881–1883). A new translation edited by Vladimir Solov'ëv shortly before his death and done from the fifth German edition appeared in 1899–1900. See Lange (1899–1900).
8 Vvedenskij (1888), 44.
9 Vvedenskij (1888), 51. He made no reference to, and thus displayed no knowledge of, Kant's "Refutation of Idealism" found in the first *Critique*.
10 Vvedenskij's terminology is most confusing, if not amateurish. Here, he wrote of consciousness without distinguishing the empirical from what must be seen from a transcendental standpoint. Additionally, he often identified consciousness with the I.
11 Vvedenskij (1888), 79.

of a philosophical physics. Such a physics would isolate or abstract (*otvlechenie*) the *a posteriori* elements in experience from the *a priori*. Vvedenskij saw this as an ongoing project. What we are unable to recognize as an *a priori* element today may be recognized as such tomorrow. Presumably, then, we cannot at present fully distinguish the *a priori* elements from the *a posteriori* elements owing to the former being tightly intermeshed with the latter rather than the former being a set of relative *a priori* principles.

Instead of the first task, Vvedenskij took up the second, as Kant earlier proposed doing and for the same reason.[12] However, Vvedenskij faulted Kant for not providing a preliminary investigation before announcing that matter is that which fills space. Whereas the former suspected a lingering influence from dogmatism on Kant in this, he also – but more forcefully – saw dogmatic influences on Kant's discussion of impenetrability. For these reasons and his resulting dissatisfaction with Kant's foray into the metaphysical foundations of natural science, Vvedenskij saw the need to redo Kant's "critical" analysis, taking not some *a priori concept* of matter, as Kant did, but what results from a careful, critical analysis of all the empirical *data* concerning matter. The resulting concept would not be connected with others that influence its content, such as force and inertia. Kant provided no preliminary analysis of these other concepts – only the respective definitions. In other words, Kant first took a philosophical understanding of the possibility of mathematical physics to be dependent on an elucidation of the principles for the construction of concepts involved in the possibility of matter.[13] Vvedenskij believed Kant did not rigorously adhere to this project, but he, Vvedenskij, intended to do so in light of the physics of his day. Russian neo-Kantianism started with this pronouncement of undertaking anew a Kantian-inspired inquiry into the foundations of natural science, and not as some historians have portrayed with reflections by disgruntled social activists.

Like Kant, Vvedenskij concentrated on motion, and like Kant, he initiated his inquiry with the claim that mass is the movable in space. He will later in his work attempt to justify this assertion as well as his claim that all physical concepts, such as force and energy, must be explained by means of motion. For now, though, if we assume its isolation, a moving body will continue in motion unless there is a cause to alter that motion. This is Kant's "Second Law of Mechanics,"

---

12 Vvedenskij (1888), 130.
13 See Kant (2002), 187 (Ak 4, 472). References to Kant's works in German are to the standard edition published under the auspices of the Königlich Preussischen Akademie der Wissenschaften and cited throughout as "Ak.").

which Kant admittedly recognized as the law of inertia.[14] Vvedenskij followed this train of thought in order to look at an *a priori* derivation of inertia. The conclusion to which he arrived, though, is that inertia is a condition of the reality of the causal law. Descartes, Leibniz, and others attempted to prove the conservation of inertia, but all of their proofs rested on presuppositions of the essence of matter. A successful *a priori* proof, however, must proceed from the conditions of the possibility of consciousness. Vvedenskij, even more forcefully than Kant, attempted a transcendental deduction of motion, via the causality of a body, rather than the Newtonian appeal to experience, which therefore more readily allowed for the motion's quantification.

A central issue in the development of physics as a discipline from Descartes onward at least to Kant and into the nineteenth century concerned impenetrability and the role, if any, of dynamic forces. Whereas Kant thought impenetrability to be an obscure idea, but explicable in terms of the primitive concept of force, many did not share his position.[15] Euler, to mention just one, took impenetrability to be a primitive and rejected the notion of action at a distance. Vvedenskij acknowledged the seeming incomprehensibility of dynamic forces and that they act independently of mechanical contact, but the idea of action at a distance does not contradict any of the known facts of nature.[16] Certainly, there are objections to the dynamism proposed by Critical Philosophy, but many of the same objections can be leveled as well against mechanism. Vvedenskij expressed the hope that in a planned second part of his study – which was never realized – he would show that at the basis of mechanism lies an incorrect view of *a priori* concepts. Vvedenskij, oddly for us today, attempted to connect the mechanist position with the problem of free will, namely, that the assumption of mechanism presents an obstacle to belief in a free will. If matter can act and be moved to action only by means of mechanical pushes or impulses, how can we explain intentional actions? How can we account for the interaction of the mind, which is not material, with the body? The only possible approach is through dynamism, i.e., by viewing matter in terms of dynamic forces. Even were we to disregard both the conservation laws and the law of inertia but take into account only the concept of substance along with *a priori* principles, we would still obtain the concept of matter but only as relatively impenetrable, not as absolutely impenetrable.

---

[14] Kant (2002), 251 (Ak 4,543–444).
[15] Kant wrote, "Absolute impenetrability is in fact nothing more nor less than an occult quality." Kant (2002), 214 (Ak 4, 502).
[16] Even in Vvedenskij's own day, not all physicists accepted the notion of action at a distance, and not all affirmed a dynamic model. Among the most prominent of these was Heinrich Hertz.

Vvedenskij certainly had much more to say regarding dynamism. Often enough, his arguments were merely of the form that mechanism cannot explain a certain phenomenon without contradicting conservation laws, which as we saw follow from *a priori* principles. For example, the undeniable phenomenon of gravity can neither be satisfactorily explained on a mechanistic basis nor can it explain the conservation laws themselves. On the other hand, dynamism can speak of the conversion of kinetic to potential energy. The same carries over to the possibility of action at a distance. Either we recognize its possibility, which mechanism does not, or we reject the conservation of energy.[17]

We see that Vvedenskij closely followed Kant's views on the metaphysics of nature and decisively sided with him on the possibility of action at a distance. We can put the case even more strongly, saying that the competing view, viz., mechanism, must be eliminated *a priori*, for it runs counter to established *a priori* principles. It may appear to us that with this Vvedenskij – and Kant too – transgressed the boundaries of Critical Philosophy. Instead of limiting the inquiry to asking how established theories are possible, we are telling nature how it must be. Instead of asking for the conditions for the possibility of a scientific theory, taken as valid, was Vvedenskij hypocritically attempting "to deduce" how nature must specifically operate on the basis of conservation laws derived *a priori*? Vvedenskij asserted that Kant himself raised the possibility that we can know *a priori* the laws governing forces. Without referring specifically to Kant's *Metaphysical Foundations*, Vvedenskij wrote,

> Are there any grounds for claiming that some force must be subject to a particular law but not to another? In other words, can we know *a priori* the laws of physics? Kant thought this to be possible and derived that the attractive force depended on the square of the distance and that the repulsive force depended on the cube of the distance. Robert Grassman deduced *a priori* the dependence on the square of the distance for all forces without exception. Cauchy, by means of a mathematical analysis of empirical data, concluded that the force of ether particles must be inversely proportional to the fourth power of the distance. ... Why these differences? If the law of forces can be determined *a priori*, this should not be so.[18]

In light of these differing conclusions from purportedly *a priori* "deductions," Vvedenskij himself concluded that apart from their existence and their necessary causality, we know nothing about the intrinsic nature of the attractive and repulsive forces. We must learn from empirical investigations, i.e., scientific experiments, the quantitative determination of the dependency of a force on distance.

---

17 Vvedenskij (1888), 252, 272.
18 Vvedenskij (1888), 291. See Kant (2002), 232 (Ak 4, 522).

At the end of what he regarded as but the "First Part" of his work, Vvedenskij wrote that he hoped to provide the only possible means to deduce *a priori* the laws of force in the continuation of his project. He would do this by examining the connection between these laws and the possibility of consciousness, presumably by means of an analysis of the *a priori* forms and intuitions involved in cognition. In the physical sciences of his day, Vvedenskij could not see an empirical basis by which mechanism could be decisively rejected. However, an appeal – a philosophical appeal – to the conditions of scientific experience would provide the avenue, and this is what he sought.

Needless to say, Vvedenskij neither had an inkling of the relativistic *a priori* that would come to the fore with Cassirer's neo-Kantianism on the philosophical side nor did he acknowledge talk of inertial frames of reference, which we today most often associate with Mach. But by seeking the fundamental physical laws through their connection to *a priori* conditions of cognition, meaning presumably the forms of space and time and the concepts of causality and substance, Vvedenskij was hoping to elaborate what Cassirer would later call "the universal invariant theory of experience."[19]

Vvedenskij's thesis was not his final word on the physical concept of mass. In March of the following year (1889), he published another long – and little recognized – article on the topic, "An Analysis of Mass from the Standpoint of Critical Philosophy and the Connection of the Higher Laws of Matter to the Law of Proportionality." Explicitly acknowledging his allegiance to Kantian Critical Philosophy, Vvedenskij alleged that from its point of view matter "consists of the regular connection of objectivized sensations and their changes. Thus, every particle of matter is empirically real only to the extent to which it can become an object of our senses (of sight, of hearing, etc.). In this way, for our experience, it can consist only of what we perceive in it with our senses."[20] The remarkable feature of these words was his choice of terminology, which certainly does not reflect what we would expect from someone inquiring into the possibility of mathematical physics, in particular at a time when physical science was advancing in step with enormous technological advances. Yet if we emphasize and thereby further develop Vvedenskij's definition with regard to what is involved in establishing a "regular connection," i.e., law-like behavior that can, then, be subject to mathematical expression, we approach more closely a functional

---

19 See Cassirer (1923), 268.
20 Vvedenskij (1889), 1.

representation of objectivity.²¹ With an ever-increasing recognition of the regularity, and therefore possible mathematization, of phenomena, its objectivity increases, making its dependence on being actually observed, heard, etc. less necessary.

Still attempting to come to grips with the physical concept of mass, Vvedenskij repeated his Kantian declaration that absolute impenetrability is impossible. Matter, thus, consists of dynamic forces, exhibited within the bounds of experience as the laws of the convergence and dispersal of particles. However, what is mass? If it is something independent of forces, and thereby independent of motion, Kant would be fundamentally wrong. Should his view be correct, however, then in principle we must be able to express mass in terms of motion, and thus deal with it mathematically, although Vvedenskij, unlike Kant, showed no concern with this all-important consequence. Evincing familiarity with much of the recent scientific literature on the topic, Vvedenskij wrote that Wilhelm Weber and Gustav Fechner attempted to demonstrate that the mass given in experience is not tied to extension, that mass is given in experience quantitatively in connection with motion. Mass is a constant coefficient inversely proportional to a body's acceleration and directly proportional to that body's force. The concern of a philosophical investigation of basic physical concepts is with a clarification of what is involved in the scientific phenomenon, the connection between the elements of that involvement and their connection to consciousness. This will determine what is real. A correct understanding of a physical concept will utilize only what is accessible to observation. In writing this, Vvedenskij gave every indication that he was referring only to direct sense experience. Thus, we can fault him for not realizing that his concepts of observation and experience must be broadened so as to include observation by way of instrumentation. Nevertheless, he wished to avoid the slightest hint of a metaphysical ontology while adhering strictly to law-governed behavior. Such a wish lay behind his refusal to admit the concept of weight, which he saw as a variable quantity.²²

The mathematical expression of Newton's laws in which mass is a variable tells us what mass is. Apart from that expression, the concept of mass has no sense. The physical law that mass is proportional to the applied force, other pos-

---

21 Vvedenskij here approached the Marburg understanding of objectification and objectivity. Natorp in 1912 would write that cognition "refers to an object and finally to the unity of the object. Cognition of an object is based on cognition of the law." Natorp (1965), 154.
22 Unfortunately, Vvedenskij wrote, "Mass is considered to be one of the factors that determine the velocity [*skorost'*] of a body: the latter is directly proportional to the force and inversely proportional to the mass." Vvedenskij (1889), 18. He, of course, should have written that the rate of change of the velocity (i.e., acceleration) is directly proportional to the force.

sible variables remaining constant, is what really exists (*real'no sushchestvuet*). "The number indicating this proportionality is the mass, which obviously would have no sense in the absence of proportional quantities. Mass is real only in them."[23] For this reason, it is impossible to ascribe existence to mass independent of forces. If there were no forces associated with matter, there would be no mass. In Vvedenskij's understanding, the typical view that mass does indeed exist in itself, independent of forces, is a remnant of philosophical dogmatism. And in one of his few explicit criticisms of Kant, he wrote,

> Kant considered the law of the equality of action and counteraction to be *a priori*. If this were correct, then the law of the conservation of energy would also be *a priori*. But Kant, in this case, is obviously mistaken. He looks on mass from the ordinary point of view, as some mysterious property that is independent of forces, a consequence of which is that in each body the velocity is the inverse of some definite coefficient. Having recognized mass as something that comparatively precedes the forces, he, of course, derived from this the required law, since with his understanding of mass he already presupposed it, which is immediately apparent from his argument.[24]

In short, Vvedenskij saw Kant as expanding, without grounds, the scope of the *a priori*, but in doing so he, Kant, also by implication compromised the very mathematization of physics that he proclaimed is the hallmark of scientificity. Vvedenskij accepted the form of the causal connection and of substance, i.e., in every conscious object or experience there must be something that appears as unchangeable or persists despite the changes.[25] In this way, he saw the persistence of substance as closely related to the principle of the conservation of matter.

Vvedenskij also turned to the fourth proposition in Kant's discussion of mechanics in his *Metaphysical Foundations*. Although Vvedenskij did not comment specifically therein on Kant's *a priori* proof of the law of the equality of action and reaction, Vvedenskij did allude to a difference of opinion concerning it, finding it to be all-too-concise. Even if Newton's third law of motion could be derived *a priori*, it would have to be by a more complex route than the one Kant presented. That is, it would have to invoke more than merely the concepts of causality, substance, and interaction. Its derivation would have to take into account several factors that *thus far* we can know only through experience. These are that: (1) the results of a push or impulse (*tolchok*), or of some other interaction, are inversely proportional to the weights (*vesa*) involved; (2) all bodies fall with the same

---

23 Vvedenskij (1889), 23–24.
24 Vvedenskij (1889), 38.
25 Kant refers to the persistence of substance as standing "at the head of the pure and completely *a priori* laws of nature." Kant (1997), 301 (A184/B227).

speed [*skorost'*]; and (3) not only the atoms but the entire system has the inertia of being at rest. By applying the concepts of kinetic and potential energy to the above, we obtain the law of the conservation of energy. As Vvedenskij hoped to remind us, "This path is not entirely *a priori*. The pure *a priori* is that without which there can be no experience (i.e., the necessary forms of the cognizing process) and what follows from it. However, what is based on at least one thing given *a posteriori* (on what is not necessary for the cognitive process) is only a relative *a priori*."[26]

Vvedenskij wished to remind us that Friedrich Mohr, who is generally credited with being at least among the first to state the conservation of energy, regarded that law as purely *a priori*, i.e., a proposition of pure reason, whereas the conservation of matter is a purely empirical concept.[27] Vvedenskij pointed out that of course if we take the atom to be a substantial particle and not a bundle of dynamic forces, there is no possibility of empirically proving the conservation of matter. Vvedenskij's intention with this analysis of mass was to confirm the Kantian understanding of matter. Mass does not exist apart from forces but exists as an expression of the forces themselves. If we take the predominant competing theory of matter, viewing atoms as primitive with an irreducible and non-dynamic property of solidity, the concept of mass, the law of the conservation of energy, and the law of the equality of action and reaction are not interconnected. On the other hand, if we adopt the Kantian viewpoint, the two laws and the concept of mass stand in a tight, organic connection with each other.

Vvedenskij penned the next year one more piece on the philosophy of science. This two-part article, "On the Question of the Structure of Matter," broke no new ground. However, it is a cautionary tale of the consequences that ensue in Vvedenskij's opinion from a hasty conclusion based on scant data concerning the reality of atoms. He intended the article to be his contribution to the argument against mechanistic materialism. As envisaged by its proponents, all natural phenomena ultimately are a result of the movements of atoms operating in accordance with strictly mechanical laws. All chemical reactions as well as all natural processes would be a result of these atomic movements. However, this view of the world is, in Vvedenskij's eyes, entirely speculative with no solid evi-

---

26 Vvedenskij (1889), 43. We leave aside here whether Vvedenskij was correct in his attempted correction of Kant. The point is simply to illustrate his *intended* correction of Kant while still maintaining an allegiance to Kantian principles, and thereby his neo-Kantianism.

27 For Mohr's view on the conservation of energy, see Mohr (1869), 40. On the conservation of matter, Mohr wrote, "The indestructibility of matter is a purely empirical concept, which has emerged from a thousand chemical experiments. No philosopher could have developed this concept through pure speculation." Mohr (1869), 31.

dence to substantiate it. The arguments introduced in its favor contain metaphysical presuppositions that implicitly presuppose atomism. It is no surprise, then, that these arguments culminate in that very assumption. "Only a disregard for a philosophical analysis of our theories and a blind faith in our preconceived metaphysical views can spawn a conviction in the reality of chemical atomism."[28] Yet after his examination of a number of contemporary positions on the matter Vvedenskij came to a rather agnostic conclusion, viz., that we are unable to determine whether or not atoms exist in the way its proponents depict them. We cannot know the fundamental nature of the substances involved in physical or chemical phenomena, but only the events that take place between substances. The fundamental principle of knowledge, viz., that all events have a cause, can be applied only to events. Thus, from the perspective of Critical Philosophy the aim of chemistry lies not in studying the structure of chemical substances, but in investigating the operative laws in chemical events.

The second part of Vvedenskij's article, which appeared the following month, was an examination of Dalton's atomic theory. The details of Vvedenskij's presentation are of little concern to us today. However, he wished to show that Dalton's argument in support of the existence of fundamental atoms consists of a number of auxiliary hypotheses, each of which was chosen for adapting the facts to support a predetermined conclusion. Indeed, the fundamental intent of the atomistic worldview is to present a mechanistic explanation of phenomena, but this "mechanism" does not stand as a rejection of dynamic forces. Rather, according to Vvedenskij it is an interpretation of the corporeal properties and events that arise as a result of the spatial movements of the particles from which the bodies are constructed. Such, at least, is how he understood the fundamental hypothesis of Dalton's atomism and the defense of chemical atomism by Alexander Naumann.[29] However, Vvedenskij did not believe he had demonstrated the untenability of atomism. It simply remained unproven. The concepts of matter and atoms as found in the natural sciences need not be banished, but the question whether atoms empirically exist should be put aside. Instead, other avenues should be pursued, which will, in turn, provoke new questions. In Vvedenskij's mind, the mentioned concepts may prove to be useful for scientific purposes. The hypothesis of mechanistic atomism, though metaphysical and thus unprovable, may yet have value owing to its utility.

---

[28] Vvedenskij (1890a), 42.
[29] Vvedenskij (1890b), 202.

1.2 A Short Argument for Kantian Criticism

Unfortunately, Vvedenskij's two-part article from 1890 largely ended his excursions into philosophy of science. It is even more regrettable that none of the other figures who will feature prominently in a discussion of Russian neo-Kantianism ventured into the philosophical foundations of natural science, which thematically played a prominent role in German neo-Kantianism. Vvedenskij himself became increasingly absorbed with traditional issues in philosophy, such as that of free will versus determinism. Nonetheless, he went on to attempt an outline and defense of a distinctive, albeit bare-boned, variety of neo-Kantianism largely, though not exclusively, in textbook presentations for class use at St. Petersburg University. Before turning to them, let us look, albeit briefly, at a popular presentation Vvedenskij offered in 1909 of just what he considered Kant's core teaching to be and which he steadfastly upheld.[30] Above all – and this we cannot stress enough – we see that Vvedenskij demonstrated both here and throughout his long professional career little patience or regard for the many technicalities and intricate arguments in Kant's first *Critique*. With such omissions, Vvedenskij joined a long list, going back at least to Karl L. Reinhold, of those who sought a "short argument" for idealism. Like Reinhold, Vvedenskij too would stress the distinction between representability and conceivability.[31]

In his essay "What is Philosophical Criticism?" Vvedenskij gave every indication that he saw himself as a proponent of a philosophical direction that stemmed from Kant and which in his day was called "Criticism."[32] Vvedenskij left unclear whether he thought Criticism was more than merely a minimalist epistemology. In any case, he subsequently employed both expressions "Critical theory of cognition" and "Criticism," thereby leaving not even a hint as to his position regarding the complexities found in the three *Critique*s. Nevertheless, Vvedenskij posed the question what is required of cognition under the assumption that mathematics and the laws of nature established thus far are indisputable knowledge. This promising start, of course, is reminiscent of Kant's regres-

---

[30] For those familiar with the present author's previous expositions of Vvedenskij's philosophical position, the following presentation is to be read as a summary of them with an emphasis on his *neo*-Kantianism.
[31] This conception of a "short argument" for idealism plays a major role in Ameriks (2000). It is hardly likely that Kant himself would have been satisfied with the proposal. As Ameriks remarks, for Kant "there is no such short argument." Ameriks (2000), 164. It is unclear to what extent Vvedenskij was familiar with Reinhold's attempt.
[32] Vvedenskij (1909a), 23.

sive or "analytic method" in the *Prolegomena*.³³ Unfortunately, having raised the issue Vvedenskij failed to follow through with a detailed account in response to the task he himself introduced. Instead, he immediately took up the question whether metaphysics, understood as a concern with transcendent entities, can be considered knowledge. In other words, can we have knowledge of transcendent entities, i.e., of objects that in principle cannot be perceived empirically? Whereas the objects of interest to the natural sciences are empirical, those of mathematics are not. Again, it is unfortunate that Vvedenskij did not raise the question, let alone investigate, whether the objects of interest in mathematics are transcendent. One will also notice here, particularly in light of Vvedenskij's claim that "metaphysics as knowledge is impossible,"³⁴ that his usage of the term is not isomorphic with Kant's. Certainly, both agreed that all metaphysical judgments are *a priori*, but Vvedenskij was oddly silent concerning the Kantian analytic/synthetic distinction. In any case, since his concern – judging from his silence – is solely with empirically transcendent entities, such as God and the human free will, he averred that metaphysics is possible but only as a form of faith, not as knowledge. That is, we cannot *know* whether God exists, but we can *believe*. Such faith is both irrefutable and, to believers, certain, but also unprovable.

It is understandable that in a short, popular article outlining his own philosophical position Vvedenskij could not delve into details but could offer only general conclusions. But his complete omission of those details must have left any careful reader coming without a knowledge of Kant's writings baffled. Vvedenskij took for granted, for example, that anyone who accepts mathematics and the natural sciences as yielding indisputable knowledge must then dismiss the very possibility of metaphysical knowledge. "Either we must not recognize the validity of mathematics and knowledge of the laws of nature (even of the laws of nature, with the help of which we make astonishingly precise astronomical predictions) or we abandon any hope of ever having metaphysics in the form of knowledge."³⁵ Someone without a background in Kantian philosophy may grant that knowledge of the supernatural is impossible, since the supernatural is, by definition, not open to empirical scrutiny. However, why did Vvedenskij mention mathematics in this connection? Its objects are not empirical, but *a priori*. Why would a belief that mathematics contains valid propositions entail the rejection of the utter possibility of metaphysics as knowledge? Vvedenskij pro-

---

33 Cf. Kant (2002), 70 (Ak 4, 275).
34 Vvedenskij (1909a), 24.
35 Vvedenskij (1909a), 24.

vided no answer to this question. But he noted that that limitation of metaphysics is the fundamental conclusion of philosophical Criticism. Despite this denial of metaphysics as knowledge, Vvedenskij added that metaphysical or transcendent hypotheses are not to be dismissed from scientific investigations provided they are recognized and utilized merely as auxiliary tools, i.e., as working hypotheses. They are to be used to help orient ourselves with the facts, to help make new discoveries, and to help raise new questions. We can even employ, within bounds, heuristic fictions, provided they are useful. Understandably in this context, Vvedenskij was far more circumspect than was Kant on the limits of their usefulness. Nevertheless, a difference in their respective illustrations is clearly discernible. Whereas Kant abjured the use of transcendental hypotheses that allegedly would explain natural phenomena,[36] Vvedenskij – alert to his academic standing and to the watchfulness of governmental censors – wrote, "if it is easier to work this way, the psychiatrist has a right to deny the existence of the soul and not be troubled by the fact that it is easier to work this way than by assuming the existence of the soul or by the fact that almost all of humanity believes in the existence of the soul."[37] The only caveat that Vvedenskij added to the working psychiatrist's hypothesis is that he/she not impose it on others or on other scientific endeavors. In contrast, Kant was chiefly concerned not with what can be legitimately denied, but with the legitimate employment of working hypotheses. For example, we cannot conjure hypotheses for our work that do not meet the conditions of possible experience. Such hypotheses, while they may be free of contradiction, would lack any possible object.[38] They could not aid in any explanation.

Vvedenskij did not restrict his understanding of "Criticism" to epistemology. He envisaged it as an integral worldview, and as such it must say something about morality, even though we must take anything "Criticism" has to say in this sphere on faith, and not as knowledge. Nonetheless, unlike Kant, Vvedenskij was noticeably short on specifics. But he did add that any metaphysical tenets that should enter our worldview must be morally grounded, and not, say, aesthetically grounded. Although Kant provided no reason why our choice of metaphysics should be based on morality, the answer is not hard to find: Not every-

---

36 Kant, in this regard, wrote, "Order and purposiveness in nature must in turn be explained from natural grounds and in accordance with laws of nature, and here even the wildest hypotheses, as long as they are physical, are more tolerable than a hyperphysical hypothesis, i.e., the appeal to a divine author, which one presupposes to this end." Kant (1997), 660 (A772/B800 – A773/B801).
37 Vvedenskij (1909a), 25.
38 Kant (1997), 659 (A771/B799).

one has an aesthetic sense.³⁹ "Hegel and Solov'ëv had no taste for music, but no one would dare blame them for this. However, it is indisputable that one cannot be without any moral principles. ... Therefore, a metaphysics, which is part of a worldview, must be based on moral views and not on any other ones."⁴⁰ Whether and to what extent Kant in his ethical writings proceeded by way of a regressive argument from an *a priori* given moral sense, a "fact of reason," need not concern us here.⁴¹ Unlike Kant, Vvedenskij failed to follow up this idea by developing a far more complete ethical theory. He never delved into ethical theory in any way comparable to Kant or, for that matter, to Hermann Cohen. Most importantly, Vvedenskij never answered how to derive a detailed normative ethics from an allegedly given moral sense, another contention of his that he failed to elucidate but that would play a prominent role in his later writings whenever ethics came to the fore. This is quite far from Kant's position.⁴²

Vvedenskij also published in 1909 a technical presentation of the ideas we have just seen.⁴³ In this "New and Easy Proof of Philosophical Criticism," Vvedenskij hoped to establish that: (1) the objects of mathematics and natural science are appearances, not things in themselves; and (2) metaphysics, as concerned with things in themselves, cannot yield knowledge. Vvedenskij claimed that proceeding from the assumption that mathematics and natural science provide us with knowledge "Philosophical Criticism" reveals the conditions by which we have the right to accept this assumption and also the two conclusions just mentioned.⁴⁴ He asserted that Kant did this too, but the latter's procedure involved difficult and involved investigations. Vvedenskij provided no reason why Kant took the harder, convoluted path if he knew an easier one. In any

---

39 Kant would hardly have agreed here with Vvedenskij's characterization of the reason to base metaphysics on morality.
40 Vvedenskij (1909a), 26.
41 See Kant (1996), 164 (Ak 5, 31). Kant also wrote in the second *Critique*, "But how is consciousness of that moral law possible? We can become aware of pure practical laws just as we are aware of pure theoretical principles, by attending to the necessity with which reason prescribes them to us and to the setting aside of all empirical conditions to which reason directs us."
42 Vvedenskij's stance here stands in stark contrast to that of another figure in Russian philosophy, Vladimir Solov'ëv, who recognized that Kant "gave flawless and definitive formulations of the moral principle and created a pure or formal ethics, a science as valid as pure mathematics." Solov'ëv (1988), 478.
43 See Vvedenskij (1909b). Vvedenskij's presentation was delivered first at a meeting of the St. Petersburg Philosophical Society in November 1908. He also published a German translation in 1910. See Wedenskij (1910).
44 Cohen in his *Kants Begründung der Ethik* from 1877 expressed his understanding of the transcendental method in much the same way. See Cohen (1877), 24.

case, Vvedenskij believed he could do all of the essentials independently of the actual path Kant took and do so extremely easily through an examination of our logical laws of thought alone. Thus, Vvedenskij hoped to employ Kant's regressive method to achieve the same ends as Kant did, but use that method differently than did Kant. Such a project would surely be a new and original form of Kantianism – a neo-Kantianism.[45] Unlike some varieties of neo-Kantianism, which would tie philosophy to contemporary mathematics and science as they evolved and make philosophy a perpetual handmaiden to science, Vvedenskij now in 1909 would have philosophy essentially terminate with the determination of logical laws of thought. Gone was his talk from twenty years earlier of seeking fundamental physical laws in accordance with our *a priori* intuitions of space and time.

Vvedenskij singled out certain elementary laws of logic, that of (non-)contradiction being one. He stated that these laws are applicable to objects if and only if they are representable. Thus, "inferences are logically valid and admissible only with regard to representations. ... But inferences are obviously completely inadmissible and logically invalid concerning what by their essence cannot be taken to be representations, i.e., concerning things in themselves."[46] In other words, our knowledge rests on making inferences. Since we cannot know whether the laws of logic apply to things as they truly are, things "beyond" or apart from our possible representation of them, we cannot make any logically assured inference concerning them, which also means we cannot know things in themselves. "In short, one can believe what one wants about things in themselves, but it is quite impossible to know anything about them. For it is logically impermissible to make inferences about them."[47]

By the same reasoning that we cannot extend the elementary laws of logic to things in themselves, we cannot extend the law of causality and the other Kantian categories beyond representations. Thus, we must leave open the question whether the causal law and the other categories are applicable to things in themselves. Vvedenskij believed this simple piece of reasoning in this way independ-

---

45 We cannot, however, help but have misgivings from the outset. The conclusions to which we are to arrive are synthetic judgments, but we are to analyze only the laws of logic, which are analytic. See Kant (1997), 130–131 (A7-A8). In short, it appears that we are to get something from nothing. Cohen would have looked askance at Vvedenskij's approach. The former wrote, "The goal is the clarification of the possibility of synthetic propositions *a priori*. They form the genuine and entire content of experience. And this content, given in mathematics and pure natural science ... is to be explained by its possibility." Cohen (1871), 206.
46 Vvedenskij (1909b), 134.
47 Vvedenskij (1909b), 135.

ently reinforces the thrust and consequences of Kantian epistemology. "If all the fundamental points of philosophical criticism have now been proven, independently of Kant, then at the same time everything that follows from them has been proven independently of Kant, namely our right to construct metaphysics as a morally grounded *faith* – as a faith that knowledge can neither refute nor prove."[48]

Vvedenskij had no wish to eliminate the concept of the thing in itself from his neo-Kantianism. He held that were we to do so, we would resume the direction inaugurated by Kant's immediate successors, such as Maimon and Fichte, who saw the concept of a thing in itself as an anomaly. We would be referring, as it were against our own wishes, to appearances as if they were things in themselves. The contradiction that some of Kant's critics found in that very concept lies, according to Vvedenskij, not in the concept itself, but in the opposition between how we should conceive the thing in itself and how we do conceive it. This or any contradiction in the concept of the thing in itself does not mean that we should banish it from philosophy and science. Were we actively to do so, we would have to banish as well the idea of a triune God and of a God who creates something out of nothing.

In the face of their unknowability – even whether there are any – does the concept of the thing in itself serve any utility? Vvedenskij recalled Kant's position that this concept serves as a limit or border to the extension of cognition. But, of course, Kant had in mind the *a priori* forms of intuition, not the laws of logic. For Vvedenskij, the concept in question serves as a limit or border of the logically permissible use of inference, which in turn is what can be a possible object of representation in mathematics or natural science. What determines these objects is left unspecified. The question how Vvedenskij's approach averts psychologism – assuming he wished to do so – remains open.

1.3 The Evolution of Vvedenskij's Neo-Kantianism

Vvedenskij's most detailed exposition of his epistemological standpoint is, undoubtedly, his textbook *Logic, as Part of the Theory of Cognition*, which went

---

[48] Vvedenskij (1909b), 136. Vvedenskij dismissed the possibility of metaphysics by way of a direct intuition of the thing in itself, saying that even if there were such an intuition its object would amount to a single assertion that something in itself exists. Any further claims concerning this object would involve logical deductions, which we assuredly cannot make.

through four editions during his lifetime.⁴⁹ He reaffirmed in the second edition from 1912 that Kant's epistemology needed "certain additions," drawn from the presupposition that mathematics and the natural sciences certainly yield knowledge. This confidence does not initially stem from proofs, but from the facts.⁵⁰ However, whether metaphysics can yield knowledge as well is yet to be determined. We already know Vvedenskij's answer, and we need neither repeat nor linger on it. But in line with his position, he criticized Kant for the view that *a priori* judgments certainly cannot be applied to things in themselves. Kant's claim stemmed from his erroneous conclusion that experiential space and time are merely subjective and therefore "things in themselves are definitely located outside *any sort* of space and time."⁵¹ Owing perhaps to the nature of the 1912 work, viz., as a student textbook, Vvedenskij did not further elaborate his view of time, which considerably differed from Kant's. However, increasingly during the 1890s culminating in a public lecture in 1900, Vvedenskij emphasized that the temporal element in our representations both of externality and internality is conceivably merely a representation of an actual time, a time in itself. If there are things in themselves, which he believed we must concede, then there is a possibility that they are "in" such an actual time. Vvedenskij speculated that this possibility would resolve many otherwise perplexing philosophical problems, such as that of freedom versus determinism and that of the immortality of the soul.⁵²

We saw that Vvedenskij, based on the law of contradiction as well as the other laws of logical thought, concluded that we can say nothing with confidence about things in themselves. He saw another difference with Kant concerning the applicability of those laws. The former wrote, "Kant does not consider the

---

49 These four editions, from 1909, 1912, 1917, and 1922, were all published with the same title and were labeled as successive editions. Prior to the first edition, Vvedenskij prepared texts with abbreviated titles (e. g., *Lectures on Logic*) covering introductory logic for student use. The fourth edition from 1922 is essentially a reprinting of the third. Vvedenskij remarked in the later "Preface" that the third edition needed no significant changes, since he had already done so much work on that third edition. Vvedenskij (1922), 3.
50 Vvedenskij (1912), 273.
51 Vvedenskij (1912), 379. Closer to our own day, Guyer, citing Kant's statement that "space represents no property whatever of any things in themselves" (A26/B42) concludes, like Vvedenskij, that "transcendental idealism is nothing other than the thesis that things in themselves, whatever else they may be, *are not* spatial and temporal." Guyer (1987), 333. Guyer calls Kant's claim "harshly dogmatic." Kant, nevertheless, wrote that space and time "cannot exist at all outside our mind." Kant (1997), 511 (A492/B520).
52 Vvedenskij (1924), 99. Vvedenskij's conjecture is hardly compatible with Kantianism, particularly in the points elaborated in Kant's "Second Analogy" in the first *Critique*.

logical laws of thought from the epistemological viewpoint, and therefore together with everyday thinking, he assumed the law of contradiction to be valid everywhere, even to things in themselves, to true being."[53] Vvedenskij averred that his application of Kant's investigative method to the logical laws of thought had as a consequence an expansion and deepening of Kantian Criticism, a supplement, not a substitution, to the Kantian theory of cognition.

We have mentioned that Vvedenskij posed the legitimacy of starting our quest with the presupposition that mathematics and the natural sciences provide knowledge (*znanie*). However, his silence, unlike Kant's, on space and time is both troubling and bewildering. He acknowledged that mathematics consists of *a priori* judgments, but in his unique understanding of the matter they are unprovable (*nedokazuemyja*) and therefore belong to faith.[54] Yet Kant held that, for example, geometry, which is certainly a branch of mathematics, "is a science that determines the properties of space synthetically and yet *a priori*."[55] This is possible because our representation of space is an *a priori* intuition. Vvedenskij failed to address the Kantian apriority of space and time directly and definitively, maintaining an agnosticism with regard to whether space and time are *merely* subjective.[56] In the absence of an understanding of what this judgmental abstention means for the "critical theory of cognition," it is not surprising that Vvedenskij could not see mathematical propositions as provable.

Vvedenskij held that everyone needs a worldview. A scientifically-elaborated worldview must consist not just of knowledge, but also of metaphysical hypotheses based on faith in addition to faith-based views without metaphysical content.[57] These non-metaphysical views constitute our morality, and the metaphysical hypotheses must be consistent with our morality. The certainty we ascribe to our moral position is based on a moral feeling, and the appropriate metaphysical hypotheses form a morally grounded faith. Vvedenskij saw this framework as "the metaphysics recommended by Kant for the construction of an integral, sci-

---

53 Vvedenskij (1912), 274.
54 Vvedenskij (1912), 421.
55 Kant (1997), 176 (B40).
56 For the "classic" statement of this so-called neglected alternative, see Kemp Smith (1962), 113. For one reply to this possible argument, see Allison (2004), 128–132. Many others have expressed the view that Kant did not neglect this alternative. Make no mistake, though, Vvedenskij did see space as subjective. The question is whether space is *merely* subjective. In his 1915 *Psychology* text, he specifically relied on Kant's second argument in the Metaphysical Exposition of space (A24/B39), writing "We actually cannot perceive space, taken by itself, apart from all things and sensations. We always perceive the various qualities of sensations as having extension and as situated in space." Vvedenskij (1915), 124.
57 Vvedenskij (1912), 434.

entifically-elaborated worldview."[58] In order to determine just what these metaphysical hypotheses are Kant employed the analytic or regressive method. In Vvedenskij's narration, Kant asked what metaphysical positions we must take to be true if we believe in the obligatory nature of moral duty. Vvedenskij reaffirmed Kant's three postulates of practical reason but added another postulate, viz., psychic activity in others similar to that in myself, which, in Vvedenskij's eyes, is to the cognizing subject a thing in itself.[59]

Despite some deviations from Kantianism that he found necessary even under a broadly understood interpretation of the original doctrine, Vvedenskij in 1912 still saw himself as adhering to it and as propagandizing it. He conceived the changes and supplementations he proposed as merely adding depth, consistency, and completeness to Kant's thought. Nevertheless, one must not forget that Vvedenskij ended his work with the statement that none of his points were in the least dependent on Kant's theory of judgments and the conclusions drawn from it. Vvedenskij thought he had adequately established the impossibility of metaphysics as knowledge and that he had done so without Kant's help.[60]

The third edition of Vvedenskij's *Logic* textbook appeared in 1917 while Russia was immersed in a prolonged war with Germany. Any sign of sympathy with the enemy was intellectually, if not politically, risky. No wonder, then, that Vvedenskij sought to deemphasize his own philosophical debt to the original formulator of Critical Philosophy. In the Preface to this 1917 edition, Vvedenskij remarked that the principal difference with the second edition is to be found in the elaboration of what he now called "logicism" as a particular direction within epistemology. He stressed that his additions and corrections to Kant's work were of such an extensive nature that they amounted to a "new proof of the unrealizability of metaphysical knowledge quite independently of Criticism."[61] Given the unique providence of "logicism," viz., in Russia, Vvedenskij believed it also could be termed the Russian proof of the impossibility of metaphysics as a body of knowledge. He thereby wished to distinguish his proof from other philosophical directions that also shared the same intent but that stemmed from other countries, such as French Positivism, English Empiricism, and German Criticism. Moreover, logicism sought its goal without any assistance from these other philosophies and, thus, completely independently of them.[62]

---

58 Vvedenskij (1912), 436.
59 For more on this see Nemeth (2017), 233–236 and especially Nemeth (1995).
60 Vvedenskij (1912), 445.
61 Vvedenskij (1917), iii.
62 Vvedenskij (1917), 307.

Vvedenskij also claimed that although his logicism fulfilled the intention of Kantian Criticism, it was neither the same as the latter nor a mere supplement to the latter. Kant had concluded that things in themselves were not in space and time. This conviction is erroneous in that it contradicts the entire thrust of his epistemology, which asserts a complete agnosticism in the matter. We cannot even confidently affirm the existence of things in themselves. Therefore, Kantian Criticism could uphold the *possibility* of metaphysical knowledge. Logicism corrects Kant's error, thereby completing the project.

Thus far, Vvedenskij's approach to neo-Kantianism was to see it solely in terms of a limitation on any pretension by metaphysicians to knowledge of what lies beyond the bounds of experience. As such, his position includes little that commentators on Kant have found to be of particular interest. Vvedenskij had little to say about space and time, nothing substantial on the analytic/synthetic dichotomy, nothing on the distinction between judgments of experience and judgments of perception, nothing toward an understanding of the transcendental deduction of the categories. The list could include many additional topics. Putting aside the tenability of this quite anemic variety of neo-Kantianism, its sheer philosophical poverty understandably attracted few followers despite Vvedenskij's long teaching career. There was, however, one topic to which Vvedenskij devoted considerable attention and with which his only philosophical successor, as we shall see, also concerned himself, viz., the problem of other minds.

Vvedenskij in the 1890s engaged in a lively discussion with domestic colleagues concerning the issue of whether one can know, in the rigorous sense, that other human beings have mental processes similar to my own. The subject itself played only a very small role in Kant's thought. We find in the *Prolegomena* §18 that all judgments are first judgments of perception, i.e., have subjective validity, and "only afterward" do we relate what is being expressed therein to an object, intending that the judgment be valid always and for everyone, not just myself. These are, then, judgments of experience.[63] Kant, however, neither refuted nor even sought to refute solipsism. He largely took other minds for granted. His argument in the *Prolegomena* was merely that judgments of experience are intended to be always valid and for everyone *if* there are even conceivably more rational beings than myself.

In the chapter on the Paralogisms in the first *Critique*, Kant expressed surprise that the condition under which I think also holds as a condition for all thinking beings. Nevertheless, each of us can represent others only by analogy with my own self-consciousness, through "the transference of this consciousness

---

63 Kant (2002), 92 (Ak 4, 298).

of mine to other things."[64] Further on, in the Criticism of the Second Paralogism, Kant again reiterated that the representation of the other's mind requires that one "put oneself" in the shoes of the other and "demand absolute unity for the subject of a thought,"[65] so that the "I think" can be said of it. It remains a mystery what justifies this analogical argument, this transference of my formal mental processes to another so that I can say others are mentally like me.[66] Kant also did not mention that we do *in fact* hold others to have minds and mental processes similar to our own and proceed from that fact to ask the transcendental question how that fact is possible.

Although without specifically referring to Kant's writings, Vvedenskij in his first examination of the issue accepted without qualification that our access to the minds of others is only by way of behavior that we see as analogous to our own: "Since the other's mental processes can be neither observed nor presented, this substitution is the only means for penetrating into the other's mind."[67] In fact, my belief that others have minds similar to my own arises on the basis of an analogical argument, since the minds of others is, in effect, a thing in itself. Thus, we have every reason to think that at this time Vvedenskij thought he was merely developing and defending Kant's own position.[68]

Such, at least, is how Vvedenskij approached the problem of other minds from the "theoretical" perspective. However, already in 1892, he conjectured that the issue could be approached from another – a "practical" – perspective. It is with this incorporation of an additional viewpoint to correct Kant's oversight that Vvedenskij passed from being a "Kantian" to a "neo-Kantian." Vvedenskij conjectured that since the belief in other minds is so firm and ubiquitous among human beings and since the analogical argument cannot account for this belief, we must have in addition to our five physical senses a "metaphysical," or practical, sense that furnishes us with information about what others are thinking or feeling. To be sure, there is no direct evidence for this sense, but its very activity may obscure its presence. Its existence is explained by the results of its employment. Of course, the sensitivity of this sense can vary

---

64 Kant (1997), 415 (A347/B405).
65 Kant (1997), 418 (A354).
66 Beiser writes, "Kant's dismissive treatment of the problem only raised more questions than it answered." Beiser (2002), 335. We can concur with Beiser and add that Kant's treatment leads directly to Vvedenskij's equally poor treatment and its attempt to resolve the problem. But at least Vvedenskij realized that there is a problem here awaiting a solution.
67 Vvedenskij (1892), 108.
68 Vvedenskij opened his initial inquiry, exclaiming "The present investigation belongs to the sphere of critical philosophy." Vvedenskij (1892), 73.

from individual to individual just as some have better vision or hearing than others. Some people are deaf or blind, and some may have trouble recognizing what others are thinking or feeling. However, Vvedenskij, approaching the issue from the practical perspective, saw this metaphysical sense as connected with morality, i.e., with a sense of moral obligation, which is neither empirical nor present in everyone to the same degree. In fact, in the course of his argument, he affirmed the identity of this metaphysical sense with our moral sense. Moral obligation would be impossible were it not supported by our absolute confidence that others have minds similar to my own.[69] Kant had postulated immortality, a free will and the existence of God as necessary for the observance of the moral law. Those postulates are to be considered extensions of pure reason for practical purposes. Vvedenskij viewed the existence of other minds as a fourth postulate just as necessary as the three Kant mentioned and just as theoretically unprovable.

Vvedenskij held to his position in a footnote in his 1912 *Logic*, calling the existence of other minds a postulate of practical reason. He had nothing more to say about it there.[70] However, in the various editions of his psychology text, Vvedenskij did treat the issue in detail, though without adding anything substantial to it. Although there are no objective indications of the existence of other minds, there is also nothing speaking against it. From a scientific viewpoint, then, we can accept it as a working hypothesis[71] – a position, as we saw he had already enunciated years earlier. He remained silent, however, on any talk of a postulate of practical reason.

Vvedenskij was undoubtedly a neo-Kantian in that he believed he needed to correct, expand, and supplement what Kant had originally pronounced. He was not interested in simply defending Kantian positions.[72] Yet, he veered away from even mentioning, let alone defending and amplifying, most of his contentious claims. His interest in engaging with domestic critics was largely at the expense of any involvement with the predominant concerns of Western philosophy at the

---

[69] Vvedenskij's contention here arose from his view of practical reason, which while deeply indebted to Kant is not above reproach. His view that we can speak of moral obligations only to others, not to ourselves, sharply distinguishes him from Vladimir Solov'ëv, who believed we can meaningfully speak of obligations also to ourselves.
[70] Vvedenskij (1912), 438 f.
[71] Vvedenskij (1915), 75.
[72] Dmitrieva doubts the justification for characterizing Vvedenskij as a neo-Kantian. If one chooses as one's criterion a resemblance to either of the two major branches of *German* neo-Kantianism, then, of course, Vvedenskij, offering a distinctive variety of his own, was not a neo-Kantian. However, that criterion is unduly restrictive with nothing speaking for it. On Dmitrieva's grounds, neither Alois Riehl nor Leonard Nelson, to name just two, were neo-Kantians. See Dmitrieva (2007b), 134.

time. Arguably, as a result, he received scant recognition during his lifetime in the West despite his efforts to make his position known in Germany. His view of other minds offered little new to the discussion. His abandonment of philosophy of science at an early date left Imperial Russia with no neo-Kantian investigators – indeed no philosophical investigators at all – into the conceptual apparati of the rapidly developing natural sciences. His advocacy of a fifth sense, a metaphysical sense, as a means to solve otherwise seemingly intractable problems, appeared too convenient to be taken seriously. Moreover, its affinities with intuitivism, which was being developed in greater depth by his former student Nikolaj Losskij made Vvedenskij, in effect, the pupil. Finally, the claims made for his short proof of Criticism could not and would not be taken seriously by those with further philosophical training. Notwithstanding a long career and many students, few acknowledged more than a pedagogical debt to him. Today, Vvedenskij's musings must appear antiquated. Moreover, unlike Cohen, he never offered an extended commentary on any of the three *Critiques* of the sort that Cohen did. Vvedenskij's neo-Kantianism quietly died with him in Soviet Russia.

Chapter 2
The Problem of Other Minds: Lapshin, Khvostov, Lappo-Danilevskij

The problem of whether and how we can know other minds is not a traditional one handed down through the ages. Whether others have a mind, thought processes, and feelings quite similar to my own formed one of the central problems of the first Russian neo-Kantians. That it rose to such prominence within neo-Kantianism is itself unusual in that none of the various German schools gave it much consideration. Whereas it did receive attention in Germany by some unassociated with neo-Kantianism, such as Theodor Lipps and Edmund Husserl, it played an extraordinary – and surprising – role in the theoretical constructions of some unlikely social scientists.

2.1 Some Biographical Data on Lapshin

There have been only a few commentaries on Lapshin's philosophical thought, but the characterizations of it have sharply differed. Opinions have varied wildly even on whether to characterize him as a "neo-Kantian." Whereas Vvedenskij's affiliation with Kant – whether simply as a Kantian or as a neo-Kantian – has never seriously been in doubt, the same cannot be said of Lapshin. Although he was a loyal student, colleague, and Vvedenskij's eventual successor in the chair of philosophy at St. Petersburg University, even those who wished to categorize Lapshin as a "neo-Kantian" have often done so with caveats or simply passed over the objections of others in silence.[1] Before proceeding further in this matter, let us look briefly at his biography, which has also been largely neglected but is of compelling interest. To the best of our knowledge, apart from the basic facts, he left no autobiographical information that would shed light on his innermost philosophical and personal reflections. One thing that seems eminently clear is that he stoically lived through tumultuous events apparently without

---

[1] The noted Russian philosopher Nicholas Lossky, who was a colleague of Lapshin's and was certainly in a position to know Lapshin's views quite well, characterized him as a neo-Kantian. A contemporary Russian scholar, however, has offered a different opinion, writing "Lapshin never considered himself a neo-Kantian, and the Russian neo-Kantians did not consider him one of their own." Shitov (2015), 268. Nevertheless, we should add that Lapshin did characterize his position as representing a "Critical Philosophy." Pustarnakov (1999), 344.

succumbing to them psychologically and lived a surely lonely existence as a bachelor for a number of years in a foreign land, from which most of his exiled compatriots gradually dispersed and which for a time had a hostile political regime. It was there in a country then called Czechoslovakia that he would die, although to the end of his days he longed to return to a homeland that had banished him and that, officially at least, had no particular love for those such as himself.

Ivan I. Lapshin was born in Moscow[2] in 1870 to parents who were deeply interested in the fashionable topic at the time of spiritualism. Believing in the authenticity of the phenomena associated with that movement, Lapshin's parents organized a circle to observe these manifestations, a circle that came to include such notables as Pamfil Jurkevich, Vladimir Solov'ëv, and Aleksej Kozlov – three eminent nineteenth-century Russian philosophers.[3] Solov'ëv was, in fact, a frequent guest in the Lapshin residence, and therefore knew Lapshin from his childhood.[4] Upon finishing his secondary education, Lapshin studied at St. Petersburg University (1889 – 1893), where he came under the influence of Vvedenskij. He stayed on at the University to prepare for a position there, and in 1898/99 he went to England to study that country's early reaction to Kant's philosophy. Lapshin's decision to spend his time at the British Museum – where Solov'ëv had earlier also spent his "study" time abroad – rather than at a German university, as did most Russian philosophers of that era, may account in part for his decidedly independent form of "neo-Kantianism." Unlike the early Vvedenskij, Lapshin exhibited no special interest at any time in philosophical reflection on natural science – another characteristic that sets him apart from many of the German neo-Kantians.

Already in 1897, Lapshin was appointed *privat-docent* at St. Petersburg University. He supplemented his income by teaching at other institutions as well – a not uncommon practice at the time. As Vvedenskij's junior colleague, they developed not only a close professional relationship but also a personal one that last-

---

2 Apparently, even the city of Lapshin's birth is disputed. Lossky, for example, gives it as Moscow, whereas Tikhonova gives it as St. Petersburg. See Lossky (1972), 166 as against Tikhonova (2010), 154.
3 Psychobiographers would surely enjoy establishing a link between the mature Lapshin's abhorrence of metaphysics and his parents' intoxication with spiritualism if only more biographical information were available.
4 After the death of Lapshin's father's in 1883, his mother re-married, and Solov'ëv's close relationship with the family, understandably, cooled somewhat. However, Solov'ëv and Lapshin remained in contact until the former's death, with Lapshin visiting Solov'ëv on occasion at the hotel where he often stayed when in the capital. Tikhonova (2010), 155.

ed through the years ahead despite Vvedenskij's overt efforts to forestall Lapshin's academic advancement. Nonetheless, Lapshin did finally receive the rank of professor in 1913. In late 1921 with the consolidation of the Bolshevik regime, the philosophy department was disbanded, Vvedenskij alone being retained owing to his advanced age. The final blow, of course, came in August 1922 when Lapshin along with many others, such as Nicholas Lossky, Lev Karsavin, Semion Frank, and Ivan Il'in, were forcibly placed on a one-way boat trip out of Soviet Russia. The reason for the expulsion of Lapshin, who never publicly demonstrated an interest in politics, has to this day never been clarified.[5] Whatever the basis might have been, Lapshin took up an offer by the relatively new Czech government, headed by Tomaš Masaryk, a student of Brentano's and a close friend of Husserl's, to come to Prague to support and to teach in budding Russian-language institutions there. Lapshin remained in the Czech capital through the years even after most of his Russian friends departed for other lands and opportunities. Despite an acclimation to the new culture, he apparently longed to return to Russia, for in 1946 with the conclusion of hostilities Lapshin sent a request to the Soviet consulate in Prague to reinstate his citizenship. The petition went unanswered. Lapshin died in the Czech capital on 17 November 1952 at the age of 82.[6]

Lapshin's first publication was connected with his translation of William James's 1892 abridged text on psychology (*Psychology Briefer Course*). Lapshin added an introductory essay in which he turned also to Kant. He granted that Kant had established that there are *a priori*, i.e., universal and necessary, forms of cognition and that these forms are entirely subjective. Kant, however, never ventured into an investigation of their origin and never considered them to be innate, i.e., temporally antecedent to our experience. Lapshin faulted James for his psychologistic interpretation of the *a priori* and for characterizing the Kantian transcendental I as an active agent that imposes the categories of the understanding onto the sensible manifold. Perhaps unaware that James was American, Lapshin wrote, "The English, accustomed to analyze the psychological genesis of consciousness, take the methodological techniques of Kant for a description of real psychological processes."[7] Lapshin also therein regarded as

---

[5] Lesley Chamberlain writes that Lapshin may have been suspect owing to his Anglophilism, for having "garnered some very un-Leninist views from having an Anglo-Swiss mother, and from reading William James while studying in England." Chamberlain (2007), 153. Such an explanation hardly sounds plausible, but in irrational times the irrational just might become normal. We may never know the basis of Lapshin's expulsion.
[6] Tikhonova (2010), 169.
[7] Lapshin (1896), 19.

unproven such attempts, as that by James and Stumpf, to derive psychologically the idea of space from non-spatial elements of sensation.[8] Rather, for Kant, empty space, space devoid of sense elements that fill it, was thought to be a simple logical function, i.e., a "thing in itself."[9] The important points here are that the early Lapshin: (1) dismissed a psychological understanding of the Kantian *a priori* in favor of viewing its epistemic role in cognition, (2) upheld that the "thing in itself" itself serves a merely epistemic (or logical) role in cognition rather than an ontological role as an independent existent causally "responsible" for our sensations.

Lapshin wrote several pieces in the immediate subsequent years, but they hardly shed much light on his emerging philosophical stand. In a long – and tediously verbose – essay from 1900, which he, in a few years, added as an appendix to his thesis, Lapshin charged that "cowardice in thinking" is not to be understood as an open fear of people or things, but as a fear of intellectual inconsistency when it leads to conclusions that clash with one's other cherished beliefs or values.[10] The result is a hesitancy to draw the unmistakable conclusions. Although some of Lapshin's observations are somewhat interesting, perhaps even intriguing, they reveal, above all, his pronounced penchant for psychological pronouncements, rather than in technical philosophy, a penchant that will become increasingly obvious as the decades pass. He held, in particular, that metaphysical beliefs can have a tenacious hold on the human mind and often serve to obstruct our thinking even in the face of the objective absurdity of those views.

2.2 Space and Time as *a priori* Categories

Lapshin's major philosophical work was his thesis *The Laws of Thought and The Forms of Cognition* submitted in 1906 for a *magister*'s degree but which in the opinion of his examiners already qualified as a doctoral dissertation. As a result, he was awarded the higher degree. Lapshin, in commenting on Kant's text, advanced his own position vis-à-vis Kant's epistemology. Not unlike so many other neo-Kantians, Lapshin rejected Kant's apparent division of cognitive faculties. There are no pure sensations; the categories of quality and quantity are just as inseparable from sensations as are the forms of intuition from the manifold

---

8 Lapshin accused Chelpanov, to whom we shall turn later, as another who advocated a psychological understanding.
9 Lapshin (1896), 25f.
10 Lapshin (1906a), 275–276.

of intuition. Importantly, this recognition as well as that of our own self, i.e., empirical self-consciousness, and of the forms and laws of thought requires reflective attention on these cognitive elements. In other words, despite his earlier dismissal of viewing the *a priori* psychologically, Lapshin understood Kant's systematization of consciousness not as the result of a logical, but of a psychological analysis!

Lapshin also raised another question in the wake of the inseparability of the sensible manifold from the categories and forms of intuition. A crucial question in his eyes concerned whether an intuition of space and of time could be had that is "unconditionally abstracted from the entire content of the senses."[11] Kant, as we know, answered in the affirmative.[12] Lapshin answered in the negative. That is to say, we cannot represent either a truly empty space or time. If we try, as it were in an Einsteinian thought experiment, to remove from perception all of its properties, space will indeed remain as the necessary *form* of any sensible intuition. But such a purification of perception would not yield some "pure space." For in fact, we cannot completely rid our representation of space of properties. Lapshin did not offer a list of these supposed features, in the absence of which we must conclude that his argument is exceedingly weak at best.[13] Lapshin also added in support of his position that if the world had a temporal beginning, an empty time would have to be conceivable preceding that beginning, and so on.[14] The quest for pure or empty space is comparable to a quest for pure being, which, as Hegel showed, is nothing. Lapshin charged that if an absolutely empty space were representable, our experience would not be continuous. If the flow of experiences were discontinuous, it would not necessarily be subject to the laws of causality and substantiality (laws of inertia and conservation). Lapshin wrote, "the law of causality acquires the significance of a necessary and universal form of cognition only in the eyes of a person who understands that an absolutely empty space and time cannot be conceived (*ne mogut byt' myslimy*) in any experience."[15] This, as a conclusion, was of the greatest importance for

---

[11] Lapshin (1906b), 46.
[12] Kant (1997), 158 (A24/B38–39).
[13] The reader, hopefully, will realize that a dismissal of Lapshin's argument does not necessarily imply that Kant's argument is valid – at least certainly not with the intention Kant had in mind.
[14] Lapshin, of course, took this argument – as he did acknowledge – from Kant's first "Antinomy of Pure Reason" (A429/B457), though without acknowledging Kant's corresponding "Remark" (A431–433/B459–461).
[15] Lapshin (1906b), 49. This, the third argument, for his position is quite confusing to say the least. Lapshin assumed in it that space and time are distinct features and not *forms* of sensible intuition.

2.2 Space and Time as *a priori* Categories — 43

Lapshin. He held that were we to admit the representability (*predstavimost'*)[16] of empty space and time, most of the essential points of Kantian philosophy would be jeopardized. Moreover, Kant himself, Lapshin charged, recognized this![17] If we were to uphold the apriority of the categories, we would have to affirm the unrepresentability of empty space and time.

Kant in the "Transcendental Aesthetic" explicitly asserted that space is a pure intuition and not a discursive concept. Kant drew this conclusion from his claim, which Lapshin rejected, that although we cannot represent that there is no space, we can conceive that there are no objects to be encountered in it (A24–25/B38–39). Lapshin recognized Kant's position in this matter and rejected it. The former agreed with Hegel that space and time are concepts. Lapshin wrote, "in fact, if we recognize that space and time, as pure intuitions, are inconceivable by themselves without a sensible substrate in the form of sensations and reproduced representations, then what makes us call them intuitions and not concepts?"[18] In a specific instance of a visual intuition, for example, I find an *illustration* of a concept, but never an intuition of a concept. The same can be said with regard to space and time. Every intuition of space and of time is an illustration of the respective concept. Thus, Lapshin wrote, "it seems to me that Kant in the given case became the victim of an illusion, as if a pure intuition of space and time were something absolutely *sui generis* in comparison to the simple conceivability of concepts."[19] The space of which I have a representation when I look out my window is a particular "illustration" of the general concept of space.[20] From this claim, from the alleged psychological fact that pure space and time cannot be represented as an object of experience and that this fact "forms the basis of the *Critique*," Lapshin, without further ado, wrote, "we must admit that space and time are in essence just as much concepts as are

---

16 Lapshin made no attempt to distinguish "representability" from "conceivability" (*myslimost'*).
17 Based on the references Lapshin supplied, he concluded that Kant too acknowledged the unrepresentability of empty space and time from a reading of the first "Antinomy" and certain of Kant's pre-Critical writings, but not from the "Transcendental Aesthetic."
18 Lapshin (1906b), 51.
19 Lapshin (1906b), 51–52.
20 Lapshin's points were already rejected by Kant in his "Metaphysical Exposition" of space. There is no need for us here to reply from an orthodox Kantian standpoint. That space and time, for Kant, are *a priori* forms of intuition and *not* general concepts is the subject of an enormous amount of secondary literature and discussion.

quantity, quality, causality, and substance, i.e., concepts *of a higher sort*, *categories* of cognition and not *pure intuitions*."[21]

However bizarre we may view Lapshin's contention that space and time have as much right to be considered Kantian *categories* as those of causality and substantiality – bizarre, that is, from a strict Kantian standpoint – Lapshin believed that his position was a *correction* of Kant's teaching, and not a radical departure or break from it.[22] Kant was deceived by what we today might call his "faculty psychology," by his sharply drawn dichotomy between sensibility and understanding. Owing to that dichotomy, he had to find a means to bridge the two faculties, and for this purpose, he turned to a schematism of the categories, an intermediary between the pure concepts and the objects of cognition. This need for the schematism was a problem of his own making. That is to say, what Kant did not recognize, according to Lapshin, was that the relation of a spatial intuition to the ideal concept of space is just more clearly recognizable than anything about or in an intuition to, say, the ideal concept of quality. This failure on Kant's part to see the categories in intuition coupled with his rigid dichotomization of sensibility and understanding led to his need for a schematism.

The question remains, though, whether Lapshin was committed to Kantian transcendental idealism. If, as at least some commentators today affirm, transcendental idealism rests on an affirmation of the apriority of space and time, *and* if, as Lapshin claimed, space and time are not pure, *a priori* intuitions, would this entail an affirmation of or a retreat from transcendental idealism?

---

21 Lapshin (1906b), 52. The reader will recognize that quantity and quality are not as such among the twelve Kantian categories, but are two of the four principal logical functions of judgments. Under each of these two, we find three categories. However, Lapshin was not always consistent in this. He also on occasion wrote of the categories of quality, which leads us to think he did recognize the distinction. See, for example, Lapshin (1906b), 61. We must add as another curious fact that in a twenty-two point summary of his dissertation that appeared in a 1909 issue of *Kant-Studien*, Lapshin did not so much as mention this thesis of space and time as categories. See Lapschin (1909), Although Lapshin demonstrates some acquaintance with Cohen's 1902 *Logik*, in which Cohen writes, "Space is a category," Lapshin does not mention Cohen in this connection. Cohen (1902), 162. Lapshin does note, however, Cohen's earlier statement, "Whoever takes space and time as categories destroys the entire foundation of the transcendental system." Cohen (1885), 211; Lapshin (1906b), 58.

22 Although understandable, it is regrettable that Lapshin came to his conclusion concerning the conceptuality of space and time quite independently of Émile Durkheim, who in his 1912 work *Les Formes élémentaires de la vie religieuse* wrote, "I call time and space categories because there is no difference between the role these notions play in intellectual life and that which falls to notions of kind and cause." Durkheim (1995), 8. Of course, Durkheim's "move" enabled him to investigate sociologically the different conceptions of space and time held by various societies. Such a project remained only latent in Lapshin's thesis.

2.2 Space and Time as *a priori* Categories — 45

The answer is not obvious, and there are various possibilities. Whether we view Kant *coming to* transcendental idealism based on his proofs that space and time are *a priori* conditions of cognition or dogmatically *proceeding from* it, we have either way the bald assertion that although we cannot cognize things in themselves we can think of the objects of experience as things in themselves. These things in themselves are, in some manner, responsible for appearances (Bxxvi-xxvii). Regardless of whether we accept the two-worlds interpretation or the two-aspects interpretation of the thing in itself, we still find in either case an affirmation of the thing in itself. The categories, such as causality and subsistence, taken in isolation from sensible intuition, are applicable merely to appearances, but not to things in themselves. For, as is the case with space and time, the categories in isolation become unrepresentable fictions.[23] This is not to say that we know things in themselves to be unextended, extra-temporal, or to lack any quality. Lapshin simply wished to emphasize that we cannot conceive a non-sensible experience that lacks space, time, a quality, and a quantity without falling into logical absurdities. In this, then, Lapshin was more heavily indebted to Kant's "Antinomies" in the Transcendental Dialectic than to the Transcendental Analytic, the traditional source for the argument for transcendental idealism.[24] However, we have not yet addressed the issue of whether for Lapshin in 1906 there are Kantian things in themselves.

Make no mistake, there is nothing comparable in Lapshin's work to Kant's metaphysical deduction of the categories. Nor did Lapshin provide a complete accounting of the categories. He focused, understandably, on substantiality and causality in his treatment of things in themselves. Clearly bearing in mind the veritable avalanche of criticism Kant's own treatment received over the years and decades, Lapshin refused to recognize that those two categories can legitimately be applied beyond the bounds of what our senses provide. The conclusion he drew from those criticisms is that we, as cognizing agents, cannot infer that things in themselves exist. We must remain agnostic. Such is the core idea of what Lapshin called "critical idealism," which, I submit, was his ex-

---

[23] Lapshin (1906b), 62 – "Qualities are known to us only as conscious states. A quality in general is just as much an unrepresentable fiction as is an empty space."
[24] Guyer finds the same reasoning, such as that advanced by Lapshin here, to be precisely Kant's in the "Antinomy of Pure Reason" and thus serving as an indirect proof of transcendental idealism. Guyer writes, much like Lapshin, "On the assumption that things as they are in themselves really are spatial and temporal, he [Kant] argues, reason is necessarily ensnared in paradoxes…. But on the assumption that space and time are merely features of our representations of objects but not of the objects themselves, Kant claims, and only on this assumption, these paradoxes can be avoided." Guyer (1987), 385.

pression for what Kant called "transcendental idealism" (A369), though suitably modified to reflect his own position.

> The idea that the category of *substantiality* extends only to sense data and is inapplicable to things in themselves is the purest expression of *critical idealism*.
> ... We can view all of experience as the interaction of substances without assuming thereby that the law of substantiality does or does not extend to things in themselves and even without touching on the question of whether these things in themselves exist or not.[25]

Lapshin's conception can be easily compared and contrasted with that presented in one passage from Kant's "Paralogisms," which states that transcendental idealism is the doctrine that all appearances are merely representations and not things in themselves. Thus far, Lapshin is in agreement. However, Kant then continued: "and accordingly that space and time are only sensible forms of our intuition, but not determinations given for themselves or conditions of objects as things in themselves" (A369). If we take Kant's word "accordingly" [*gemäß*] to mean that having already correctly established that all appearances are representations and consequently (in the logical sense) space and time are merely forms of intuition, Lapshin would also be in agreement. But if the word "accordingly" is used in some looser sense, Lapshin would not be in agreement. The fact that not just space and time are inseparable from sense data, but so are also the other categories shows that all appearances are representations. In short, Lapshin broadened Kant's doctrine but by way, again, of the Transcendental Dialectic, rather than by the Transcendental Aesthetic.

Lapshin agreed that the unity of consciousness is the bedrock of all cognition. Since animals, in addition to humans, have an awareness, a consciousness, we must postulate that they too have a unity of consciousness. But insofar as animals lack self-consciousness, they lack the *idea* of the unity of consciousness and its associated laws. For Lapshin, these laws included, for example, the law of causality. Thus, animals, along with humans, can form, albeit implicitly,

---

[25] Lapshin (1906b), 63. Lapshin's agnosticism, as with so many other aspects of his discussion of the "thing in itself," is problematic. Kant was not agnostic in this matter. To speak of appearances is to say that those appearances are *of* something. If, on the other hand, these same "appearances" are *not* of something, they, as objects, are, then, not appearances, but things in themselves. Since we certainly do apply, say, causality to "appearances," we would face again the Humean problem. The simple logic involved here was recognized by Friedrich Paulsen, whom Lapshin quotes concerning this specific point in a Russian translation. Lapshin presented no reply, writing that Paulsen "did not suspect that there is a possibility of a *completely agnostic* point of view, according to which the category of causality is seen as applicable only within appearances." Lapshin (1906b), 70; see Paulsen (1902), 154.

what Kant called "judgments of perception," in which they notice the uniformity of nature within their empirical intuitions. However, beings with self-consciousness, such as humans, are capable of conceiving the law of causality as a law. They are capable of subsuming empirical events under the concept of a necessary law, thus forming "judgments of experience." Self-consciousness arises when we notice (*podmechaem*) a unity among the elements of our consciousness and begin to conceive this unity conceptually.[26] The "I" of the "I think" is to me no more than the simple logical unity of my consciousness. Being a simple concept, the "I" cannot be a factor in experience apart from the sensible manifold and its forms. The "I" is neither an intuition nor a substance.

The original unification of the sense manifold of which we are conscious by way of the abstract concept of "I" lies at the basis of all logical thought, the most elementary expression of which is the law of identity. "Therefore, the identity of our consciousness at each moment of our thought is a condition for the very concept of identity."[27] With such reasoning, Lapshin believed he could conclude that the laws of thought rest on a synthesis of identity and difference. In other words, they rest on a categorial synthesis, namely, on the category of quality.[28] In this manner, even Kant's so-called analytic judgments are grounded on a logically prior synthesis of the manifold. Lapshin, continuing with this train of thought, also claimed that owing to the inseparability of the categories and the intuitions of space and time from all logical thought, analytic judgments rest on them as well.[29] Presumably, then, simple arithmetical equations, such as $2 + 2 = 4$, are "ultimately" dependent on an entire assortment of categorial syntheses, one of which inherently involves temporality. An elementary consequence of this position – and there are others – is that a being for whom objects are directly given –

---

26 Lapshin (1906b), 72; cf. Kant (1997), 309 (A197/B242). Lapshin's explanation of self-consciousness is quite circular. The act of noticing is itself an act of reflection. Moreover, the act of conceiving a unity at least on Kantian grounds logically requires the very transcendental unity that Lapshin holds to arise from this conceiving, unless, that is, he has in mind here empirical self-consciousness. But then who, looking in a mirror, thinks he/she sees a unity of conscious elements?

27 Lapshin (1906b), 90. He, in an arguably more cryptic fashion, reaffirmed this position in an undated and until recently unpublished manuscript entitled "What is Truth?" He wrote, "Let us start with the relationship of our 'I' to *time*. Kant grounds the idea of identity on the *identity* of the cognizing subject in the temporal process of the change of states of the subject's consciousness. The worldly process, the change of temporal states, is logically connected for me with the consciousness of an unchanging identity of the cognizing subject." Lapshin (2006), 286.

28 Lapshin clearly wrote "category of quality" even though Kant did not speak of such a single category.

29 Lapshin (1906b), 94 and 176.

a "divine understanding" – could conceivably obviate the laws of logic and, as a corollary, obtain starkly different arithmetical formulations than we do.[30] However, Lapshin concluded from this that those who reject *his* understanding and interpretation of "Critical" epistemology have no *right*, i.e., epistemologically grounded justification, to employ the laws of formal logic![31]

Another conclusion Lapshin drew from his deliberations on the mechanism of cognition – one that he reiterated repeatedly – is that the laws of logic and of thought are inapplicable beyond the empirically given. Thus, they are inapplicable to things in themselves, presuming, of course, that there is a set of numerically distinct things (transcendental things in themselves), which are separate from the objects of everyday experience, although Lapshin also repeatedly and incautiously omitted this proviso.[32] Thus, metaphysics, conceived as concerned with objects beyond the sphere of possible experience, loses its legitimacy as a branch of knowledge.[33] In a similar vein, Lapshin was dismissive of the criticism that Kant, for one, did not provide a logical deduction of the completeness of his table of categories. The former averred that no such deduction is logically possible without introducing a vicious circle. Any such proof would presuppose the applicability of logical laws, which that very proof is meant to establish.[34]

2.3 On Other Minds

Lapshin's hasty exploration into epistemology without detailing the ramifications of its conclusions provided little room for continuing philosophical work.

---

30 Cf. Kant (1997), 253 (B145).

31 Lapshin's remark here may not be audacious as it may appear at first if understood from a *strict* Kantian viewpoint. For Kant writes in the Transcendental Deduction that the "whole of logic" obtains its necessary character from its relation to the synthetic unity of apperception. Kant (1997), 247 (B134 f). See Longuenesse (1998), 75.

32 See, for example, Lapshin (1906b), 152. If neither the laws of logic nor the *a priori* categories, understood to include as Lapshin did space and time, are inapplicable to things in themselves, that concept is vacuous. The concept of a thing in itself would be equivalent, as Hegel noted, to non-being. Therefore, contra Kant (A29–30/B45), the distinction between the transcendental thing in itself and the empirical thing in itself collapses, and we are back again to Hume.

33 Many a mystic would hardly feel seriously threatened by Lapshin's position, for he did not take into account mystical experience. Its defenders contend such that experience can be as "real," if not more so, than other forms of experience.

34 Lapshin (1906b), 184. Again, Lapshin misunderstood the criticism and the difference between the Metaphysical and the Transcendental Deductions, the task of the latter being to address the issue of the validity of the categories (*quid juris*). In our own day, Allison has argued against the sort of circularity that Lapshin saw. See Allison (2004), 152–153.

Unlike his mentor Vvedenskij, for whose own epistemology Lapshin's thesis acted as a support and bulwark, Lapshin showed little technical interest in other philosophical subdisciplines apart from aesthetics.[35] Vvedenskij tried to accommodate religious belief within his neo-Kantianism; Lapshin would have none of it. Even when dealing with issues related to art, however, he demonstrated little to no interest in the concerns that riveted Kant in his third *Critique*. Nonetheless, there was one topic that apparently held his interest through the years, one that he shared with Vvedenskij, namely that of the problem of other minds.

In a treatise from 1910, *The Problem of the "Other I" in Modern Philosophy*, Lapshin lamented that so little attention had thus far been accorded to the problem of the other I in the history of philosophy. Whereas Kant in the first *Critique* undoubtedly sought to overcome skepticism, he had completely neglected the question of knowing that others have minds similar to my own. To be sure, Fichte raised the issue, but he took the existence of other minds as a moral postulate and not one based on theoretical considerations. Nor did any of the founders of the German neo-Kantian movement (Lange, Liebmann, and Cohen) pay much attention to the question. Hermann Cohen, in particular, simply assumed that all others had a formal cognitive organization homogeneous with my own.[36] Likewise, the Baden neo-Kantians simply postulated that all others had a similarly organized consciousness allowing for the possibility to speak of a "consciousness in general." We can add, however, that it, apparently, did not occur to Lapshin that Cohen's concern was not with the cognitive apparatus of the individual human mind, but with scientific experience as manifested in the publicly available theories of natural science, particularly mathematical physics.[37] Cohen, thereby, circumvented the entire issue of other minds without addressing the assumption Lapshin ascribed to him.[38]

Lapshin concluded his historical survey, writing that he hoped to return at an unspecified date to the problem and present his own thoughts on the

---

35 Lapshin did write a number of works other than his thesis prior to his banishment from Soviet Russia as well as during his days in Prague. The latter are, for the most part, of little interest to Western philosophers today and, as Abramov wrote, "they almost all fall outside the framework of the Kantian problematic." Abramov (1994), 230.
36 Lapshin (1910), 164.
37 Cohen wrote, "The conditions of the possibility of experience are to be discovered. ... These conditions are to be designated as the constituting features of the concept of experience. ... This is the entire concern of transcendental philosophy. Experience is, therefore, given in mathematics and pure natural science." Cohen (1877), 24–25. See also Cohen (1885), 501.
38 It is with this circumvention, and therefore lacuna, that Lapshin could have faulted Cohen.

topic. He ventured to say, though, in 1910 that he believed the problem *could* be solved in theory on the basis of Critical Philosophy. While he did not yet have the solution, its direction was clear enough. It took years before he would again return in earnest to the topic. Lapshin resumed his reflections in 1923 while in Prague. He tells us that his interest is purely epistemological. He is unconcerned with how we as individuals acquire or develop psychologically the view that others have mental activity similar to my own. His characterization, however, of his concern was quite imprecise. For, on the one hand, he wrote that he was confronting the issue of solipsism, whereas the actual argument that he presented was aimed not at solipsism, i.e., the position that one's own self solely exists, but only at establishing that the other's mind is similar to my own. Lapshin's mentor, Vvedenskij, had no interest in combatting the former but attended only to the latter. Vvedenskij had no doubt that he confronted others with minds, but what is our basis for concluding that the minds of others are similar to my own? Their minds, i.e., their thoughts, feelings, impressions, etc. are not given to me directly as is their hair coloring or their physical dimensions.

It would be wrong for us, however, to think that Lapshin committed an elementary error at the very start, confusing solipsism with the problem of other minds. In his view, the three problems of personal identity, that of other minds, and that of all objective existence apart from my self are coordinated with each other. All three must be solved together and solved without appealing to any sort of metaphysical transcendence, which would, in effect, presuppose what is in contention.[39] Those who believe that these and other such problems are fundamentally of no philosophical interest are deluded by an illusion. The world and, in particular, the minds of others are not given to me immediately, as the intuitivists believe, although owing to the close and integral connection between corporeal movements and the presumed mental states of others the illusion is unavoidable. Rather, the other I, taken as an individual reality in isolation, is – and here Lapshin employs an expression closely associated with Cohen – a limit-concept [*Grenzbegriff*], a mental construction that necessarily arises if we lose sight of the correlation of the other I to our own consciousness and the infinite inexhaustibility of psychic reality.[40] Surely, the mental states of others are, in some sense, transcendent, but, by the same token, so are factually inaccessible locations in space and places in time. Coherent cognitions of exter-

---

[39] Lapshin (1923), 26.
[40] Lapshin (1923), 56. Gustav Shpet at roughly the same time saw the necessary correlation between consciousness and the essence of the object of consciousness, i.e., the essence of intentionality, as the phenomenological reduction. However, he did not venture to say that the thing in itself results from a diremption of this connection.

nality are impossible without the assistance of the imagination and of inductive conclusions that fill the lacunas of immediate perception. Similarly, coherent cognitions of mental activity are impossible without the assistance of the imagination to complete my personal experience with the mental states of others along similar inductive principles. In order for my cognition of nature to be coherent, an agreement between my judgments of it with those of others plays an enormous role. These statements imply, however, that the imagination and a certain creative mental faculty are factually necessary for my consistent representation of the external world. That I take the other I as, in general, having mental states similar to my own and ascribe an importance to them in the other similar to the importance I ascribe to my own mental states is necessary for my consistent rendering of externality. A doubt that another person has a similar mental state as I do on a particular occasion is no more alarming than a doubt concerning the cause of some particular event.

Vvedenskij proposed that we have "faith" in the very existence of the other's mental activity. Such faith enables us to speak of morality, e. g., that others have rights, that we should treat others always as ends, not as means toward an end. Lapshin was more restrained. For him, we can speak of faith as immanent or transcendent. We already saw that Lapshin urged that the other I be taken as a limit concept or a limit fiction. There is no need, then, to speak of taking the existence of the other I as a matter of belief. The other I is always cognized by me as the set of my representations of him/her, although such representations do not and cannot exhaust everything about the other's mental activity. There is always more just as there is always more to the world than I experience at this moment. On the other hand, a transcendent faith in the other I is not only unnecessary but is philosophically unjustified – unjustified, that is, along Lapshin's principles. In other words, the mental activity of the other I is not a metaphysically transcendent thing in itself, but a thing in itself conceived in a properly neo-Kantian sense, a Kantian idea, as an inexhaustible phenomenon to which we asymptotically approach through ever widening experiences. Lapshin's solution to the problem of the other I bore little semblance to a traditional understanding of a solution. We cannot prove the existence of anything, let alone that of the other I. Moreover, we can be wrong in our determinations with regard to the other just as we can be wrong in anything. But we can speak of the other's immanent reality and of its states as they are presented, and we can understand them "as if" they were our own.

Lapshin failed to develop *in extenso* his own philosophical position or his relationship to his predecessors. He did go on to write additional tracts, but for the most part, they were both ignored by his contemporaries and had little to do with the technical philosophy emerging in Western Europe, of which he displayed little knowledge. The phenomenological movement in which he, as a resident of Prague, could have participated to some degree utterly and amazingly passed him by. The linguistic turn afforded by analytic philosophers with their prominent background in mathematics and formal logic went entirely unnoticed. Unlike Solov'ëv, who had a deep interest in moral philosophy but little if any in music, Lapshin had a deep interest in music but little if any in moral philosophy. True, already in his 1911 extended essay "Universal Feeling," Lapshin wrote of such a feeling, an immanent feeling, the recognition of which all people manifest in the thought of highest values. However, this remained unelaborated apart from the assertion that humanity is indefinitely approaching, though asymptotically, an absolute moral good. Certainly, we can see this assertion as connected with his own distinct variety of neo-Kantianism. However, the idea remained stillborn. For Lapshin, the ideal state to which humanity is asymptotically approaching is a conclusion drawn from concrete biological, psychological, and historical analyses. As such, he saw it not as an *a priori* truth, but merely as an *a posteriori*, though probable, one.[41]

Lapshin's uncompromising hostility toward metaphysics surely also set him apart from many of his contemporary compatriots who moved toward an open embrace of the Orthodox Christian religion. Since he remained largely isolated both physically and intellectually from the world philosophical community, we cannot offer any firm judgment concerning any continuing commitment while in exile to the neo-Kantianism of his Petersburg period. Nevertheless, another particularly noteworthy figure whom we shall look at later, declared that Lapshin's work represented "the most valuable and important manifestation of the critical way of thinking."[42]

2.4 A Historian Weighs in on Other Minds

We have seen that for both Vvedenskij and Lapshin the problem of other minds, i.e., whether others have minds similar to my own, loomed as a distinctive feature of their own variety of neo-Kantianism.[43] Both thought they had found a sol-

---

[41] Lapshin (1911), 88.
[42] Jakovenko (2003), 353.
[43] One scholar goes so far as to write that Vvedenskij's grappling with the issue of other minds "in many respects defines the specific Russian version of neo-Kantianism." Rumjanceva (2012), 133. Such a view is exaggerated. It is unclear why this specific problem should be considered a neo-Kantian one. Moreover, most of the figures discussed in the pages that follow did not concern themselves with other minds. If we were to accept a concern with it to be a necessary con-

ution, but the respective solutions pointed in different directions. Remarkably, another who found the topic of interest with regard to his own pursuits was Aleksandr S. Lappo-Danilevskij (1863–1919), a highly regarded historian who also taught at St. Petersburg University and at other institutions in the capital.[44] Writing with knowledge of Theodor Lipps' classic 1907 treatment as well as of Vvedenskij's earlier controversial work and Lapshin's 1910 historical survey of the problem, Lappo-Danilevskij in his massive *Methodology of History* offered a reason why even a historian should care about whether the other I can be known. With a deep appreciation for the Baden School's version of neo-Kantianism, Lappo-Danilevskij believed the search for universal and invariable laws in society and human behavior at the expense of the individual peculiarities would fail to provide insight into events. All social scientists including the historian must seek an understanding of the changes in the state or condition of one's object of study, be it an individual or some collective, in order to understand the events that ensue from such changes. The social scientist needs to understand the other's psyche or mental attitude, which is "essentially inaccessible to empirical observation."[45]

In an arguably veiled reference to Vvedenskij, Lappo-Danilevskij wrote that he was uninterested in metaphysical formulations that see the other I as, for example, an unknowable thing in itself or as the soul's essence. In the first case, the solution is found in faith, and in the second an appeal to some metaphysical I. Lappo-Danilevskij also dismissed the easy solution of the intuitivist who is satisfied with the virtually instinctive belief that other minds are given directly via a non-sensible intuition together with our sensible perceptions. The Critical epistemologist, however, seeks the reasons for our everyday ascription of an objective sense to our experience of the "other." The other I is not given to me immediately through some non-sensible intuition, but I have a firm conviction about this other I apparently based on my observations of his/her corporeal move-

---

dition of Russian neo-Kantianism, the present investigation would have to conclude with this chapter.
44 Lappo-Danilevskij also dealt extensively with historical methodology informed by a considerable familiarity with the relevant writings of the Baden School of neo-Kantianism. We will return again at greater length and with additional biographical information to Lappo-Danilevskij later in this work when our study turns to others influenced by the Baden School.
45 Lappo-Danilevskij (2006), 237. In her quest to disentangle him from the Baden School, Rumjanceva emphasizes Lappo-Danilevskij's discussion of the problem, which she finds as corresponding to the "philosophical quests of Russian neo-Kantianism, whose founder was Vvedenskij." She calls that discussion the backbone (*sistemoobrazujushchij*) of Lappo-Danilevskij's methodology of history. Rumjanceva (2014), 8. We shall return to this interpretation of Lappo-Danilevskij later.

ments. What, then, Lappo-Danilevskij asked, is the link between the unobservable and the observable? Some (German) neo-Kantians skirted the entire problem, invoking the concept of a consciousness in general that certainly appears to include a tacit recognition that others have mental activity comparable to my own. When speaking of universal categories of cognition employed in unifying representations, we clearly are thereby affirming that those categories are a condition of all cognition by finite beings, not just my own.[46] However, a consciousness in general, being an epistemological concept, provides no basis for the recognition of the actual existence of the other, concrete individual I.

If we grant that no theoretical proof is possible that others have mental activity similar to my own, but that we as a matter of fact have no doubt in that mental activity, we must seek the basis of this conviction somewhere other than in a theoretical proof. Lappo-Danilevskij stated that this conviction is more firmly established in a regulatory-teleological sense.[47] In his view, it can be used as a scientific hypothesis in connection with an explanation of some observable behavior. It can also be used as a moral postulate with which we take the other I as an end in itself. Our behavior toward others, thereby, takes on a moral character.[48] Lappo-Danilevskij wrote,

> In fact, the moral consciousness of my "I" demands that I recognize the other's mental activity. The morally free activity of my "I" cannot be an activity without an object. Such an object, however, should encourage me to engage in morally free activity and not mechanically "bind" my action in a cause-effect sense. In other words, the other I should be conceived by me as morally free.[49]

---

[46] Kant affirms so much in the *Prolegomena*, §22, writing that judgments are objective if the representations "are united in a consciousness in general, i.e., are united necessarily therein." Kant (2002), 98.

[47] In a manuscript stored in the St. Petersburg branch of the Russian Academy of Sciences designated for a course in 1902–1903, Lappo-Danilevskij already at this early date wrote, "Although the assumption of the existence of the other's mental activity can be critically *grounded*, it cannot be proven. By virtue of an inference made on the basis of analogous objective features, the assumption is only probable. In any case, I constantly use the principle of the other's mental activity as a regulative principle to explain, for example, the expediency of the other's actions, etc." Lappo-Danilevskij (2013), 97.

[48] Thereby, whereas, our behavior toward inanimate things is widely – though not universally – recognized as amoral, our behavior toward rational beings like ourselves is worthy of moral judgment. Lappo-Danilevskij here differed from Vvedenskij. The former stated that the existence of the other I can be *postulated* for moral purposes, whereas Vvedenskij took that existence at one time as an object of faith with epistemological import and at another time as the intentional object of a moral sense.

[49] Lappo-Danilevskij (2006), 242–243.

Additionally, social scientists – psychologists, sociologists, and historians – accept that others have mental activity similar to my own as a necessary working hypothesis in order to coordinate and to unify one's own knowledge of the observed actions of others. In the absence of contradictory evidence, the hypothesis is more likely than its opposite. True, we, as it were, instinctively have trust in the hypothesis, but the apparent efficacy of it adds further support to the trust we have initially placed in it.

We can look at the issue of how I know the mental activity of others not just as a philosophical problem, inquiring into the rational grounds for our recognition of the other I, but also in terms of how that recognition psychogenetically arises in consciousness. One of the instinctual expressions of our recognition of the mental activity of others is through sympathy. Through the experiencing of that feeling, our recognition of the other I actually develops. Lappo-Danilevskij realized that an account of sympathy still awaits its full elucidation.[50] He himself briefly suggested several concrete ways in which the feeling arises. The details of these examples need not concern us here, particularly since in any case he offered them merely as tentative. Nevertheless, one basis for the emergence within consciousness of our conviction concerning the other is an association of symbols, including meaningful language and corporeal expressions and signs. Bertrand Russell, writing on the topic much later, offered an analogical argument to "justify" an inference to other minds that common sense makes unreflectively.[51] But it is interesting that Lappo-Danilevskij warned against putting too much weight on the analogical inference writing,

> only if I substitute the concept of my "I" with the concept of "a certain I" can I by analogy conclude concerning the "other I," i. e., pass my representation of a "certain I" (in this case, identified with my "I") to the "other I." In such a case, however, I already proceed from a recognition of the other's mental activity, for it underlies my concept of "a certain I" as an example by which I can judge another example of the same sort through an analogical inference.[52]

From this viewpoint, the analogical argument as a piece of philosophy presupposes what it seeks to establish. If we start with the presupposition that a certain person X has a mental state and displays a certain physical behavior we can by

---

50 We should note that Lappo-Danilevskij, in a footnote, remarks that his term "sympathy" (*sochuvstvennoe perezhivanie*) "comes close" to Lipps' term *Einfühlung*, which the former translates as *"vchuvstvovanie."* Lappo-Danilevskij (2006), 244 f. Whereas, then, the meaning of the two terms are close, they are not the same in his mind.
51 See Russell (1948), 482–486.
52 Lappo-Danilevskij (2006), 245.

way of analogy associate the same behavior in another person Y with having that same mental state as X. But Lappo-Danilevskij's point is that it is unclear how the very conception of a transference of *my* personal, subjective mental state to another can occur, as Wittgenstein would later also point out.[53]

A factor in the psychological development of my conviction in the other's mind is that I constantly ask for and refer to the opinion of others regarding the condition of some object. That others have the same opinion as I do about, say, the taste of some food, about the temperature outside, etc., reinforces my conviction that they sense or think as I do. The more often this happens, the stronger my conviction that they are "rational" beings and the more certain I am that some state of affairs is an objective truth. Lappo-Danilevskij holds that the notion of self, my self-consciousness, develops in tandem as I distinguish myself from others. That I am aware of my own self in conceiving the other prevents my identification with the other. Even when I have a sympathetic experience of another person, I cannot help being this I.

For the most part, the social scientist operates under the assumption that the human mind, the psyche, is uniform across humanity.[54] The historian assumes that he can understand the mind of some historical agent despite the passage of time. This presupposes that the agent operated in a calculating manner that would be understandable if only we would perform all due research. Lappo-Danilevskij was willing to entertain the logical possibility of psychophysical parallelism. But he saw as its consequence that the historian cannot confidently conclude anything about the mental state of a historical individual based on that individual's behavior. The historian could regard that behavior as an indication of the individual's state of mind, but not as an effect caused by the latter. The entire issue comes down to one of scientificity: Can the historian hold with assurance that a particular figure operated with a specific mental attitude and intention? Such may be the theoretical concern facing the historian, but he/she in practice tacitly at least employs a recognition of the other's mental activity in order to construct the concept of a qualitative change in the psyche of the individual or collective being studied.[55] The historian, after all, judges his/her sub-

---

[53] "Look at a stone and imagine it having sensations. – One says to oneself: How could one so much as get the idea of ascribing a *sensation* to a *thing*?" Wittgenstein (1953), 98 (§284).

[54] Whereas many of the German neo-Kantians simply assumed this uniformity of human nature, Lappo-Danilevskij again already in the first years of the twentieth century raised it as an issue. "The assertion of a similarity between my mental activity and that of others is in need, however, of an additional principle of the uniformity of human nature." Lappo-Danilevskij (2013), 97.

[55] Lappo-Danilevskij (2006), 250.

ject by the latter's actions and results. Such a superficial statement of the historian's actual practice does not dispense with the topic of other minds, however odd it may seem to others that an eminent historian would linger on a supposed problem with such a seemingly obvious solution.[56] For even the practicing historian, who accepts that the other I exists must ask to what extent the individual is a product of the surroundings and to what extent the historian must look at the individual as unique. We will look at such a concern later when we investigate the influence of the Baden neo-Kantians. The point is that Lappo-Danilevskij's treatment of the other I, while not common, was not outlandish and without its own logic within either a Baden-influenced neo-Kantianism or a distinct Russian variety of neo-Kantianism.

2.5 An Eclectic Neo-Kantianism

Gustav Shpet penned in 1918 a lengthy review of a book on Hegel by Ivan Il'in, a remarkable scholar in his own right and who, like Shpet, also briefly studied under Husserl. Shpet noted the peculiar status of philosophy within Russian schools of law and the distinctive philosophical works that emerged from the philosopher-lawyers there. The first thing that strikes the eye, he thought, was the unique rhetorical style of their works and their "great philosophical dilettantism." They exhibit "a tendency to solve purely philosophical problems from the perspective of practical life" as though philosophy were in need of practical justification and only in this way does it acquire a supreme value.[57] Although not mentioned by name, one such "philosopher-lawyer" would, surely, have been Venjamin M. Khvostov (1868–1920), a graduate of Moscow University's law school, where he also became a professor in 1889 with a specialty in Roman law.[58] He resigned from the university in 1911 in protest against government re-

---

[56] Such is how it surely appeared to Shpet in 1918. Shpet also gave every indication that the problem of other minds, at least as Lappo-Danilevskij had presented it, had been satisfactorily resolved by contemporary philosophy. Since so much of Lappo-Danilevskij's discussion contains elements that Husserl himself at the time was considering, it is hard to believe Shpet, an early phenomenologist, would have looked favorably on Husserl's treatment. See Shpet (2019), 112. Shpet's dismissal of the issue does not speak favorably of his philosophical acumen, perspicacity, and inquisitiveness. In any case, of course, Shpet did not know Husserl's treatment of the issue.
[57] Shpet (2004), 281.
[58] Andrej Belyj, in his memoir *The Beginning of the Century*, wrote, "Lopatin knew of Cohen and Rickert only from the exposition of Khvostov, a professor of law. The two lived for years in the same neighborhood. The animated Khvostov, a dilettante in philosophy, passed the sum of his

strictions on university autonomy instituted by the Minister of Education Lev A. Kasso. In succeeding years, Khvostov taught at several other higher educational institutions. In short, Shpet's statement, taken as an overall assessment, was surely overly harsh. Khvostov served for a time as co-chairman of the Moscow Psychological Society and worked on Russia's premier philosophy journal *Voprosy filosofii i psikhologii*.[59] He also helped create the law journal *Voprosy prava*. During his virtually de rigueur study period abroad, Khvostov attended Georg Simmel's lectures in Berlin and became familiar with Wilhelm Wundt's thought. Khvostov's philosophical works demonstrate a broad familiarity with not only the ideas of the mentioned Germans but also with other German and notably French scholars. During the last years of his life, Khvostov clearly came to appreciate Bergson and became deeply interested in sociology along with the work of Durkheim. Tragically, whatever the circumstances may have been, Khvostov committed suicide in 1920.[60]

In his principal philosophical writing, *The Theory of the Historical Process*, Khvostov proposed providing far more than a "mere" philosophy of history or even of historical methodology. He held that any such narrowly focused philosophy must also furnish a philosophically grounded worldview. Although a trained lawyer, albeit in effect a professor of the history of law, Khvostov, like Lappo-Danilevskij, believed that the historian must clarify and ground the fundamental presuppositions from which his further work would proceed. He, again like Lappo-Danilevskij, devoted considerable space to problems of the philosophy of the social sciences from a neo-Kantian perspective, which we will examine in due course. Let us now, however, look at how he too addressed that most curious issue for a historian of law – the problem of other minds.

Khvostov was clearly far less indebted to the St. Petersburg theoreticians (Vvedenskij, Lapshin) than was Lappo-Danilevskij, but he could not simply and indeed glibly dismiss the problem, as Shpet did, or even the specter of solipsism. Like Lappo-Danilevskij, Khvostov maintained that whether and how we know the minds of others is of fundamental significance for historical method-

---

reading to the experienced old 'goat' on walks." Belyj (1990b), 382. Lev Lopatin (1855–1920) was a professor of philosophy at Moscow University and who upheld a neo-Leibnizian standpoint. He displayed no interest in recent Western ideas. For more on his position vis-à-vis Kant, see Nemeth (2017), 242–246.

59 This title henceforth in this study will be abbreviated simply as *Voprosy filosofii*.
60 In his editorial notes to a collection of Sergej Hessen's letters, Vadim Sapov remarks, "Unable to bear the hardships and horrors of life under the Bolsheviks, Khvostov hanged himself." Hessen (1993), 532.

ology.⁶¹ Undoubtedly, we routinely encounter others in our everyday experience similar to ourselves. On what basis, though, do we ascribe to them mental activity in all essential respects similar to my own when that activity appears to be essentially concealed from my five senses? Clearly, the corporeal movements of others are much like my own, but their thoughts and emotions cannot be perceived. As Husserl remarked at almost the same time, if I could experience the other's psyche, as I can experience my own, it would cease to be the other's and would be my own.⁶² Of course, this does not establish that the other has a psyche, let alone one similar to mine, but if there is one in the other, it is not mine.

Unlike Lappo-Danilevskij, Khvostov had few qualms concerning the efficacy of the analogical argument as such. We start from the fact that we know through reflection on our corporeal movements that they correspond to certain mental states within us. Observing essentially the same movements in others we infer that they too have the same mental states as we have when we move in that particular way. "Therefore," Khostov inferred, "the other's psyche is not immediately perceived by us, but is mentally constructed through analogous deduction. ... We always judge about others by analogy with ourselves."⁶³ Khvostov readily admitted that the analogical argument does not prove anything. No philosophical argument can prove that something exists that is essentially inaccessible to the senses. That others have mental activity similar to my own amounts to a very probable hypothesis, but we can say nothing stronger than that. Nonetheless, virtually everyone accedes that others have mental activity and that it is similar to one's own. In this respect, then, the theory of knowledge, owing to its inability to prove this basic conviction, is deficient. We can hardly be surprised in this, knowing that Khvostov wrote under a strong influence from Kant. However, the former added that in practice we routinely acknowledge the other to be a being like ourselves and that we stand in a practical, or moral, relation to other people. Khvostov held that this relation implies that we have a practical, or moral, knowledge of a world transcendent to our own consciousness, and that this is what Kant had in mind when he wrote of the primacy of practical over theoretical reason. We routinely acknowledge in practice the other and utilize our moral knowledge. It vouches for the reality of the other as a real moral being like myself, and through it we dispel the specter of solipsism. Were it not

---

61 Khvostov (1914), 70.
62 Husserl (2006), 5.
63 Khvostov (1914), 66.

for the complete assurance provided by moral knowledge, there could be no talk of duties toward others or of their duties toward us.

With his increasing interest in sociology during the frenzied years of World War I and revolution, Khvostov turned, as mentioned, to Bergson, whose thought, in any case, was attracting considerable attention within Russia. Bergson represented a break from the narrowly focused concern with epistemology emblematic of German neo-Kantianism and with Vvedenskij's own version of it. Khvostov along with a number of others had already noticed Bergson some years earlier.[64] Now, however, he particularly looked at Bergson's emphasis on life and the presence within it of an élan vital. In a lengthy essay from 1917, Khvostov asserted that life and the "spiritual element" – Khvostov's rendering of Bergson's key expression – form the basis for the affirmation of mental activity similar to my own in others. Was Khvostov's attraction to French vitalistic philosophy a retreat from neo-Kantianism?

Khvostov wrote that, of course, we have no need of a mediate inference to conclude that the "I," i.e., the content of consciousness and the will, exists, or is real. The "I" is given immediately. The "not-I" – note the Fichtean language – is all that appears in consciousness as standing apart from consciousness and the will. But that appearance contains hints (*nameki*) of something existing independently of the cognizing subject.[65] Khvostov did not clarify his statement or what these "hints" are. Nevertheless, the legitimacy of the hints begins to be associated with our conviction in the existence of an independent objectivity in general. He again repeated his earlier claim that objectivity, a world of things in themselves as the logical foundation of what in our consciousness appears as existing independently of our will, is only a hypothesis. Khvostov went on to reiterate the analogical argument in virtually the same terms as years earlier, leading us to conclude that the "hints" he mentioned are merely the inferences from the analogical argument. In short, then Khvostov has not surrendered his neo-Kantianism in confronting Bergson. The French philosopher's variety of vitalism is seen as supporting rather than overturning neo-Kantianism. The sphere of practical relations has certainly expanded now in light of Bergson to include not just Kantian moral duties, but also such feelings as sympathy and antipathy, which connect us to others and thereby support the hypothesis of other minds.

---

64 Even in the years prior to World War I, Bergson's works received several translations. *Matière et mémoire* appeared in Russian in 1911; *L'Évolution créatrice* appeared in 1909 and again in 1914. *Essai sur les données immédiates de la conscience* appeared in 1910 in a translation by S. Hessen. For a general overview of Bergson in Imperial Russia, see Nethercott (1995) and Tremblay (2017b).

65 Khvostov (1917), 75.

Khvostov wrote, "When we act, not for a moment do we doubt the presence in others of their own consciousness quite analogous to our own, although they are quite transcendent to us. If we did not have this firm belief in the presence of other people with mental activity, our entire life would lose all of its moral sense."[66] Our moral values provide our lives with meaning. Thus, without the former, our lives would be empty. But there is no fear in this regard: we have an unshakable practical certainty in the existence of other minds. We simply take it as a fact, and with that fact concomitantly comes an acceptance of the reality of the transcendent world in general.

We must ask, though, what is this reality? How does Khvostov conceive the transcendent world given the acceptance of essentially unknowable other minds? He explicitly tells us that our practical reason dispels any doubt we might entertain in the reality of the unknowable "things in themselves."[67] These things in themselves are not Kantian things in themselves, for the latter are atemporal. No, Khvostov merely used Kantian expressions infelicitously, for to him other minds are empirically real and unknowable but in time.

In his last philosophical publication entitled "The Social Connection," Khvostov argued for the reality of a collective or social consciousness and in doing so against the conception that consciousness had to have a bearer or owner. For Khvostov, a social consciousness is the aggregate of mental states developed in the interaction between people, and that this social consciousness then can act on the minds of the individual participants of society.[68] Since we can meaningfully speak of a social consciousness in this manner, i.e., as acting on other human individual minds, it must be real and was not either a mere metaphysical construction or a scientific abstraction. A common objection to such talk is that whereas the individual consciousness is the property of a human individual, there is no comparable bearer or owner of social consciousness. Therefore, according to this objection, there is no "real" social consciousness. Khvostov replied in a manner not dissimilar to that offered by Shpet some two years earlier but with which Khvostov demonstrated no direct familiarity despite its bearing on the topic at hand.[69] He wrote that not only does the social conscious-

---

66 Khvostov (1917), 78.
67 Khvostov (1917), 74.
68 Khvostov (1918), 55. His definition bears a striking similarity to one offered by Durkheim himself. See, e.g., Durkheim (1997), 61.
69 See Shpet's essay "Consciousness and Its Owner" in Shpet (2019), 158–205. The irony in Khvostov's omission of Shpet lies in that both rejected, albeit from different angles, an owner of consciousness. In this, Khvostov could have seen Shpet as an ally. However, whereas Shpet

ness need no owner but the individual consciousness also has no distinct owner. Khvostov elaborated, writing that "the individual consciousness is a connected process of spiritual experiences that is its own bearer. There is no need to double the process and put some special spiritual substance under it, regardless of how subtly and dynamically that is understood."[70] We find a recognition of this unity of the contents of the individual consciousness expressed in the form of self-consciousness. But self-consciousness is just one function of the normal human consciousness and not some spiritual substance. Khvostov averred that social consciousness differs from the individual consciousness in that the former "usually" (*obyknovenno*) has no self-consciousness, no awareness of itself, or at least it is expressed only in special forms, on which Khvostov did not elaborate. Nonetheless, social consciousness does exist, as we can recognize through its effects on individual consciousnesses. Unfortunately, Khvostov failed to discuss how this owner-less conception of social consciousness was compatible with his other positions. Can we not say that social consciousness is a possession of all members of society, that the Kantian unity of consciousness is a logical synthesis by all concerned individuals?

Khvostov's position within Russian philosophy has largely gone unrecognized, although we can and will see that he turned his attention to many of the central issues being discussed in the neo-Kantianism of his time. With his concern for practical matters, he failed to analyze those issues as deeply as we would expect of a professional philosopher. Nonetheless, his perceptiveness and knowledge of the central issues and current literature place him far ahead of a number of professional philosophers in his own country. Here in this chapter, we have concentrated on the problem of other minds understood as the recognition that other individuals have mental activity essentially identical in its operation with my own. Introduced and upheld by Vvedenskij, we can see it as forming a key concern of one distinctive version of neo-Kantianism, a Russian version, even though it was discussed outside Russia as well, at different times, and with no connection to neo-Kantianism. We shall see that there were many additional representatives of neo-Kantianism in Imperial Russia who did not concern themselves with this problem. These other figures, to whom we shall turn in later chapters, largely represented forms of neo-Kantianism found in Germany. Nonetheless, we can see both Lappo-Danilevskij and Khvostov as having one foot in the door of a distinctive Russian neo-Kantianism and one

---

combatted Kantianism and neo-Kantianism, Khvostov broadly allied himself with Kant and *a* neo-Kantianism.

[70] Khvostov (1918), 55.

foot in another neo-Kantianism, viz., that of the Baden School. Khvostov actually had, as it were, still another foot in the neo-Kantianism represented by Georg Simmel and still another later in life in the French neo-Kantianism presented by Durkheim. Before turning to the influence of the Baden philosophers, let us look at the closest representative of neo-Kantianism – if he was that – in the philosophy department at Moscow University.

Chapter 3
The Psychologist as a Transcendental Realist

In the last years of the nineteenth century, a central and heated debate raged whether psychology was a natural science or a branch of philosophy. Some pondered whether the introduction of experimentation in studies of the mind would impact our understanding of philosophical issues. Some thought it might even resolve many traditionally fundamental philosophical issues. Russian academics faced these concerns and prospects no less than their colleagues in western Europe and America. In Imperial Russia, the controversy took on a political and theological coloring, making an impartial adjudication of the matter difficult, if not impossible. If the human mind could be studied by natural means, would that not strike a blow against the Biblical teaching that the human being was created in the image and likeness of God? Few, perhaps even only one, torn between the two worlds of experimental psychology as science and philosophy as independent reflection could not unambiguously choose and alternated their allegiance.

3.1 The Preparatory Years in Kiev

Georgij I. Chelpanov (1862–1936) was born in Mariupol in what is now Ukraine. After completing his secondary education in 1883 at a local school and allegedly already attracted to studying psychology, he entered the university in Odessa, where Nikolaj Ja. Grot (1852–1899) had only recently assumed a professorship.[1] Chelpanov graduated in 1887 and was recommended to pursue further studies at Moscow University. In a few years, he was teaching as a privat-docent there, but in 1892 he accepted a position at Kiev University. Already by this time, Chelpanov was publishing at a furious pace, as we shall see. In the mid-1890s, he went to Germany, where he was able to attend lectures by Stumpf and Wundt. He defended a *magister*'s thesis in 1897, followed by a doctoral dissertation in 1904. Most of his writings over many years reflected his deep pedagogical concerns rather than an attempt to present and to argue for his own philosophical position. This is particularly the case with his first publications. In fact, it is surprising how little

---

[1] Chelpanov wrote years later that upon finishing his secondary education he had to decide which university to attend. With a firm desire to pursue psychology, Grot's presence in Odessa was decisive: "All of my sympathies were on the side of N. Ja. Grot." Chelpanov (1911), 188.

he clearly revealed of his own stance – assuming he had one – in so many of his own publications.² This reticence, whether deliberate or not, in the secondary literature surely lies behind the failure to reach a consensus as to Chelpanov's relation to Kant and, more specifically, to neo-Kantianism, however that may be understood.

We cannot possibly hope to survey here all of Chelpanov's extensive list of works, many of which were concerned with experimental psychology and as such have little bearing on the present topic. He also authored a number of purely expository pieces, explaining the works of others and the latest advancements in scholarship abroad. In this chapter, we shall proceed chronologically to the extent possible. In this regard, in what may well have been the first statement of his relationship to Kantian philosophy, Chelpanov took up in 1893 a comparatively brief study of inner time consciousness. With respect to the age-old problem concerning the nature of time, Chelpanov thought it most useful to study how we perceive it, i.e., its psychological nature, how it appears to each of us, rather than objective time. It is strictly from that point of view that he intended to examine the question. Thus, Chelpanov's exposition bears a greater similarity to Husserl's work on the topic – minus any "transcendental reduction" – than it does to Kant's treatment in the "Transcendental Aesthetic." Chelpanov made a sharp distinction between what he termed "mathematical time" and "psychological time." The former is abstract. Just as we distinguish the mathematical representation of a line from the representation of a line drawn on paper, so too can we distinguish between the time measured by clocks, "mathematical time," and our inner awareness of the passage of time, "psychological time." Contra Kant in the "Aesthetic," we cannot think of either space or time as completely empty of representations.³ Chelpanov wrote, "we can as little intuit time devoid of content as we can intuit space without filling it with content."⁴ Although we may conclude from this that Chelpanov's opposition to Kant in this instance is a simple proof that the former was not a Kantian (or neo-Kantian), it did not appear to Chelpanov that he was contradicting Kant. For in Chel-

---

2 In a revealing statement of his fundamental outlook presented in a public lecture in 1902, Chelpanov said, "It would seem most natural in this case to present my own idealistic system. However, since this presentation of public lectures must be brief, it would have to bear a dogmatic character. ... I will survey the philosophical constructions that were proposed in the second half of the nineteenth century in order to show that philosophical thought unavoidably had to pass from positivism to metaphysical idealism." Chelpanov (1902c), 4.
3 "In regard to appearances in general one cannot remove time, though one can very well take the appearances away from time." Kant (1997), 178–179 (A31/B46).
4 Chelpanov (1893), 48.

panov's estimation, Kant had in mind not psychological time, i.e., inner time consciousness, but mathematical time, which is infinite, continuous, and uniform.[5] Time is subjective and an a priori property of the mind, but Kant's idea of time is not what Chelpanov thought he was investigating. Based, then, on what he wrote, Chelpanov in 1893 was committed to neither an acceptance nor a rejection of Kantianism.

Chelpanov's concern with our psychological conception of space, which obviously consumed a significant measure of his attention, dated back at least to 1884 when he gave a presentation on "The Doctrine of Space in Kant and in Contemporary Psycho-Physiology" while still a student. However, Chelpanov's major work on the topic was undoubtedly his *magister*'s thesis, the defense of which took place in late October 1896, the opponents at the defense at Moscow University being Grot and Lopatin.[6] In the "Preface" to this work, Chelpanov wrote that his fundamental concern was philosophical: the structure and origin of our representation of space. The choice of topic evokes Stumpf's own earlier work from 1873 on the psychological origin of our representation of space *Über den psychologischen Ursprung der Raumvorstellung*, with which Chelpanov clearly was familiar. The similarity was unlikely coincidental. In any case, Chelpanov added that he must ultimately address additional questions: Is the representation of space original or derivative? Is it simple or compound? He hoped to address the problem of the origin and the structure of space by means of an analysis of psychological facts and laws. However, the fundamental task remained a philosophical problem. The turn to psychology serves merely as a preliminary, though necessary, investigation in order to prepare for a confrontation with the philosophical problem.

It is not always easy to follow Chelpanov's presentation with its many distinctions that, in effect, obfuscate rather than help clarify whether he belonged at an early date to the neo-Kantian camp. It is even more difficult, as a consequence, simply to determine Chelpanov's viewpoint. Kant's theory of space, he acknowledged, is not in the rigorous sense a psychological theory, but an epistemological one. Since Chelpanov's investigation in this work was intentionally of a psychological nature, we should not be surprised if we were unable to find any textual evidence to support one position or another in the matter. Fortunately, however, this is not entirely the case. Chelpanov contrasted genetic or empiricist theories of spatial representation with nativist theories. Genetic theo-

---

5 This, of course, flies in the face of several arguments in Kant's metaphysical expositions of space and of time, which Chelpanov, however odd it may appear, did not acknowledge.
6 Many sources place the date of the defense in 1897, without providing a source. Rubcov (2012), 97 gives it as occurring on 25 October 1896, the year of the publication of the thesis itself.

ries hold that the operation of separate mental processes form the mental representation of space from non-spatial elements in sensation, whereas nativist theories are defined as those that view spatial extension as sensed just as are other properties of things, such as color.[7] Chelpanov contended that Kant's position belongs among the geneticists, since "he could not recognize the immediacy of our perception of the spatial quality as the nativists did."[8]

Chelpanov, indeed, saw Stumpf as criticizing Kant. Space, for Stumpf, is inseparable from sensation. Criticizing Kant's abstraction argument, a space without qualities, Stumpf wrote, "One cannot represent space without some quality, e. g., not without color through our sense of sight, without feelings of contact through our sense of touch. ... Anyone who actually attempts to carry out Kant's experiment by eliminating all qualities, particularly all colors, even black and gray, will be left not with space, but with nothingness."[9] We see here a coincidence with Chelpanov's earlier statement to the same effect. In fact, he stated that although he had objections to some parts of Stumpf's statement he could not help but admit that Stumpf's resolution of the problem had indubitable merits. More than others, Stumpf consistently developed the nativist principle.[10] Since Stumpf explicitly distinguished his position from Kant's, and Chelpanov, albeit with qualifications, endorsed Stumpf's, it is becoming unclear on what basis we could characterize Chelpanov as a neo-Kantian at least in the 1890s. Can someone who holds different views than that of Kant as expressed in the Transcendental Aesthetic but who is silent on the Kantian tenets be labeled meaningfully as a Kant or a neo-Kantian? Or is such a judgment, however tentative, premature?

---

7 Although he ascribed this position to others, Chelpanov omitted that Kant did not view the representation of space as comparable to our representation of color. Indeed, several years earlier Chelpanov wrote, "According to Kant, the intuition of space is a subjective form of intuition similar to qualities of sensation, to red, to sweetness, to cold." Chelpanov (1891), 43. Kant, in contrast, wrote, taste and colors "are not *a priori* representations, but are grounded on sensation." Kant (1997), 161 (A29/B44).
8 Chelpanov (1896), 9f. He returned again and again to this topic in subsequent years. For example, he wrote in 1903, "Many think that Kant was a nativist in his theory of space, but this is incorrect. Kant was not interested in whether the representation of space is acquired gradually or whether it forms an innate property." Chelpanov (1903), 174. This assertion is not quite correct. In his 1790 response to Eberhard, Kant wrote that the formal ground of the possibility of intuiting space is innate, but not the spatial representation itself. Kant (2002), 312 (Ak 8, 222). See also his 1770 *Dissertation*, Kant (1992), 400 (§15, Ak 2, 406).
9 Stumpf (1873), 19–20. Cf. Kant (1997), 156 (A20–21/B35).
10 Chelpanov (1896), 121.

Let us proceed further in Chelpanov's first publications. Following an earlier path carefully treaded by Stumpf, Chelpanov characterized the representation of space, considered quantitatively, as consisting of parts, i.e., that it can be conceptually broken down or reduced, although we do not consciously see it as such. More importantly, he again affirmed that spatial extension is as much an aspect of a sensation as is its intensity. As an object, extension has no meaning apart from what is sensed.[11] He wrote,

> I hold the position that extension is just as much a moment of sensation as is intensity. In general, by 'moment' I mean something that in a given appearance has no independent meaning. For example, the quality and the intensity of a sensation are moments, because neither of them can exist independently. They can be conceived separately from other moments only in abstracto.[12]

It is not clear in this, though, that Chelpanov departed significantly from Kant. The latter in the second argument of his Metaphysical Exposition asserted the psychological impossibility of imaging objects apart from space.[13] Chelpanov, however, was concerned, above all, with the perceiving or sensing of objects. He agreed with Kant that we can separate spatial extension in thought, i.e., abstractly, from the particular contents of sensation. Kant's concern, however, was different from Chelpanov's, and as a result, we can view Chelpanov not as differing with Kant, but as viewing the matter from a different perspective. Chelpanov did not disagree, though, with Kant's affirmation that the universality and necessity associated with our representation of space affirms its apriority as a condition of cognition.

Kant's third argument in the second, or B-edition, of the first *Critique*, holds that "one can only represent a single space, and if one speaks of many spaces, one understands by that only parts of one and the same unique space."[14] Chelpanov alluded to the same point. He asked what do people see when they open their eyes for the first time. The original object of perception is a general spatial field populated with separate elements that subdivide, as it were, the spatial field and that the objects "determine the interrelation of the separate parts of space, their relative positions, etc."[15] He added that it does not follow from

---

11 Stumpf had earlier made the same point in asking, "Cannot both contents of sensation – space and quality – be given in an immediate sensation in the same way?" Stumpf (1873), 15.
12 Chelpanov (1896), 150–151.
13 Kant (1997), 158 (A24/B38–39). It is admittedly not uncontested that the second argument asserts a *psychological* impossibility.
14 Kant (1997), 158 (A25/B39).
15 Chelpanov (1896), 168.

this – although it may seem to do so at first – that our awareness of space, i.e., our representation of it, is constructed from these separate elements. Chelpanov ever so guardedly opined, writing that "in all probability" the formation of our representation of space is not from parts to whole, but rather just the reverse. He, thereby, added his qualified support for Kant's stand that space is not discursive, but a pure intuition. Whereas Kant remained content without inquiring into the physiological basis for our intuitive representation of space, others later were not. None, though, could successfully explain space. Chelpanov held that neither Johann Herbart, Kant's successor in Königsberg, nor Wundt could demonstrate how the representation of extension could arise from non-spatial elements. The representation of extension cannot be produced from what does not possess extension any more than can the representation of color be assembled from color-less elements. In Chelpanov's eyes, in short, space is underived. "It is impossible to prove that space is composed of non-spatial elements."[16]

However, even should we concede the non-reductionist position, one may argue that the argument establishes only that our representation of space is an *a priori* condition of our representation of empirical objects. The argument alone, as Guyer has more recently pointed out, does not entail the transcendental ideality of space itself – what space is – without an additional argument that apriority entails such ideality. Unlike Kant, Chelpanov was usually careful not to confuse our (psychological) representation of space with ("objective") space itself, though as we just saw and will continue to see he did not always do so.[17]

Chelpanov's primary interest in the emerging field of experimental psychology during the 1890s did not totally exclude him from venturing into the philosophical arena. One such incursion was his contribution to the popular monthly

---

16 Chelpanov (1896), 378.
17 Chelpanov's book received a rather stinging review in the hands of N. Ja. Grot, who wrote that by counting Kant as a geneticist in the theory of space Chelpanov obviously revealed his untenable starting point and the inadequacy of his method as demonstrated by his subjugation of epistemology to psychology. Kant's standpoint was, contrary to Chelpanov's reading, purely epistemological, not psychological. Grot (1896), 653. However, Chelpanov refused to concede. As we saw, Chelpanov acknowledged that Kant's standpoint was essentially epistemological, and in a reply to Grot, he pointed this out. See Chelpanov (1897a), 278. Chelpanov also pointed out in this reply that he did not explicitly subordinate epistemology to psychology. Rather, "the philosophical problem of space must be examined jointly from both the psychological and the epistemological viewpoints. ... It seems to me that the elaboration of empirical psychology is, to a certain degree, just as independent of epistemology as it is of physics and chemistry." Chelpanov (1897a), 277.

journal *Mir Bozhij* [*God's World*][18] on the perennial issue of free will/determinism. Although he spent far more time discussing Mill than Kant, Chelpanov largely concurred with Kant's position, albeit without entering into the complexities of or even assenting to Kantian idealism. Chelpanov conceded that Kant's stand was not without contradictions and imprecisions. Nevertheless, its kernel contains an indubitable truth that can be understood with effort and recognized as correct, although with even greater effort. Here lies the "kernel" for regarding Chelpanov to be, at least in part, a neo-Kantian. Thus, let us look a little closer at his view.

Chelpanov in his popular exposition "On Freedom of the Will" discussed what for him are the three types of proof in support of free will: (1) the psychological proof, the thesis of which is that we "feel" the will acts without causes; (2) the metaphysical proof, the thesis of which is that the will can act originally on the body and, therefore, can violate universal causality; and (3) the moral proof, the thesis of which is that our sense of responsibility for our actions presupposes a free will. We indubitably have such a sense of responsibility, therefore we have a free will.[19] Chelpanov had no need to invoke the physicist's notion of universal causality. As a psychologist, he could see that every willful action has a definite motive, and as an experimentalist, he would be unable to engage in objective investigations of behavior if he denied the link between the will and human action. That we "can more or less precisely predict human actions" would be impossible if causality did not operate in the psychic sphere.[20] If contingency reigned there, if certain causes did not regularly and consistently evoke the same actions, psychology would be a useless enterprise. Chelpanov recognized that this attitude did not amount to a definitive refutation of the psychological and the metaphysical proofs. However, he stressed that he merely wished to defend free will by way of the third proof, i.e., the moral proof, and thus independently of first making a physicist's determination of whether the will is uniformly caused or not.

That we have a sense of responsibility for our actions testifies to our free will for all practical purposes. Our awareness that we are the true cause of our actions follows from that sense. Chelpanov held, though, that this is an ethical, not a metaphysical, argument for the existence within us of a free will. At the same time, responsibility also entails that our actions are subordinate to the

---

[18] The journal billed itself as a popular journal intended for self-education. That Chelpanov published often in it lends credence to the view that he saw himself primarily as an educator, rather than as a philosopher.
[19] Chelpanov (1897b), 18.
[20] Chelpanov (1897b), 46.

law of causality. If our actions were uncaused, punishment for incorrect behavior would be meaningless. On the other hand, that we are punished for incorrect behavior shows that society recognizes that individuals can and should control their actions. Chelpanov, in short, argued for a soft determinism quite consistent with Kant's teaching even though it is bereft of much of Kant's metaphysics and the architectonic of Kant's system.

Before turning to his philosophical writings intended for a professional audience, let us turn to one published at the beginning of the new century that Chelpanov intended to introduce Kant to a broad, educated public. Much that he said in this four-part piece is a simplistic but uncontroversial elaboration of the essential points Kant enunciated in his three *Critiques*. Nevertheless, Chelpanov made the rather odd claim – from a Kantian standpoint, that is – that for Kant there is objectively nothing that corresponds to the concepts of space and time, that "space and time, as we conceive them, have only a subjective existence."[21] Such a statement, of course, makes us question how he understood Kant's transcendental idealism. In Chelpanov's eyes, that is, Kant argued for space and time as having not an absolute, but merely a subjective, existence, that "we perceive things not as they are in reality."[22] Chelpanov did acknowledge that mathematics is an objective science in that its objects are the subjective forms of space and time. He concluded from this that Kant's theory is, consequently, a subjective idealism, a phenomenalism. Although today few would so characterize Kant's position, we must recognize that Chelpanov here was writing for a lay audience and that apart from whether he "correctly" depicted Kant's positions we cannot conclude whether he personally agreed or disagreed with the positions he attributed to Kant. For this reason, we must proceed cautiously – though that is difficult – concerning what we ascribe either to Chelpanov's understanding of Kant or to his own personal standpoint.

Chelpanov was largely silent on the complexities of the synthesizing process in Kant's first *Critique*. He characterized the twelve Kantian categories, i.e., those listed in §10 (A80/B108), as forms added by consciousness in a manner similar to the way in which space and time are added to our raw sensations. However simplistic this may sound, Chelpanov at least expressed his own opinion that this additive process is not psychological, but logical. He expressly noted that in writ-

---

[21] Chelpanov (1901a), 18. However we may chafe at some of Chelpanov's wording, we should not forget that Kant at times expressed himself in words that conjure a similar impression. For example, he wrote that space and time are representations, "which dwell in us as forms of our sensible intuition before any real object has even determined our inner sense through sensation...." Kant (1997), 428 (A373).
[22] Chelpanov (1901a), 22.

ing of the apriority of space and time Kant's interest was in the objectivity of mathematics.[23]

Chelpanov also in this popular survey of Kant's thought discussed his ethics and aesthetics. As we saw above, Chelpanov saw Kant proceeding from a recognition of the real necessity of the moral law to its presupposition in a free will. "Clearly in order to submit to the moral law freedom is necessary. Freedom is a necessary condition for the existence of the moral law itself."[24] The ultimate message, though, was religious. The important point contemporary philosophy was to learn from Kant is that faith occupies an equal position in our worldview with knowledge. The world of science and mathematics is not reality in itself, but only an appearance, whereas the world of religion is the supersensual reality itself.[25]

3.2 The *a priori* Understood Psychologically

At the start of the 20th century, Chelpanov published several philosophical articles aimed at a technically-sophisticated audience and that alluded to what would become the central topic of his later thesis. The goal of the first of these essays was to determine whether there are *a priori* cognitions or *a priori* elements of cognition. He alleged that the simplest approach to the issue is by way of considering the concept of number. Chelpanov sought to show that that concept is *a priori*, and not obtained from sense experience, this contrary to Mill's version of empiricism.[26] We must admit that Chelpanov's exposition is admirably lucid, even though his arguments hardly break new ground. We need not even summarize them here; they are only too familiar. Few in either his own day sought – or in our day seek – to uphold Mill's philosophy of arithmetic. Of somewhat greater interest, in any case, is that Chelpanov, having dispensed with Mill's account, persisted, stating that "we must seek the origin of the idea of number not in changes in the physical facts, but in some feature of consciousness itself."[27]

For Chelpanov, the concept of number can be traced to an abstraction from pauses in conscious processes. We obtain the concept through an implicit reflection on our inner observation of those processes, and not as the empiricists

---

23 Chelpanov (1901a), 183.
24 Chelpanov (1901a), 66.
25 Chelpanov (1901a), 70–71.
26 Chelpanov (1901c), 537.
27 Chelpanov (1901c), 548.

claim through observing objects and their qualities. In apprehending two temporally separated events or items, we apprehend not only those events but also the pauses between the two apprehensions. Chelpanov held, in short, that we attend to these pauses: "I notice that in order to perceive two peals of a bell, I have to do something different than when I perceive one peal. ... This difference in the number of pauses of consciousness becomes the source of the idea of number."[28] I can perceive these pauses in consciousness regardless of whether the intentional object is empirical or imaginary, really present or only in my memory. Thus our concept of number has its origin in an act of reflection. Arithmetical laws have their seat not in the empirical world, but express only laws of consciousness. Not being, strictly speaking, of empirical origin, arithmetic and the concept of number are *a priori* and cannot be either proven or refuted by external sense experience. Since the concept of number is not derived from something material, it is formal. We also see that conscious processes play an active role in the formation of arithmetic and that these pauses in conscious acts are inherently temporal cessations of mental acts. In this way, an inner awareness of time is involved in the formation of arithmetical concepts and laws. Lastly, if we consider concepts that are logical conditions of experience but that are not derived from experience to be *a priori*, then the concept of number is, by definition, an *a priori* concept. Its content is not obtained from experience, but at the same time, we ascribe numerical properties to experiential objects.[29]

In summary, Chelpanov's view of the concept of number shares certain distinct similarities to Kant's. Both held it to be *a priori*, formal, based in inner time-consciousness, and a condition of experience. For both, counting is a temporal process.[30] However, Kant – thankfully – failed to emphasize the apparent psychological process involved in the emergence of number that is so prominent

---

[28] Chelpanov (1901c), 549. One could possibly reply to Chelpanov noting that in his depiction of the origin of the concept of number the concept is already presupposed. How could someone "count" the pauses unless that person already has the concept of a number? Additionally, we must leave aside here a Kantian technical objection to Chelpanov's depiction of the origin of number. Namely, by counting the pauses in reflective consciousness Chelpanov would have us come to an image of a specific number, but the thought of a number in general "is more the representation of a method for representing a multitude ... than the image itself." Kant (1997), 273 (A140/B179).

[29] The reader, hopefully, will realize this author simply wishes to determine Chelpanov's position vis-à-vis Kant, not to defend Chelpanov's theory of the origin of numbers by means of purely subjective mental operations.

[30] Kant, in the *Prolegomena*, wrote, "Even arithmetic forms its concepts of numbers through successive addition of units of time, but above all pure mechanics can form its concepts of motion only by means of the representation of time." Kant (2002), 79 (Ak 4, 283).

in Chelpanov; there is also no mention of a synthesizing process. On this basis, we can consider Chelpanov's philosophy of arithmetic to be in the spirit of Kant, i.e., neo-Kantian, were it not for his obvious omission of invoking Kant in any manner therein.

In the second of his two-part study, "On the *a priori* Elements of Cognition," subtitled "The Concepts of Time, Causality, and Space," Chelpanov expressed his own position far more audaciously than previously. He claimed that the concept of time, along with that of number, is *a priori*. In other words, there is nothing objective in the external world that corresponds to time in the way that we can say of a particular color that there is something, such as a chemical, in an externally existing object that absorbs all wavelengths of light except that which corresponds to that color. "There is no doubt that time really exists in our consciousness, but it is also indubitable that it does not exist in nature, objectively. ... In nature, there objectively are events or phenomena, the perception of which is merely the occasion for the emergence of the concept of time."[31] Thus, time is a product of a distinctive activity of consciousness itself. Yet a mere alteration of events or phenomena would be insufficient for us to have a concept of time. Chelpanov held that the changes must recur in a regular order, for we must be able to establish a definite connection between the representations of the events. "Therefore, the most characteristic feature of temporal relations is the representation of succession, i.e., that we conceive events as following one after another. But the condition just given must be fulfilled in order for there to be such a representation."[32] Our representation of temporal succession, while not empirically given, would not arise in consciousness were it not for what Hume called "impressions." Objectively, there are only sounds, color, etc., but no succession. We ourselves create it by connecting individual impressions. The job of psychology is to furnish more information about this connecting process, which, in any case, is subjective, i.e., an activity of consciousness. The concept of time is a product of reflection.[33] The singularity of time, of which Kant wrote, led him to view time as an intuition.[34] In this, Chelpanov believed Kant had inferred incorrectly. We conceive time as singular owing to the fact that all of our representations refer to a single consciousness. The concept of time is *a priori* in that, unlike *a posteriori* concepts, it is formed on the basis of reflection

---

31 Chelpanov (1901d), 699.
32 Chelpanov (1901d), 703.
33 Of course, if psychology can provide such information, then the object of that study cannot be transcendental.
34 Kant (1997), 162 (A31–32/B47). Chelpanov did not supply a reference to the first *Critique*.

on inner, conscious states alone, whereas empirical concepts are formed on the basis of reflection on externally given objects.[35] The various other properties of time, such as its homogeneity, continuity, and infinitude, are explicable from the properties of consciousness itself. Chelpanov believed that although he had shown time to be subjective, and therefore by his explicit definition psychologically *a priori*, it is also epistemologically *a priori* in the sense that time is a logical condition of cognition. Cognition, whether of the external or the internal world, is possible only as having a temporal form. In other words, time is a necessary condition for all cognition.[36]

The concept of causality, likewise for Chelpanov, is *a priori*. It certainly is not of empirical origin, since otherwise it would not be necessary and universal. Not being empirically derived, it cannot be proven empirically to be a law. The law of causality must be assumed in order for cognition to be possible. Remarkably, Chelpanov had little to say about its genesis in human consciousness. The same cannot be said concerning the concept of space. The manifold of sensation is not given in any particular order; it is simply given. As with temporality, spatial order is created and introduced into the manifold by consciousness itself.

> Spatiality is created with the help of the synthetic activity of the mind, thanks to which there is a binding and an arranging of elements given from without into one definite whole. Therefore, we can say that spatial order or, what amounts to the same thing, extension in general is not affected by the external world, but is a product of the activity of our consciousness. In this sense, it represents an *a priori* element of our cognition.[37]

Chelpanov explicitly differed with Kant when the latter wrote that space is represented as an infinitely given magnitude.[38] For the former, "it would hardly be correct simply to say that we represent space as infinite."[39] He added that we certainly cannot represent a particular extension without representing it as part of some other general and greater spatial container. In this sense, the representation of an infinite space is for us something potential. That is to say, we cannot

---

35 Chelpanov, in a 1903 essay "Psychology and Theory of Cognition," wrote "Kant did not recognize space and time to be innate representations. Both are acquired by means of reflection on inner processes." Chelpanov (1903), 116.
36 Among the many lacunae in Chelpanov's position here was his collapsing of his own distinction between the psychological and the epistemological *a priori*. He did not show that time is a necessary condition of the cognition of externality. Moreover, if time is obtained subjectively, how can it be a condition for subjective events? Criticisms aside, Chelpanov, at least, believed he was upholding Kant's conception of time, albeit interpreted in a non-transcendental vein.
37 Chelpanov (1901d), 729–730.
38 Kant (1997), 159 (A25/B41).
39 Chelpanov (1901d), 734.

represent some finite extension without implicitly recognizing as well the possibility of connecting additional extensions without end. Such is the sole correct sense of Kant's statement. Just as we can mentally extend a numerical series at will, so too can we do the same with spatial extensions. Our representation of space with the attributes Kant recognized is not given to us from externality. It is a subjective construction, and just as with temporality, the formation of our representation of space is a matter for psychological investigation.

We, surely, can now fault Chelpanov for failing to distinguish our subjective conceptions of space and time from the space measured by meter sticks and the time measured by clocks. In particular, Chelpanov egregiously erred in concluding without argument that space and time were *purely* subjective with nothing corresponding to them in objectivity. Apart from possible objections, however, we could ask of Chelpanov how, if geometry is based on a psychologically generated space, there can be more than one geometry. It is to this topic that Chelpanov turned early in the following year.

Much of Chelpanov's long four-part essay, "Neo-Geometry and Its Significance for the Theory of Cognition," is devoted to explaining what non-Euclidean geometry – "neo-geometry" in the parlance of Chelpanov's day – is and its philosophical importance as seen by others including, in particular, Helmholtz. Only at the end of his third installment and in the fourth did Chelpanov proceed to some of his own reflections on the topic. As in his previous essay, he held that the *a priori* is:

(1) that which is not obtained directly by the senses and
(2) a necessary presupposition of experience, understood as perception or knowledge.[40]

That which meets the first condition is psychologically *a priori*, whereas that which meets the second condition is epistemologically *a priori*. Chelpanov maintained there are some elements in our cognition that are *a priori* in both senses. But what is the status of geometric axioms?

Geometric axioms are assertions about the properties of spatial forms that, considered to be obviously true, are thought not to need formal proof and cannot be proven. In Kant's opinion, the apodicticity of geometric axioms is connected to their involvement with spatiality as a subjective form, i.e., as a form of consciousness.[41] If our representation of space were different – so the charge goes

---

[40] Chelpanov in another psychology text wrote, "concepts that contain elements not obtained from sense experience are called *a priori*." Chelpanov (1913), 133.
[41] Chelpanov (1902b), 1382.

– our geometry would be different. Euclidean geometry would, then, not describe the external world. If we would confine ourselves to the surface of a sphere, we would find that there are an infinite number of lines of the same length through the two points located at each end of the diameter of that sphere. The sum of the angles of a triangle drawn on a spherical surface is greater than 180°, etc. From this, someone might conclude that Euclidean geometry is based on one particular possible space and that the various non-Euclidean geometries are based on other "forms" of space. That we view the world in Euclidean terms shows that we make our choice of axioms based on experience, that geometry is empirical, and that its validity is, therefore, no greater than any physical science. Helmholtz, in an often cited paper from 1876, wrote, "But if we can imagine such spaces of other sorts, it cannot be maintained that the axioms of geometry are necessary consequences of an *a priori* transcendental form of intuition, as Kant thought."[42] In other words, Helmholtz held that neither geometry nor space is *a priori* in Kant's sense of that term.

Chelpanov, confronted with advances in mathematics, sought to disentangle our intuition of space from Euclidean geometry. Our representation of space is a form, and as such need not be conceived as having the specific features that would bind it to a particular set of geometric relations. Whereas in our (Euclidean) space a straight line is the shortest distance between two points, in another space the analogous spatial relations, e.g., on the surface of a pseudosphere, would possess other characteristics.[43] Even three-dimensionality is not necessarily tied to our representation of space. "We can conceive [*myslit'*] a space that would retain its basic properties, e.g., externality, homogeneity, continuity, and that would yet not be three-dimensional."[44] However, if we cannot say with certainty that space has a particular character, we cannot say that our geometric axioms – or, for that matter, any other set – have absolute validity.

Chelpanov contended that some proponents of empiricism (e.g., Helmholtz) explain the correspondence between the Euclidean axioms and actual spatial relations by the alleged fact that the former arises under the direct affection of actual space. He replied that all geometrical forms are produced in consciousness and are idealizations. There are no truly straight lines in reality, only in our thoughts. If our geometrical formations were dependent on empirically given

---

42 Helmholtz (1876), 314.
43 Chelpanov (1902b), 1388. The obvious objection here is that there is no dispute as to the logical admissibility of other geometric relations. The issue, instead, concerns our representation of space. Do we intuit space as Euclidean? Can we mentally picture a straight line as *not* being the shortest distance between two points?
44 Chelpanov (1902b), 1388.

space, they would have an empirical character. That a geometry can be constructed with absolute consistency demonstrates its apriority, that its mental construction is not indebted to experience and to a particular space with certain features. The construction of internally consistent geometries shows that the concept of space is epistemologically *a priori*, a condition of cognition and of perception in general. Although we can acquire the content of geometric axioms from empirical observations of spatial relations, we need not do so. What is common to all geometries is that they all invoke an *a priori* conception of space. If we should seek more information about this conception, we must determine the characteristics of space that remain invariable throughout all geometries. While space is essentially dimensional, as already mentioned, it need not be three-dimensional. The other essential properties of space – exteriority, homogeneity, and continuity – are ideal and necessary, since space is a form of thought. "If we were to strip space of the property of exteriority, it would cease to be space. Whereas if we strip space, for example, of its three-dimensionality or its zero curvature, space does not cease to be space but would only be another type of space."[45] Chelpanov concluded his study, saying that although the development of intrinsically consistent non-Euclidean geometries is commonly thought as proving the non-empirical character of geometrical axioms, he believed they show that space must be considered *a priori* in the proper sense. What this proper sense is, though, is not entirely clear. We can be sure, though, that he saw his exercise as affirming a Kantian conception of space.

Chelpanov also published a two-part contribution to the popular journal *Mir Bozhij* during the same time period as his long essay on non-Euclidean geometry. This piece, "The Evolutionary and the Critical Method in the Theory of Cognition," set out to contrast the naturalistic approach of Herbert Spencer to Kant's Critical method. Although it sheds little, if any, additional light on Chelpanov's philosophical stance, it does demonstrate his growing realization of a distinction between the naturalistically considered psychological *a priori* and the transcendental *a priori*. Now, more forcefully than previously, he found Kant's theory of cognition as having nothing to do with the origin of the *a priori* forms of scientific consciousness. Kant's concern is with epistemology, unlike Spencer whose interest was with the genesis and development of some concept. Epistemology accepts a scientific concept, such as causality, and asks whether and to what extent it can justifiably be utilized. Psychology, which was Spencer's concern, on the other hand, has no interest in justification. No doubt Kant rec-

---

[45] Chelpanov (1902b), 1408.

ognized that the *a priori* forms and categories of cognition had a history, but this was of no interest to him.⁴⁶

It is quite regrettable that Chelpanov, for the most part, did not provide a basis by which we could state with confidence his own attitude toward what he outlined in such simple and clear language. He stated that three elements are necessary for cognition: (1) sensations simpliciter; (2) the forms of space and time; and (3) categories, which are synthesizing forms.⁴⁷ While we can clearly see Kant behind this assertion, was this Chelpanov's view as well? No unequivocal answer is given. Yet, he gave every indication further on in his discussion that he accepted the Kantian schema just mentioned. Some who accept a naturalistic interpretation of Kant may argue that owing to changes in our physiological conditions, the exact nature of our *a priori* forms may change in direct response. Chelpanov found such a view to be wrong in that it presupposes a direct and naturalistic connection between physiological processes and our epistemological forms of consciousness. In the absence of evidence in support of this presupposition, "we cannot draw any conclusions concerning the variability of the forms of consciousness."⁴⁸ Whereas the manner of perceiving space and time may vary, that we conceive space and time as forms does not change.

3.3 A Defense of Transcendental Realism

Chelpanov published and defended as a doctoral dissertation in 1904 his long-promised work on our perception of space considered philosophically. In it, he reaffirmed yet again that our conception of space is *a priori*, ideal, and a subjective form of cognition. He devoted considerable attention in it to Kant's theory and in fact repeated much that he had stated in his two-part article from the previous year. Indeed, a large portion of the dissertation itself was taken over from the extensive articles he had previously published and which we have already examined.⁴⁹ For this reason, we need not belabor our probe into it. Let us look, though, at some of Chelpanov's discussions that are new or at least amplify

---

46 Chelpanov (1902a), 98.
47 Chelpanov (1902a), 101.
48 Chelpanov (1902a), 113. In this matter, he explicitly invoked the neo-Kantian Friedrich Paulsen in support of his position.
49 The ideas presented in the dissertation's Chapter 3 stem from Chelpanov (1901a); Chapter 4 is largely a reprint of Chelpanov (1903); the ideas in Chapter 5 stem from Chelpanov (1902a); most of Chapter 6 is a reprint with insignificant changes from Chelpanov (1901c) and Chelpanov (1901d); and, finally, Chapter 7 is largely a reprint of Chelpanov (1902b).

what we have already seen. For example, he now, in his dissertation, explicitly contrasted many of the positions he had supported in the past to those offered by others including contemporaries. One instance of this concerns Ivan Lapshin, who, in the "Preface" to his Russian translation of William James's *Psychology Briefer Course*, wrote that the psychological theory of the underived nature of the *a priori* is indistinguishable from the epistemological view of the same. Lapshin contended that those who uphold Kant's epistemological view logically must also hold the psychological view.[50] As we would expect from what we have seen, Chelpanov disagreed. For him, the aim of psychology in this matter is to show that extension cannot be derived from unextended elements, whereas the epistemological enterprise is to show the impossibility of thinking of unextended objects.[51] The viewpoints of psychology and epistemology also differ with respect to the concept of necessity. Necessity in psychology is a statement concerning the constant conjunction of representations, whereas in epistemology – Kant's standpoint – necessity has an objective character. The objects of cognition are seen as bearers of some necessary character. In Kant's case, this necessity is introduced by cognition itself.[52]

We have seen that to a surprising degree Chelpanov agreed with much of Kant with respect to space and time. A question that remains and that has haunted Kant-scholarship since its first days is whether the acceptance of Kant's *arguments* concerning space and time necessarily entail his *conclusion* that space and time are *nothing* other than the forms of appearances, i.e., transcendental idealism.[53] Is it possible to accept Kant's arguments and yet hold that although our representations of space and time are necessarily subjective, there is still something objective, something independent of us, that corresponds to them? This is the topic of the final chapter in his dissertation, Chapter 8.

There are some, Chelpanov charged, who reject the existence of things independent of consciousness. One such idealist, to use Chelpanov's terminology, was Berkeley, who had to turn to God, Who interacts constantly with human individuals to provide us with representations. If we decline to proceed down such a road, we have no way to explain specific representations that arise in the human mind and often enough contrary to our wishes without conceding the existence of a transcendent or at least a transsubjective world.[54]

---

50 Lapshin (1896), 25.
51 Chelpanov (1904), 122 f.
52 Chelpanov (1904), 128; Chelpanov (1903), 178.
53 Kant (1997), 159 (A26/B42) and 163 (A33/B49).
54 Chelpanov (1904) 397. Such a stand is quite compatible with Kant's own form of transcendental idealism as against Berkeley's empirical idealism. In Kant's eyes, transcendental idealism

Chelpanov recognized as Kant's response to empirical idealism that things do exist independently of us, but these things provide only the material, or manifold, for sensations whereas time and space come from our cognitive faculty as forms. The cause of our sensations, then, is, in Chelpanov's understanding of Kant, the thing in itself, which exists independently of our representations. Since the forms of space, time, and causality are imposed by consciousness onto the material provided by things, these things themselves can have none of those forms. Thus, Chelpanov alleged that Kant and his followers err in recognizing "transcendent" things in themselves as the cause of sensations:

> The recognition of something transcendent, i.e., something existing independently of our consciousness, to which the forms of our mind cannot be applied but which nevertheless must exert an affection on our consciousness, seems quite incomprehensible. ... This is why Kantians, in order to retain Kant's point of view, i.e., in order to recognize transcendent 'things in themselves' as the cause of sensations and at the same time explain the origin of our cognition, introduce the concept of an 'empirical affection' in contrast to 'transcendent' affection.[55]

Thus, based on these words alone we see that Chelpanov set himself apart from the "Kantians." He did not see himself in 1904 as a Kantian transcendental *idealist*. This does not mean, however, that Chelpanov must be stricken from the ranks of neo-Kantianism. Chelpanov, next, launched into an exposition of the "double affection" theory, which proposes to eliminate the apparent contradiction concerning how a thing in itself can "cause" my empirical representations by having things in themselves causally affect the transcendental ego (or "I") and empirical objects causally affect the empirical subject. We need not pursue this issue further, for Chelpanov again shied away from controversy proclaiming he will not attempt to adjudicate the merits of the theory.[56]

Chelpanov found that problems regarding the existence of the thing in itself in no way are resolved with Kant's "Refutation of Idealism," as found in the second edition of the first *Critique*.[57] Whereas for Chelpanov the true correlate of my

---

is an empirical realism, whereas Berkeley's empirical idealism is a transcendental realism. For more on this, see Kant (1997), 426 (A369).

55 Chelpanov (1904), 399–400. As his very terminology indicates, Chelpanov confused the transcendent with the transcendental. In doing so, he fell victim to an illusion utilizing the category of causality beyond its limited immanent use. See Kant (1997), 385 (A295–296/B352).

56 Chelpanov did not shy away, though, from writing that in any case Kant himself did not hold to any conception of an epistemic noumenal affection, whether it be encased in a theory of double affection or not. Kant wrote only of "causality through freedom." Chelpanov (1904), 402.

57 Kant (1997), 326–329 (B274–279).

representations and intuitions should be actual things, things in themselves,[58] Kant, in the "Refutation" confused a proof for the existence of things in themselves with one for the existence of a thing in space. Although he refuted the "problematic idealism" of Descartes, he should have undertaken to refute idealism in general. If his interest were sincere, he would have shown that there are things in themselves independent of our representations of them, which are responsible for things in space. Kant's "error in this case is obvious."[59]

No doubt some neo-Kantians were – and are – aware of Kant's problems and of his failure to prove the existence of independent things in themselves. In order to uphold Kant's overall position while avoiding its problems, they re-interpret the thing in itself not as a distinct entity existing independently of cognition (under the "two-worlds interpretation"), but as an asymptotic, limiting concept, the result of lawful sensible connections produced by consciousness. In their eyes, the synthetic activity of consciousness itself creates its own objects.[60] Chelpanov commented that on this basis these neo-Kantians find they have no need to admit the existence of some transsubjective world. Chelpanov found such a position insupportable both philosophically as well as based on Kant's own texts. He affirmed, on the contrary, that we must recognize the existence of a transcendent reality that provides the material of sensation. We must also account for why in a particular instance causality is invoked and not some other category or none at all. That causality is invoked on some occasions and not on others cannot be explained relying solely on subjective factors. If the connections of sensations were due exclusively to subjective factors, there would be no compulsory connections and certainly no consistent agreement between people. "If we did not recognize the existence of a transsubjective reality with its specific law-governed regularity, it would be impossible to understand why we must apply in this or that case this or that specific category."[61] Chelpanov saw this as a proof, a transcendental proof, of the existence of a transsubjective world. However, this by no means entailed the elimination of a transcendental apperception connecting representations.[62] Our cognition is the product of an interac-

---

[58] Chelpanov provided no textual support in Kant for this claim. Instead, he referred the reader to secondary sources, such as Zeller, Volkert, and Riehl.
[59] Chelpanov (1904), 403.
[60] Chelpanov in this discussion mentioned Cohen (1877) and Cohen (1885) in a footnote, but surprisingly not Cohen (1902).
[61] Chelpanov (1904), 407–408.
[62] Chelpanov's imprecision here and elsewhere should not blind us to the essential Kantian point he is making here.

tion between the regularity of things and the regularity that consciousness introduces into things. The thing in itself is a correlate of the unity of apperception. Chelpanov likened the application of categories to the perception of colors, which is also not entirely subjective. I see the rose as red owing, for one thing, to some complex set of chemicals in the object, etc., but also owing to something in my visual apparatus that regularly responds to the reception of the same wavelength of light from the rose. One or the other could conceivably not be regular, in which case I would not see that same rose as consistently red.[63]

Chelpanov read Kant along these lines and deemed him to be correct. The former wrote, "For me, it is indubitable that Kant must have recognized *the existence of an objective foundation of space.*"[64] Therefore, space is the product of the interaction of two factors. On the one hand, we have consciousness with certain properties, and on the other hand, we have an objective foundation in the intentional object with definite properties. Conceivably, this objective foundation in the object is not itself space, but only an analog of space comparable to what is responsible for us seeing something as a particular color. Chelpanov added that we must remain ultimately agnostic in this. All we can say is that our representation of space corresponds – much like a symbol – to something existing in exteriority. He denominated this position as "transcendental realism" as against "transcendental idealism," which takes space to be something merely subjective. In transcendental idealism, there is no accounting for the regular order of a particular given spatial object. There is no explanation why you see my particular books arranged in the same order that I do. "Therefore, we must recognize the existence of an objective correlate, of a certain objective foundation of space, in relation to which our concept of space is, in a certain sense, a reflection."[65]

Chelpanov's own position was transcendental realism, or to use the expression he evidently preferred "ideal realism," owing to its allowance for ideal elements within a general realist ontology. But he also believed it was fully consistent with and can be found in Kant's texts. Although Kant wrote that space and time are subjective forms, Chelpanov believed this must not be understood as implying that they are found in consciousness in advance [*zaranee*] of the sensible manifold. He absolutely rejected the position that space is exclusively sub-

---

[63] Although this particular example of the color red is taken from Kant, Chelpanov does mention sound and color. He was especially indebted in this matter to Riehl, whose discussion is quite similar. See Riehl (1879), 50–65.
[64] Chelpanov (1904), 411.
[65] Chelpanov (1904), 415.

jective.⁶⁶ If he were to do so, it would mean transgressing the bounds of empirical reality. He did not concern himself with what transcendent reality is. He recognized its existence and therefore could not deny the objective foundation of space. Such are the positions concerning space that Chelpanov believed Kant espoused and that Chelpanov also held in 1904.

3.4 A Neo-Kantian *malgré lui*

Sergej Trubeckoj, a well-regarded professor of philosophy at Moscow University, suffered a fatal stroke while waiting for a meeting with the minister of education in St. Petersburg in September 1905. His sudden death created a vacancy at the University. The academic committee there in the following year invited Chelpanov to fill the opening, finding him to be the uniquely qualified candidate for the professorship in philosophy and psychology.⁶⁷ Chelpanov took up his duties in Moscow in 1907 and as customary gave an inaugural lecture in September "On the Relation of Psychology to Philosophy." Unfortunately for us, the talk covered little new ground. It largely summarized his view that each discipline needed the other, a heated topic at the time.

In the years subsequent to his appointment at Moscow University, Chelpanov devoted himself far more to psychology and in particular to the promotion of experimental psychology than to philosophy. His treatise in 1908 "On the Mind of Animals" examined to what extent animals can be held to have minds comparable to those of humans. Do they have mental concepts as we do? Do they have faculties of reason and understanding? Regrettably, Chelpanov failed to address these questions in a manner that would allow us to draw any additional conclusions that would comfortably allow us to characterize him as a neo-Kantian. Kant is not so much as mentioned in the published papers stemming from this line of inquiry. And also disappointingly, the issue of whether animals perceive space and/or time as we do received no attention.

In 1909, Chelpanov published a defense of experimental psychology vis-à-vis the more traditional reliance on introspection as a psychological technique. He defended both objective and subjective methods, finding that psychology, in

---

66 "Could Kant look on space as something exclusively subjective? It seems to me absolutely not. Kant thought that the existence of things in space is immediately obvious. This required no proof." Chelpanov (1904), 411 f.
67 Rubcov (2012), 98. The elation which Chelpanov naturally would have felt with the news of the appointment was tempered if not made bittersweet by the unexpected death of his wife that same year leaving him with three young children.

general, will always rest on self-observation even while relying on objective techniques. This piece hardly dealt with the issues confronting us here, but it remains interesting to see Chelpanov's acknowledgment of Husserl at this early date even though that name did not figure prominently. Evidently consumed with his new duties, his family, and a newly created experimental psychology laboratory, Chelpanov devoted little time to pure philosophy. Psychology, after all, was his chief interest. However, let us look at two works, both of which went through multiple editions during his lifetime.

Chelpanov's large, sustained attack on materialism, *Brain and Mind*, was based originally on public lectures presented in Kiev in 1898–1899. The first edition of the collected lectures appeared in 1900 and the last edition, the sixth, appeared in 1918. Its masterfully clear and thorough critique of materialism notwithstanding, much of its concern falls outside the bounds of our concern here. However, Chelpanov did turn in one of the book's chapters to space and in another to time. Inasmuch as he reiterated his stand without substantial modification over the various editions and in more than one publication, we can conclude that his statements represented his settled conviction. As in his dissertation, Chelpanov likened our representations of space and time to our sensation of sound and our representation of color. Indeed, they are completely analogous, and like them, space and time have a subjective character.[68] Just as other earthly creatures may conceive colors and hear sounds differently than we do, the very possibility of the construction of non-Euclidean geometries demonstrates that at least with regard to space, it may have different properties than we represent it as having in everyday life. That we cannot represent space in some manner different than we do is itself proof that our representation of it is connected in the strongest way possible with our psycho-physical organization.

> If our organization conditions our representation of space, it is *subjective* in the sense in which the sensation of a color or a sound is subjective. Consequently, we must not think that our space is absolute, an existence independent of our subject. ... In this sense, our space exists only for us, only for our subject. For another consciousness our space would perhaps not exist.[69]

And Chelpanov again reiterated that our representations of space and time, while subjective, certainly correspond to something in the objective world just as colors do in the perceived object. This "something," though, is quite different

---

68 Chelpanov (1906), 197; Chelpanov (1912), 176.
69 Chelpanov (1906), 208; Chelpanov (1912), 185.

from our mental picture of it.[70] Time, as we represent it, exists only in our consciousness. Objectively, i.e., outside the human consciousness, there is no time. If the process of transmission by which the objective world conveys to us our representation of time were to cease functioning, time would be eliminated. In such a fairytale world where all living beings otherwise cognizant of time are fast asleep, time would not exist.

As with *Brain and Mind*, Chelpanov reworked his basic text *Introduction to Philosophy* numerous times during his professorial career. Indeed, it went through seven editions – the last appearing in 1918 – in his lifetime and offered wide-ranging presentations in remarkably simple language. Chelpanov in this *Introduction* reiterated yet again his stand that our concepts of space and time do *not* correspond to an objective something such that our concepts would be mere copies or reflections of that something. Therefore, our concepts are products of the functioning of our consciousness. Consciousness creates those concepts. Although we conceive space to be infinite, it, being merely subjective, is not objectively so.[71] Chelpanov also added the "form" of causality to those of space and time. Possibly leading the reader astray, however, he began mentioning Kant's name in conjunction with the position that other forms, such as substance, are created in consciousness by its functioning. In this manner, whether what Chelpanov ascribed to Kant was also his own position is ambiguous. What he did clearly state is that there are no innate ideas as such. "There can only be innate capabilities to form these or those ideas."[72] Nonetheless, Chelpanov insisted that this is not to be understood psychologically. He was making no claim about temporal origin, since such a concern would presuppose temporality. Rather, epistemology accepts a concept as it is already found in science and seeks to establish its attributes with the aim of determining why and how it is applied in experience.

Chelpanov also reiterated his position that the forms of consciousness, while *a priori*, are symbolic of something in objectivity much like our representation of a color is symbolic of a certain wavelength of light. We see from this that there is

---

70 This and what Chelpanov further said regarding time can be discussed *ad nauseam* with respect to Kant. Briefly, Kant at one time held that colors are *not* objective qualities of intuited bodies. Kant (1997), 161 (A28). He then apparently reversed himself in the second edition of the first *Critique*, saying "the predicates of appearance can be attributed to the object in itself." Kant (1997) 190 (B69). There are many other facets to the issue Chelpanov raised. What certainly appears to be indubitable is that Kant objected to comparing the ideality of space with "things like colors, taste, etc." Kant (1997) 161 (A29/B45).
71 Chelpanov (1916), 34.
72 Chelpanov (1916), 53.

an enormous difference between a representation and that which evokes it. But we can conclude that there are things in themselves, though how they "truly" are must remain beyond our cognitive ability. On the other hand, such talk appears to invoke causality beyond the bounds of its legitimate use – or at least in Kant's scheme it does. Chelpanov recognized the issue but apparently preferred to remain agnostic. He recognized that some have proposed a "double affection," but he stated that Kant would hardly have found that to be satisfactory. Whereas Kant did write of another causality, a causality through freedom, its applicability is only to human action.[73] In the end, Chelpanov claimed agreement with Windelband that the cognized object is a totality of sensations connected by certain rules.[74] Further inquiry is essentially pointless, for it would involve exiting the theory of cognition and venturing into metaphysics. We must, in the final analysis, limit ourselves to saying that our sensations and representations are indicative of transcendent somethings, which we can never know. These "somethings" in some manner induce our consciousness to apply particular *a priori* categories or forms in a particular instance and not in another. Such a standpoint, Chelpanov's standpoint, he now called "critical realism."[75] In his 1913 psychology textbook referenced above, he elaborated slightly, writing that our certainty in the constancy of elementary physical laws remains steadfast even though this stand itself remains unproven. In fact, no factual evidence could disprove the law of causality. This, to Chelpanov's mind, demonstrated that the causal law is a presupposition, i.e., a transcendental *a priori* condition, of cognition. Furthermore, our apprehension of the incessant change of appearances given to cognition would be inconceivable if we did not presuppose the persistence of some unalterable thing grounding these appearances. This ground is

---

73 According to the most technical treatment of "double affection," transcendent things in themselves affect the transcendent consciousness in itself while correspondingly empirical things affect my empirical consciousness. The development of this position took place more recently than Chelpanov's work, but he knew of its early formulations. See Chelpanov (1916), 86–87.
74 Chelpanov (1916), 118.
75 Chelpanov (1916), 120. There is no indication in Chelpanov's numerous texts that he essentially distinguished the various designations he used to characterize his own overall position. Why Chelpanov employed one particular expression at one time and another expression at another time is unclear. Certainly, the expression "critical realism" had already been attached to the positions of several philosophers whom he evidently held in high regard such as Riehl, who in effect labeled his stance as a "critical realism." See Riehl (1887), 174; Riehl (1894), 165. In particular, Wilhelm Wundt, to whom Chelpanov often positively referred, had used the expression "critical realism" in the very title of his serialized article "Über naiven und kritischen Realismus," which started to appear in 1896.

called "substance." In other words, substances are a necessary presupposition of cognition.[76] Chelpanov, in this manner, reiterated Kantian transcendental arguments for the categories of causality and substance, and thereby again demonstrated his stand that these two key categories are to be understood as Kant did.

Chelpanov did provide summaries of various ethical theories such as utilitarianism and the then-popular views of Spencer and Schopenhauer. These are little more than perfunctory expositions as required in an introductory survey text. However, he realized that any talk of morality required some account of responsibility and that in turn leads to the perennial issue of free will. Being, above all, a psychologist, Chelpanov recognized that human actions are not totally arbitrary. Not only are many of our actions determined by some external factor, but we must take into account also social causes. Our particular will is very much shaped not just by the environment, but by our past which formed our character. In general, there is a reason we act as we do. Therefore, in a sense, all of our actions are caused and, therefore, necessary in the sense of being law-governed. Yet, even though our actions are caused we are responsible for them. Chelpanov recognized the unoriginality of this conclusion. Such diverse thinkers as Kant and Mill recognized it, but so too did Paulsen and Wundt. "It is precisely the feeling of responsibility that is the cause of the fact that we consider the will to be free."[77] We routinely interpret this human feeling as meaning that externality is not the decisive factor in our choice of action. We decide on a course of action through a realization of what kind of individual we are, how we understand our own character, which requires a self-consciousness including an ability to connect the present with the past. We see from this why animals cannot be held responsible for their actions. Lacking self-consciousness, they cannot ascribe freedom to themselves.[78] Only human beings in the natural world are ethical beings.

In the final years of the Imperial Russian regime, Chelpanov devoted little attention to purely philosophical problems. If we except revisions to his introductory text and a few reviews, his writings at the time, though still significant in number, belonged to psychology and, to a lesser degree, pedagogy. The psychological institute that Chelpanov created with generous funds from a non-academic source and directed until November 1923 was expropriated from him as was his position at Moscow University. Chelpanov valiantly attempted to defend non-partisan scholarship against official assaults to no avail, of course. During a significant portion of that decade, he was attached to the State Academy for the

---

76 Chelpanov (1913), 135. Cf. Kant (1997), 300 (A182/B225–226).
77 Chelpanov (1916), 473.
78 Chelpanov (1916), 474–476.

Study of the Arts (GAKhN), where he was able to carry out some psychological investigations. However, in 1930 he and the remaining staff were let go with the liquidation of GAKhN, and Chelpanov was forced into retirement.[79]

In light of the exposition above, the question still lingers: Was Chelpanov a neo-Kantian? What is clear is that he never referred to himself in print as a neo-Kantian. Indeed, he distinctly distanced himself from those routinely characterized as "neo-Kantians" in his writings. In his own *Introduction to Philosophy*, he called F. A. Lange, the celebrated author of the three-volume *History of Materialism*, a "typical representative of neo-Kantianism."[80] The same characterization can also be found in Chelpanov's earlier 1902 book *On Contemporary Philosophical Directions*, in which he also wrote that neo-Kantianism arose in the 1860s and represented a third form of positivism.[81] One may conclude from this that Chelpanov distanced himself from neo-Kantianism owing to a difference concerning the viability of metaphysics as a philosophical discipline. However, he also listed Cohen, Natorp, Windelband, Liebmann, and even Riehl as representatives of neo-Kantianism, although these figures espoused quite different understandings of that general direction.[82] Chelpanov paid the greatest homage, based on references and direct textual expressions of agreement, to Wilhelm Wundt and Friedrich Paulsen, the latter being commonly labeled a neo-Kantian.[83]

---

79 Although arguably more information on Chelpanov as an individual and his family is not germane to our philosophical concerns here, the family story is filled with loss and tragedy. His older brother Vasilij, to whom he was close in his youth, was executed in 1920 as a representative of the bourgeoisie. The closing of GAKhN left Chelpanov in a precarious financial position. In 1930, his daughter Natal'ja, who had married a French diplomatic attaché, was forbidden entry into the Soviet Union, and Chelpanov never saw her again. In 1933, another daughter, Tat'jana, died. In 1935 his granddaughter Marina, the daughter of his son Aleksandr, also died, followed shortly afterward in the same year by the execution of Aleksandr on the fabricated and preposterous charge of conspiring for Germany against the Soviet government. Chelpanov himself, by this time in rather poor physical condition, was not informed. This and a great deal more biographical information, including the family tree, can be found in Bocharova and Borob'ev (2012).
80 Chelpanov (1916), 271.
81 Chelpanov (1902c), 59.
82 Chelpanov (1916), 283. In a logic text from his Kiev years, he wrote, "I affirm that from the logical point of view we have no right to deny the possibility of metaphysics, i.e., the science of super-experiential cognition. ... I want to show that cognition in both metaphysics and science is subject to the same laws and that truth is the same in both science and metaphysics. There certainly can be doubt only concerning the degree of validity of the cognitions attained by them." Chelpanov (1901b), 157.
83 Beiser (2014), 2.

A definitive determination of Chelpanov's philosophy vis-à-vis neo-Kantianism is impossible, particularly in light of his many inconclusive statements. On the one hand, he stated that the "inductive metaphysics" of Wundt and Paulsen represented the "latest word of contemporary philosophy." To be sure, Chelpanov admitted that Wundt's philosophy contained contradictions on specific points, but these were unimportant in comparison with the method employed, viz., that Wundt reconciled natural science with metaphysics thereby moderating the pretensions of both.[84] On the other hand, we saw that he characterized his position as a "critical realism." Moreover, we saw a repetition of key points in a number of his writings that qualify him as a realist neo-Kantian. True, even in his own day, Chelpanov's contemporaries struggled to attach a consistent label to his position. Sergej A. Askol'dov (1871–1945), who later taught at St. Petersburg University, saw Chelpanov's doctoral work as representative of a very qualified form of Kantian Criticism, certainly not of an orthodox form. On the other hand, Chelpanov, in Askol'dov's estimation, held views incompatible with the basic doctrines and most important conclusions of Kantianism. Although Chelpanov allowed for Kantian *a priori* conditions of cognition – and thus adhering to this extent to Kantianism – he also claimed that things in themselves, despite being unknowable, affect in some manner our experience and, in turn, our cognition.[85] Thus, it is impossible for him to say that things in themselves are uncognizable.[86] Askol'dov added that if Kant had accepted what Chelpanov allowed, all of Kant's conclusions in his first *Critique* would be imperiled.

Others with a greater familiarity with contemporary German philosophy and neo-Kantianism, in particular, were not so quick to expel Chelpanov from the neo-Kantian camp. We should recall, for example, that the principal basis for Askol'dov's rejection of characterizing Chelpanov as a neo-Kantian was the latter's belief in noumenal affection. Yet in addition to Paulsen, Alois Riehl, as we saw, unquestionably held a realist view, writing that "space and time are concepts that have their empirical and real basis in the manifold of sensations, their ideal foundation in the logical capacities of our mind."[87] If anything, Chelpanov, with his affirmation of the uncognizability of things in themselves, was even more of a Kantian than was Riehl, who rejected their uncognizability. Also unlike Riehl, for whom the existence of external objects was a matter of immediate knowledge, something that needed no philosophical proof, for Chelpanov the

---

84 Chelpanov (1902c), 91 and 106.
85 Askol'dov (1904), 521–522.
86 Askol'dov (1904), 546.
87 Riehl (1879), 107.

need to accept noumenal affection in order to account for appearances counted as a philosophical proof of that existence.⁸⁸ Finally, for the last word on this topic, we turn to one of the last figures we will discuss in this study, Boris Jakovenko, who in an extended essay from 1922 wrote, "Among the Kantians should also be included *G. I. Chelpanov*, who in his vast epistemological investigations devoted to the problem of space, attempted to provide a grounding for Critical Philosophy in the spirit of *transcendental realism*."⁸⁹ In conclusion, Chelpanov was a quiet neo-Kantian, who sought to avoid confrontation and controversy. That some denied him that label had much to do with their own narrower understanding of the term.

---

88 Riehl (1879), 186 and 18.
89 Jakovenko (2000), 807. For the very reason that Askol'dov gave, few would place Chelpanov among the "Kantians." Thus, Pustarnakov is correct in writing that Chelpanov cannot be classified as a Kantian and that Chelpanov, unlike Kant, held metaphysics could be a science "connected with the inductive conclusions of the concrete sciences." Pustarnakov (2003), 683. He is wrong, though, in not recognizing the common elements between Chelpanov and Kant that allow us to characterize Chelpanov as a neo-Kantian, albeit one who shied away from overt statements of his philosophical convictions.

Chapter 4
Neo-Kantian Marxism: A Curious and Unstable Blend

The first four figures discussed in this chapter are, arguably, the best known of all to be discussed in this study; the last figure among may be the least known of all. All five were commonly grouped together in the secondary literature during the brief period examined here under the label "Legal Marxists." They were labeled "Marxists" owing to their support for historical materialism to varying degrees and were committed to radical social and economic change in their home country. The eventual disillusion felt by the first four with what they took to be the orthodox, deterministic version of that doctrine and its concomitant inadequate ethical standards and grounding led them, albeit again briefly, to embrace a neo-Kantianism grafted onto their individual understandings of historical materialism. What resulted was, to use an expression coined later by one of its exproponents, a "Critical Marxism."[1] These four figures, by no means, shared a unified point of view apart from this aspiration to infuse Marxism with some element(s) of Kantianism. As individuals, they had different interests and distinct personalities. It hardly needs to be pointed out, then, that their views on occasion clashed and that they were not timid in expressing these disagreements. The final, fifth, figure found no incompatibility between Marxist social theory and Kantian morality.

4.1 From Kant to Jesus

Sergej Bulgakov (1871–1944) was the son of a village priest with a clerical lineage extending back numerous generations. He attended a local theological seminary for a time and then the local secondary school before enrolling in the law faculty at Moscow University, where in the early 1890s he learned about Marx-

---

1 In a review published in early 1903 of Petr Struve's 1902 collection of articles, S. L. Frank wrote that Struve "is in Russia the founder of a 'Critical Marxism,' a movement that again under his undoubted influence, is quickly passing from a critique of the present Marxist worldview to the construction of a completely independent socio-philosophical program." Frank (1903), 100.

ism.² In his first significant work, "O zakonomernosti social'nykh javlenij" ["On the Regularity of Social Phenomena"] published in late 1896 and while still under the banner of the "materialist philosophy of history," Bulgakov directed his youthful zeal against one version of neo-Kantianism in the form of an extended review of Rudolf Stammler's *Wirtschaft und Recht nach der materialistischen Geschichtsauffassung*. The review largely served, though, as a platform for the expression of his youthful rigid determinist outlook in both the physical and the social sciences.³ All social development, he held, is a natural process and as such is subject to natural laws. Bulgakov, at this time, had no dispute with Kant that all natural phenomena are bound by causal laws. The former simply saw causality in all matters as extending to economic relations and social development, concerns that largely escaped Kant's purview. This applied as well to human social history, which, as a natural process itself, was bound by natural laws. Bulgakov held that this rigid determinism was not just compatible with Kant's teachings but a mere extension of it to human affairs. His monistic outlook sharply departed from Stammler's dichotomizing of the natural and social sciences in that the latter have to take into account free choices in determining not just the sought ends, but also the appropriate means to secure them.

Bulgakov was particularly eager, however, in this early work to reconcile the image of a strict mechanical regularity in all matters with individual human initiative. If they are incompatible must our ultimate conclusion be a simple resignation, a quietism? Can my personal initiatives make a difference in the determined scheme of things? Bulgakov's answer was neither original nor particularly insightful. For him, the emergence and growth of our recognition that social relations are determined by economic relations is itself a result of the law of social development. But as for the individual, a recognition of determinism by no means implies resignation. Bulgakov held that quite the contrary is the case. "In fact, the idea of the regularity of human actions makes truly free, i.e., rational, expedient activity possible. It excludes indifferentism, which is the mother of

---

2 Since there is much biographical information concerning Bulgakov available in English, we can be brief here. For a clear summary, see Evtuhov's "Introduction" to her translation of Bulgakov (2000), 3–7.

3 Stammler's book clearly struck a nerve among social scientists and philosophers of law in that it drew the ire not only of the Marxists but also of such an outstanding figure as Max Weber, who devoted a lengthy tract to its second edition. See Weber (1977). Bulgakov dealt with the first edition of Stammler's work. Stammler's book appeared in three different Russian translations already in 1898–1899 and yet another in 1907.

chaos and night, to use Kant's expression.[4] The law of social development speaks not of what arises *without* our acting, but *from* our acting."[5] We will sooner act when we know that our actions will meet with success than when we do not know. Seen objectively, however, success can only be assured to those whose deliberate actions are consistent with the developmental law of a given society. Therefore, it is imperative for those who wish to effect social change to study economics, more specifically, the economy of the society they wish to change.

Stated in such stark terms as those above, Bulgakov's outlook would appear to have little to do with Kant or Kantianism. But Bulgakov framed his outlook, his "social materialism," in terms of Kantian – or at least, neo-Kantian – epistemology. For one thing, an individual's "free" will is purely a matter of subjective psychology. All natural phenomena – and that is all there is – are determined, i.e., take place in accordance with the unity of a law, which corresponds to the unity of the object, which, in turn, is dependent upon the unity of time and space. His view of social and economic relations and development, i.e., social materialism, "fully agrees with Kant's critique of cognition, which established these unities as postulates of our reason. What characterizes the theory of social materialism, as such, is the idea that this common regularity is the regularity of economic phenomena."[6] Such is the (neo-)Kantian basis of Bulgakov's claim that *all* economic relations – and correspondingly social relations – are subject to the category of causality. And this single law of causality – single in that the same law is applicable in all disciplines – is an *a priori* postulate of reason.

Whereas Stammler, as a Marburg-style neo-Kantian, sought the ground of legal right (*Recht*) in Kant's moral philosophy, specifically, connecting it with Zweckmäßigkeit (purposiveness), Bulgakov saw the "materialist understanding of history" as understanding legal right solely in terms of a causal connection with the economic structure of society.[7] Thus, social conflicts, according to Marxism, occur not because of contradictions between outmoded laws and newly emerging economic relations, as Stammler thought, but because of contradictions within those very economic relations. Changes in the economy are accompanied by changes in law. But Bulgakov even at this early date knew enough phi-

---

[4] Bulgakov did not reference Kant, but the expression comes from the "Preface" to the first edition of the first *Critique*. See Kant (1997), 100 (Ax).

[5] Bulgakov (1903), 33.

[6] Bulgakov (1903) 7. There was considerable ambiguity in Bulgakov's position. For example, what was his notion of "materialism"? Was he, at this time, endorsing a metaphysical eliminative materialism? Was this "causality" of which he wrote a *purely* cognitive one? In writing of "postulates" of reason, is it possible for cognition not to postulate them?

[7] Bulgakov (1903), 29.

losophy not to conflate questions concerning the genesis of an idea with its veracity. He remarked that every idea, regardless of whether it was true or false, could have its origin explained, but that "story" did little at most for determining whether the idea is correct or not. The important point is whether the idea is consistent with others within the framework of the laws of possible experience.

Bulgakov's fundamental philosophical objection to Stammler's neo-Kantianism was directed toward what he regarded as the latter's bifurcation of consciousness into a theoretical and a practical direction and, consequently, two ways of understanding human action. Since one of the two ways could not even in principle be proven, we have no practical means to prefer one over the other or prevent the use of one instead of the other. The practical direction views action in terms of purposiveness. It is no more and no less than ethics – "ethics *ni plus, ni moins*" – and ethics does not belong alongside epistemology as a scientific philosophical subdiscipline.[8] Its foundation is part of psychology. Bulgakov portrayed the unity of "transcendental consciousness" as a fundamental condition of the possibility of experience, but this unity would be impossible if we allowed for both a practical and a theoretical direction in the understanding of social phenomena.[9]

Without the slightest doubt or hesitation, one can say that despite mentioning Kant by name on occasion and Bulgakov's periodic use of terms commonly associated with Kantian epistemology, there was actually quite little of Kant or of any recognizable neo-Kantian elements in Bulgakov's pronouncements apart from this diaphanous terminological veneer.[10] Fortunately, he, already at this early date, allowed for the possibility of a further formal development of Marxism in terms of an elaboration and clarification of its explanatory concepts.

Bulgakov's article originally appeared in the last issue for 1896 of the philosophy journal *Voprosy filosofii*. However, two articles appeared in reply already in the next issue in 1897. The first of the two authored by the historian Nikolaj Kareev expressed the hope that Bulgakov would abandon his flirtation with "economic materialism" in favor of the Kantian Criticism that underpinned his posi-

---

8 Bulgakov (1903), 28.
9 Bulgakov (1903), 31.
10 One can hardly take seriously Evtuhov's claim that Bulgakov's "decisive argument in favor of Marxism" – if that indeed is what it was – is "fully Kantian." See Evtuhov (1997), 31. Three years later, Evtuhov would write, "He [Bulgakov] experimented with neo-Kantianism in the early 1900s, but he ultimately found in Orthodoxy a system of beliefs that could replace his Marxist creed of the 1890s." Bulgakov (2000), 5.

tions.¹¹ Bulgakov did not reply directly to Kareev, but he did respond to the second article by Petr Struve, the second figure whom we shall look at shortly. Bulgakov's reply appeared appropriately enough not in *Voprosy filosofii*, but in the somewhat more popular monthly journal, *Novoe Slovo* that had the previous month passed into the hands of Marxists.¹² Bulgakov's views had not changed in the intervening time. He reiterated even more emphatically that all human cognition stands under the condition of the applicability of the law of causation. This applies not just to our understanding of natural phenomena, but also human actions. Spinoza proclaimed that they are strictly law-abiding, and Kant had further established it despite our immediate, inner sense of freedom. The contradiction is merely apparent, but this psychological feeling serves as the source of various theoretical constructions. That Kant had accorded primacy to practical reason over theoretical was philosophically insupportable, but it did say something about human psychology. Why we as individuals adopt one theoretical point of view over another has much more to do with psychology than with the logic of the case.¹³ Nonetheless, epistemology has an undisputed *logical* priority over psychology and all other disciplines.

Bulgakov also continued to uphold at this time that the second condition of cognition was the unity of the "transcendental consciousness," on which Kant supposedly based the unity of experience.¹⁴ If there is a unity of experience – and there must be such, for otherwise, cognition would be impossible – there must be a unity of consciousness, presumably the unity of transcendental con-

---

11 Kareev (1897), 108. Kareev immediately at the time recognized the sheer vacuity of Bulgakov's invocation of Kantian terminology, remarking that Kant's critique of cognition had nothing to do with the issues at hand. Referring to the unity of space and time, if Bulgakov were correct, his reasoning could be used to uphold the unity of any fanciful object. Kareev (1897), 117.
12 The Marxist reign did not last long. Although Lenin and others managed to publish some pieces in it, the journal was banned already in December. For far more information, see Pipes (1970), 176–190.
13 Bulgakov (1903), 37f. He did not clarify how he saw the connection between the psychological basis for adopting one viewpoint over another and Kant's conception of practical reason. Yet, he gave every indication that he thought there is one.
14 As Allison remarks, Kant does use the expression "transcendental consciousness" uniquely in a note in the first edition of the first *Critique*'s Transcendental Deduction, where he says that all empirical consciousness has a necessary relation to this transcendental consciousness. He, thereby, leads us to think that the expression is a synonym for "transcendental apperception." Allison (2015), 248. See Kant (1997), 237 (A117f.). Allison has much more to say on the possibility of the attribution of various empirical consciousnesses – and thus contradictory directions – to a single transcendental consciousness. However, the technicalities involved in that discussion in relation to Kant's text would lead us considerably astray. For that reason, we must refrain from proceeding further into it and leave it for the interested reader to pursue.

sciousness. The latter is possible only if there are no contradictory directions in a single consciousness. Yet some theorists, such as Stammler, interpret Kant's theoretical and practical reason as implying both can coexist in (empirical) consciousness at a single time with one coming to the fore at the individual's discretion. Bulgakov charged that this certainly was not Kant's view and that he sided with Kant.[15] Were we to eliminate the unity of the pure or transcendental I, on what would the unity of experience be based? All cognition would become impossible. Whereas Kant did postulate the possibility of some non-experiential cognition accomplished without the help of the cognitive categories, he did not establish the possibility of two contradictory standpoints in experience.

For Bulgakov, were we to take seriously this talk of two cognitive directions, freedom and necessity, freedom would have to be an epistemological category. The concept of freedom applied to the human will yields a will with a preceding cause, a will without content. Some assert that this notion of "free will," an absolutely contingent will, is a fact of immediate inner experience. Bulgakov disagreed, finding no such concept in consciousness, nor could there be. The psychological freedom that its advocates hold to be self-evident is merely a mental state and as such is a natural phenomenon subject to natural laws including the law of causality.

What sparked Bulgakov's abandonment of his rigid determinism that accompanied his departure from Marxism in a few short years is unclear. In 1898 he still expressed support for social materialism, his rendition of a "Critical Marxism." In an essay published that year "Khozjajstvo i pravo" ["Economy and Law"], he stated, on the one hand, that the relationship between law and economy must be understood not in terms of the theory of one or the other, but as a sociological matter: "Only a discipline that studies social life on the whole is competent to answer the question."[16] Oddly, however, with hardly a pause, Bulgakov presented a Kantian-like sketch of cognitive activity. He wrote that our senses present a chaotic and discordant manifold that is brought into a particular order by our cognitive faculties with the assistance of their cognitive forms. "The unifying form of an *object* belongs to our reason, and the object is projected into space and time, which are only *forms* of our perception, a special means of coordinating sensations."[17] Although Bulgakov's terminology may be somewhat imprecise, the inspiration behind the words is clear.

---

15 Bulgakov (1903), 42.
16 Bulgakov (1903), 55. Of course, from an orthodox Marxist standpoint this statement is itself heretical, since the economy is the ultimate foundation. Social formations are superstructural upon economic relations.
17 Bulgakov (1903), 55.

Bulgakov never explained the circumstances behind his apparent philosophical epiphany. Perhaps it had to do with his time in Germany, but that would be purely conjectural. In any case, already in a public lecture delivered in Kiev in November 1901 devoted ostensibly to Dostoevsky, we notice a sharp change in his tone. Gone is the rigid Marxist determinism of three years earlier. Indeed, there is hardly a trace of it. Instead, we find a deep appreciation for the issues found in the novelist's *The Brothers Karamazov*. But most important – because most telling for Bulgakov's further intellectual evolution – is the claim, "It seems that in our day of all the philosophical problems the ethical one is advancing to the fore and is exerting a decisive influence on the development of philosophical thought."[18] Thus although he had earlier viewed epistemological issues as his foremost concern, Bulgakov had now come to see ethics as being of paramount philosophical importance. Kant was mentioned though not as an epistemologist, but for his normative ethical precept to treat everyone as an end, not as a means. Marx was nowhere to be found.

In his own introductory comments to a collection of previously published articles with just one exception, Bulgakov acknowledged that the first three pieces we discussed, i.e., those written prior to 1901, were "written in defense of Marxism." He stressed, however, that by the time of this collection (1903) he had passed to idealism.[19] He frankly stated that he had intended to graft Kantian Criticism onto Marxism in order to provide an epistemological foundation for the latter so that its principal sociological and economic doctrines would receive a Kantian formulation. To do so, he turned – or so he said – to several different "varieties" of neo-Kantianism, but Kant remained always the indisputable source even over Marx himself. Thus, he ultimately considered "it necessary to correct Marx by Kant and not vice versa."[20] Clearly, Bulgakov had by 1903 surrendered his "Critical Marxism." But in doing so did he ever adopt, however briefly, an unalloyed neo-Kantianism? He did state in his introductory comments that "for a long time" he believed Kant had almost completely closed the door to metaphysics and only owing to a personal weakness had left open a small gap for the postulates of practical reason.[21] However, Bulgakov explicitly assigned this position to the years during which he wrote the first two essays we examined, i.e., 1896–1897.

One of Bulgakov's best-known works was his contribution "Basic Problems of the Theory of Progress" to the 1902 collection *Problems of Idealism*. The

---

18 Bulgakov (1902), 836.
19 Bulgakov (1903), v.
20 Bulgakov (1903), xi.
21 Bulgakov (1903), xviii.

mere fact that he was asked to contribute to this collection indicates that in the eyes of the Russian "liberal"-minded philosophical community Bulgakov was already seen as having shed his earlier Marxism. By the time of the essay's composition, he appears to have also shed his neo-Kantianism, such as it was. The essay itself is hardly a piece of philosophy as that is understood today. Presupposing a nineteenth-century conception, even a nineteenth-century *Russian* conception, Bulgakov wrote that recent philosophy had developed in a roundabout manner, taking Kantian epistemology as its starting point, rather than the "authentic, historical Kant" for whom epistemology played only a propaedeutic role. Such a statement could easily be seen as compatible with a neo-Kantianism, but Bulgakov slams the door by then continuing, "in my view Solov'ëv's philosophy is thus far the last word in world philosophical thought, its highest synthesis."[22] And in a postscript to the reprinting of this essay in the 1903 collection *From Marxism to Idealism*, Bulgakov commenting on his earlier defense of necessity and determinism in all physical matters added that the solution to the age-old problem of freedom would require an "entire metaphysical doctrine."[23] That doctrine, he added, was for him the philosophical system of Solov'ëv.

To be sure, Bulgakov's youthful flirtation with neo-Kantianism and early immersion into German thought would leave a lasting impression on his formulations going forward. A quick perusal of his 1912 *Philosophy of Economy* shows that, in addition to lengthy discussions of Kant, et al., he framed his query in distinctly Kantian terms: "How is economy possible?"[24] And in another essay two years later entitled "The Transcendental Problem of Religion," he asked, "How is religion possible?" but gave an answer totally in line with his theistic fervor.[25] There can be no mistake that in his actual procedure and philosophical

---

22 Bulgakov (2003), 116. In the introductory comments to the 1903 collection, he wrote, "I share the fundamental epistemological and metaphysical opinions of Solov'ëv, who brought them into a living connection with the most burning issues of practical life." Bulgakov 1903: xx. We should keep in mind that at least in 1903 Kantianism and neo-Kantianism were, for Bulgakov, "subjective idealisms" whereas Solov'ëv presented an "objective idealism." Bulgakov (1903), xi, xviii.
23 Bulgakov (1903), 158.
24 Bulgakov (2000), 97. The significant intellectual value of Bulgakov's book is somewhat offset by intended captious remarks concerning Kant, accusing him yet again of subjectivism, phenomenalism, and of the fundamental sin of Protestantism, viz., anti-ecclesiastic individualism. See Bulgakov (1914), 304. In a lengthy review-article of Bulgakov's book, the neo-Kantian Nikolaj Alekseev, whom we shall look at later in detail, saw Bulgakov attempting to do for "economism" what Kant had done for mathematical physics. "Kant's Criticism gave Bulgakov a general plan for his philosophy of the economy." Alekseev (1912b), 712.
25 Bulgakov (1914), 581.

attitude he had by that time passed well beyond what could be embraced by the designation "neo-Kantian."

4.2 A Flirtation with a Realist Neo-Kantianism

Petr B. Struve (1870–1944) hardly needs an introduction to students of Russian intellectual history.[26] In the autumn of 1889, he enrolled in St. Petersburg University initially intending to study zoology, but already by the end of his first year, he decided to transfer to the law faculty. He soon became interested in the German Social Democratic movement and in Marxist theory. He immersed himself within a few short years in German economic literature and became familiar with neo-Kantian treatments that dealt with social issues. Even slightly earlier than Bulgakov, Struve proclaimed and sought to fructify Marxism through an infusion of elements from a realist neo-Kantianism. Whereas the former "ism" had announced a new vision of historical development seen in terms of socio-economic concepts, it had not yet completely elaborated that vision's philosophical foundation. In other words, Struve sought to provide a solid philosophical buttress to a doctrine he accepted on non-philosophical grounds. Moreover, other general tenets of historical materialism needed reassessment in light of new facts. This, Struve thought, could be supplied through new investigations of economic development in the years since Marx's *Capital*.

Struve's first major engagement as a neo-Kantian Marxist appeared already in 1894 with his *Critical Remarks on the Question of Russia's Economic Development*.[27] Our concern here, of course, is not with economic development, but with Struve's neo-Kantianism. In this, he relied heavily upon the work of the realist neo-Kantian philosopher Alois Riehl and the sociologist/philosopher Georg Simmel. By the expression "transcendental consciousness," which we saw Bulgakov would also employ shortly later, Struve meant the general or common human consciousness, i.e., the consciousness with the characteristics or features that all humans share.[28] Although such a conception is abstract, the transcendental consciousness is by no means transcendent. Since it is the consciousness that is

---

26 Given Struve's notoriety, we need not linger on his biography. Clearly, the best study of Struve in English is Pipes (1970).
27 Pipes remarks that Struve's book is the earliest known instance of an effort to link Marx with Kant. Pipes (1970), 107.
28 See Riehl (1894), 154–155 – "'Transcendental consciousness' is the form of the unity of consciousness abstracted from its contents, in so far as this form is thought of as the universal condition, valid not for me alone, to which the idea of every object must conform."

common to all humans, it, as Riehl wrote, "has no existence which is separate or which can be separated from the psychological consciousness."[29] Whereas our sensible manifold must be united in consciousness in some manner in order for each of us to have *a* consciousness that persists from one fleeting moment to the next, that manifold itself fluctuates simply during our days and is often quite different from person to person. Nevertheless, that we perceive a single object through our varying postures with regard to it shows that the manifold is governed or regulated by principles. These general or logical principles clearly are common to us all, insofar as we share intelligible communication about the world, etc. and serve as proof of objectivity. Struve wrote, "Transcendental consciousness is the general human consciousness for which logical principles are obligatory and which best convinces us of the real existence of the external world. ... *Logical* norms follow from the formal unity *not of one* consciousness, but of all who are in spiritual intercourse with each other."[30] Skepticism toward the possibility of objective cognition necessarily arises from a confusion of transcendental consciousness, which is, after all, social, with the individual psychological consciousness. Struve here quoted Riehl again: "to be objective means to be valid for every knowing being."[31]

The essential realism that Struve undoubtedly found in Marxism is likely to have been what attracted him to the neo-Kantianism of Riehl as opposed to that of, say, Hermann Cohen of the Marburg School. Struve accorded primacy to real life, our being, as opposed to cognition and consciousness. He found this fundamental tenet in both "economic materialism," i.e., Marxism, as well as in Riehl's writings, however far apart Marx and Riehl otherwise may have been. Consciousness is an instrument in the struggle for existence.[32]

Most startling, however, is Struve's utter rejection of individual free will, a conception he called at this time a "fiction," which Kant, in order to save it, placed within a non-empirical, non-intersubjectively cognizable, but "intelligible" realm. In our everyday perceptions of others, we see only their empirical

---

29 Riehl (1894), 155; Riehl (1887), 164.
30 Struve (1894), 34. Cf. Riehl (1894), 155; Riehl (1887), 164–165.
31 Struve (1894), 35; Riehl (1894), 155; Riehl (1887), 164.
32 "Life does not exist for consciousness, but consciousness (at least originally) for the sake of life. It is an equipment of animal life in its struggle for existence which regulates its relations to the varying conditions of the environment." Riehl (1894), 153; Riehl (1887), 161. It remains curious that, apart from Struve, Riehl exerted little discernable influence in Russian philosophy. Russian students did not flock to the universities where he taught in order to hear his lectures despite the availability of his major writing *Der philosophische Kriticismus und seine Bedeutung für die positive Wissenschaft* in Russian already in 1887.

and thus determined character. Were it the case that we perceive only empirical impressions from without, we would not conclude that we possessed a "free" will. Such a person, however, Struve argued, would also be unable to react to impressions and external, empirical sensations. Yet we have no basis to ascribe to this active side of our individuality an idealistic construction, that the will is spontaneous and uncaused. Thus, in viewing the individual human will to be entirely a product of various socio-economic factors, we find moral norms to be a representation of existing, factual norms in one's social group. A scientific view of ethics, i.e., ethics seen from the Marxist viewpoint, sees the content of morality as reducible to socially acceptable behavior. The full philosophical grounding of this position still awaits, but so does a review of the facts from which it is drawn.[33]

The reader will surely have noticed a similarity between the positions of the late-1896 Bulgakov and the 1894 Struve. The latter himself was well aware of this similarity. "I believe it necessary to note that I stand on the same ground as Bulgakov. We are both at the same time supporters of Critical Philosophy and the materialistic understanding of history."[34] Both subscribed to a distinctly non-metaphysical universal determinism, but Bulgakov more forcefully than Struve disavowed psychologism and upheld *a priori* categories as universally necessary concepts making cognition possible. However, by the end of 1896 Struve's thought was already moving away from a rigidly deterministic worldview in favor of a greater appreciation for Kant's "practical philosophy."[35] As mentioned above, Struve replied to Bulgakov's critique of Stammler in the very next issue of *Voprosy filosofii* at the beginning of 1897. Now, Struve expressed his unease with his earlier cavalier dismissal of the individual's inner sense of free will. Whereas he still recognized Marxism, or "the economic understanding of history," as the attempt to understand human history as a matter of rigorous causality, the positing of goals and our strivings for their attainment could only be done with a recognition of freedom. The fundamental epistemological task consisted in his eyes in describing and analyzing the seeming contradiction between freedom and necessity. Rather than viewing the Kantian transcendental consciousness as a general social consciousness, Struve in 1897 saw it replete with irreconcilable contradictions.

Struve was not yet willing to surrender his "materialist understanding of history," but he thought this did not necessarily entail an acceptance of metaphys-

---

[33] Struve (1894), 46.
[34] Struve (1897b), 122.
[35] Struve openly acknowledged that he had distanced himself "somewhat" from his 1894 views. Struve (1897b), 121 f.

ical materialism. He admitted that there certainly was a genetic connection between the two. Both Marx and Engels had separately shown the historical and psychological unity of the two doctrines. However, this did not imply that they are logically united, for the materialistic understanding of history belongs to the sphere of historical experience, not to that of metaphysics.

The means by which we can speak of a transcendental consciousness and concomitantly of objectivity is social communication. Lest we lapse into metaphysics, any talk of externality and objectivity is by way of intersubjectivity. Thus, to speak of some "reality" beyond the possibility of our collective consciousness is sheer metaphysics. Struve did not depart from Bulgakov on the unity of experience, but did disagree with his contention that such unity implied a unity of transcendental consciousness. Kant showed in the "Paralogisms" chapter of the first *Critique* that there are irreconcilable contradictions in transcendental consciousness. There are always present in that consciousness the ideas of freedom and necessity, corresponding to two directions of consciousness, viz., will and cognition. We now find Struve acknowledging Stammler's merit in recognizing this contradiction. Nevertheless, experientially to talk of a free will is surely a great absurdity. All events are caused. However, from the standpoint of the will, to speak of ourselves as completely determined, as non-free, is just as absurd.[36] Thus, Struve by early 1897 had nudged himself ever so slightly away from strict determinism without abandoning his commitment to a neo-Kantian Marxism. He did so in the guise of responding to Bulgakov, who had championed many of the positions Struve had earlier advocated. In doing so, Struve, several years before Lenin, answered Chernyshevskij's question "*What Is to Be Done?*" stating that Marxism alone can provide no response. The materialist understanding of history can say only how to do it. But what to do involves interests and ideals. These come from elsewhere.[37]

Bulgakov replied, as we saw, to Struve's article in the April 1897 issue of *Novoe slovo*, and Struve, in turn, replied to Bulgakov in the May issue with a piece "Again on Freedom and Necessity." Demonstrating a further drift away from his 1894 stance and even from that of a few months earlier, he now acknowledged that how we ourselves view or represent our actions is entirely necessary. Struve returned yet again to his assertion that the unity of experience is

---

[36] Vucinich writes that for Struve the "very possibility of human consciousness is based not on an identification of freedom and necessity but on a full recognition of the two as separate human orientations." Vucinich (1976), 198. This is not technically correct. Struve does not speak of these two "orientations" as conditions for the possibility of consciousness. They are simply two factually separate directions "of" human consciousness.
[37] Struve (1897b), 139.

not identical with the unity of transcendental consciousness, which, he understood Bulgakov as not accepting. Struve acceded that the unity of experience is a necessary condition of experience but not that it is identical to the unity of transcendental consciousness. He held that the root of the disagreement lies in their different understanding of the correlation between psychology and theory of cognition. Without a great deal of elaboration, Struve remarked that the theory of cognition, as a discipline, is a description of cognition.[38]

Nonetheless, as he admitted at this time, the major difference with Bulgakov is that he now saw freedom as the opposite of necessity and therefore rejected the famous expression from Engels, "freedom is the recognition of necessity." The concept of freedom has sense, which it would not have if it were reconciled with necessity. Struve feared that if freedom and necessity could at present be combined, the consequence for human activity would be a quietism; it would make no sense to attempt to change the established order. This, he could not abide. This is not to say that at some time in the future freedom and necessity will not combine in some manner despite the logical contradiction of such a union. At present, though, freedom from the viewpoint of experience or theory is an indubitable illusion, but practically it is an indubitable reality, an irrefutable fact of life.[39]

Struve was neither by training nor by temperament an armchair philosopher, and his academic interests were largely economic in nature. Nevertheless, based on the scanty evidence available he still clung in 1899 to his neo-Kantian Marxism, believing, on the one hand, that determinism ruled the experientially-given world and, on the other hand, that the further development of Marxism must be along "Critical," i.e., neo-Kantian, lines while holding to Marx's fundamentally realist outlook with its *economic* conception of history.[40] It became difficult immediately after the publication of his "Marxist Theory of Social Development" to determine precisely when Struve began to distance himself from Marxism and adopt an openly idealist position.[41] Given his high regard for theory of cognition and scarce attention to ethics and philosophical anthropology, we

---

38 Struve (1897a), 204.
39 Struve (1897a), 207. Struve added that Bulgakov's accusation that he viewed human action as theoretically free is wrong.
40 For Struve's empirical determinism, see Struve (1899), 679–681, and for his continuing adherence to a revisionist, "Critical" Marxism see Struve (1899), 704.
41 Struve's 1899 work, written in German, appeared in a Russian translation by Boris Jakovenko, himself a noted neo-Kantian, in Kiev in 1905. In an essay that originally appeared in October 1907, Struve commented, "There is a useless Russian translation of this critical essay made without my permission." Struve (1911), 578.

cannot be surprised that he severely deprecated Vladimir Solov'ëv's philosophical positions, which owed a great debt to the later Schelling and Hegel. Struve saw the post-Kantian German Idealists as not carefully considering or concerning themselves with epistemology, whereas current "critical idealism" proceeded from the theory of cognition but abruptly came to a halt before the door of metaphysics.

Struve's 1900 essay "In Memory of Vladimir Solov'ëv" unfortunately, then, offered a basis for seeing a continued commitment to neo-Kantianism, but none to Marxism. Oddly enough, though, by November of that year in an article to commemorate the seventy-fifth anniversary of Ferdinand Lassalle's birth Struve wrote approvingly of Lassalle largely at the expense of Friedrich A. Lange, the author of the much-celebrated *History of Materialism*. Struve acknowledged Lange's "critical spirit," but the latter's work provided very little to the creation of a positive worldview erected on old idealistic foundations. Instead, Struve suggested, "in a certain sense," "a return to Hegel and even further and more so to *Fichte*."[42] Where does this leave Struve's neo-Kantianism? Had he by the end of 1900 simply abandoned it in favor of Fichte's idealism? Struve would emphasize what must to us look like a wavering commitment to neo-Kantianism the following year in another piece entitled "Again on Lassalle," saying "I set therefore the idealist Lassalle against not only the materialists Marx and Engels, but also the neo-Kantian Lange."[43]

Were we limited to Struve's writings from 1900 mentioned above, we would be hard pressed to form a definitive judgment concerning his position toward neo-Kantian Marxism at the time. However, he also penned in that year a substantial "Preface" to Nikolaj Berdjaev's book *Subjectivism and Individualism in Social Philosophy*, which would appear in 1901.[44] Unlike several years earlier, Struve remarked that with respect to both "theory of cognition and metaphysics there is no difference between Marx's materialism and so-called vulgar materialism."[45] However much he appears to have irrevocably distanced himself from materialism, he did not abandon realism. Struve explicitly recognized the presence "in" consciousness of *a priori* elements and dismissed the reduction of logical laws to psychology. Such laws are natural laws and, contrary to the Baden

---

42 Struve (1902), 266. Pipes summarized Struve's philosophical path to metaphysics well, writing that his turn to Fichte was not direct, but circuitous from Bernstein to Lassalle to Fichte. Pipes (1970), 297.
43 Struve (1902), 278.
44 Struve himself dated this "Preface" as being from September-October 1900, thus placing its composition at virtually the same time as the first article on Lassalle.
45 Struve (1999), 9.

neo-Kantians, are not normative. Our thinking processes must be logical. There is in every illogical train of thought some element or content that is inadequately clear. The absolute character of mathematics stems from its basis in logic. Mathematics *is* applied logic.[46]

Struve in 1900 contrasted the laws of logic – broadly understood to include the law of causality and the conservation laws, all being natural and inviolable – to the norms of ethics and aesthetics, which can be recognized or dismissed and therefore are violable. This opposition would be simple to explain in terms of the framework that he and Bulgakov had earlier championed, viz., that epistemology concerned itself with one sphere whereas ethics belonged to another, fundamentally different sphere. Necessity reigned in the former; freedom reigned in the latter. Bulgakov, for a time, saw freedom as merely psychological. However, Struve in 1900 dismissed this stance. Freedom, he tells us, is the ability to act independently of any causal connection, and only something that is, something having substantiality, can be free in this sense.[47] We are led inevitably, Struve concluded somewhat inscrutably, to metaphysics and to viewing freedom and substance as categories. "For us, the concept of mental activity demands the concept of a mental substance. The concept of substance includes self-determination and initiative. It is impossible to deprive it of epistemological significance, to eliminate the category of substance from the general inventory of our cognitive contents by referring to its psychological character."[48] Struve clearly ventured here into metaphysics, as he himself admitted, and consequently drew away from his original Kantian inspiration. The "I think" was for him a principle, but what he has said formed in his eyes an argument in favor of a "spiritualistic metaphysics."[49]

Struve held that substance and freedom are present in cognition and play some ill-defined role therein. However, far from making experience possible, as do the other Kantian categories, their use makes "experience" – in some apparently greatly expanded sense – impossible. Thus, we must abandon the bounds of ordinary experience for metaphysics, i.e., for what is transcendent, what is *not* given in experience. Yet, he immediately added that this broader con-

---

46 Struve (1999), 28.
47 Presumably, Struve had in mind Kant's third "Antinomy" in the first *Critique*, although he did not invoke Kant at this point. See Kant (1997), 484 (A444–446/B472-B474).
48 Struve (1999), 33. Struve there remarked that he found the arguments in this regard of Lotze to be irrefutable.
49 Kant in his first paralogism in the first *Critique* treated this issue and, as hinted, came to the opposite conclusion. He wrote that the proposition "The soul is substance" is valid only if "one admits that this concept of ours [substantiality] leads no further, that it cannot teach us any of the usual conclusions of the rationalistic doctrine of the soul." Kant (1997), 417 (A350–351).

ception of experience provides material for a metaphysical synthesis that creates an integral presentation or picture (*kartina*) of the world. Surely, Struve has subtly and gradually moved further and further from a recognizably Kantian framework. He, nevertheless, remained tied to vestiges of his earlier neo-Kantianism.

Struve abruptly turned next to ethics in the third section of his "Preface," asserting that there is a universally obligatory moral law – without specifying what that law is or who first "discovered" it. Along with Kant – and others – Struve recognized that the "ought" absolutely cannot be derived from the "is." Sensory experience and logic testify to the veracity of propositions in their respective spheres. To what authority do ethical imperatives ultimately appeal? Whereas Kant repeatedly invoked *practical reason,* seeing it as concerned with the determining grounds of the will, we are hard pressed to find any such notion – or even the expression "practical reason" – in Struve.[50] And whereas Kant extensively discussed the relation of practical to theoretical reason, albeit neither sufficiently nor clearly for legions of commentators, Struve found neither a means to reconcile nor to affect a transition from the theoretical to the practical "sphere of consciousness."[51] He added that they stand side by side in sharp opposition.

Struve devoted little attention to normative ethics. He continued to maintain that epistemology should bear a descriptive character and therefore any attempt to construct a bridge between ethics, the study of what we ought to do or what should be the case, and epistemology, the study of how we know what is the case, is doomed to failure from the outset. However, this limitation of epistemology shows that it alone cannot satisfy our cognitive needs. In other words, this restriction of the theory of cognition shows why we need metaphysics. Concomitantly, metaphysics has as a task to ground the moral law, which is present in any normal human consciousness. However, Struve then took a giant leap, writing "recognizing the impossibility of an objective (experiential) solution to the moral problem, we recognize at the same time the *objectivity of morality as a problem and accordingly come to the metaphysical postulate of a moral world-order, independent of subjective consciousness.*"[52] We could conceivably reconcile Struve's talk with Kant. After all, Struve still wrote here of the moral order being a postulate, and Kant would affirm that morality, being grounded in practical

---

50 For Kant, see already the "Introduction" to the *Critique of Practical Reason.* Kant (1996), 148 (Ak 5, 15).
51 Kant (1996), 247–254 (Ak 5, 134–141); Struve (1999), 39.
52 Struve 1999: 51. Referring to these words, Pipes writes that they mark Struve's Rubicon. He now expressed his abandonment of strict positivism, i.e., of an overtly non-metaphysical neo-Kantianism, in favor of a dualistic philosophy with two parallel realms. Pipes (1970), 297.

reason, is not a matter of the individual, subjective consciousness. Struve's remarks above, however, were not his last in this "Preface." With words far more evocative of Fichte than of Kant, he added that although the existence of a personal Deity cannot be objectively proven by logical or empirical means, neither can the existence of the absolute good nor of absolute beauty be proven. Yet, we know (*my znaem*) that they are independent of us and exist objectively.[53] The question for Struve, which he did not address, is: How does he *know* these things, and are they postulates, as he also says, or mere knowledge-claims?

Struve did not answer these questions in 1900. In fact, we cannot be certain what he thought of his earlier neo-Kantian Marxism at the time. Obviously, he no longer showed any Marxist sympathies. However, he provided in 1902 a self-interpretation of his earlier positions in his contribution, entitled "Toward Characterization of Our Philosophical Development," to the collection *Problems of Idealism*. As in 1900, Struve opined that the basic error of positivism is its attempt to derive the "ought" from the "is." Instead, one cannot be deduced from the other; they lie in two separate spheres. Moreover, although "critical philosophy," meaning presumably Kantianism, has shown that experientially the existence of God cannot be either proven or refuted, it, suitably expanded, cannot reject an uncaused first mover. Although such a concept is rationally inexplicable, and thus unknowable, it cannot logically be rejected either. To do so would entail a rejection of understanding the universe.

We find here again a train of thought that can be reconciled with Kantianism. However, is the unconditional law of causality a matter of belief? Neither the early Bulgakov nor Struve previously gave any indication that they held the unconditionality of causality to be a presupposition. In 1902, Struve held causation to be a belief, and therefore apart from it we could just as well believe in an uncaused first mover, a "*creative* being."[54] Only a belief in ceaseless causation forbids acceptance of a creative being. In a moment of self-criticism, Struve admitted that he advanced the basic error of positivism of subordinating the "ought" to the "is," of "submerging" freedom into necessity, and of rejecting uncaused causality. Struve characterized his earlier stand as uncritical and dogmatic. He now in 1902 saw that his earlier attempt to meld Riehl and Simmel with Marxism yielded an incorrect understanding of critical realism and social psychologism. He also added that he ceased to be a Marxist by 1900. We, in turn, can recognize that his forthright devotion to metaphysics indicates he

---

53 Struve (1999), 59.
54 Struve (2003), 149.

held no more than a tenuous connection to any discernable form of neo-Kantianism.⁵⁵

Lastly, Struve in 1902 reprinted a negative article originally from 1897 directed against Vladimir Solov'ëv's ethical philosophy. Struve added to the reprinting a footnote dated 1901 saying that in contrast to his earlier view, which he termed "critical positivism," he now professed "metaphysical idealism."⁵⁶ The reader will surely see that although he evoked a measure of Kantianism with his qualification of positivism he did not employ the same adjective, viz., "critical," as we might expect, in the qualification of his idealism.

4.3 From Ethical Marxism to Russian Religious Idealism

In the 1890s, Nikolaj Berdjaev (1874–1948), who would become among the best-known of Russian philosophers in the West, was a student in the law faculty at Kiev University.⁵⁷ It was while there that he turned to Marxism. As a result of participation in political demonstrations in 1897, Berdjaev was sentenced to "internal exile" in Vologda for three years and expelled from the University. He never would go on to earn a university degree, and his fame in his own day, as well as today, rests largely on his literary output and his advocacy of a Christian existentialism in a time and place that was particularly receptive to such a message.

In his engaging autobiography written toward the end of his life, Berdjaev tells us that he read both Kant's first *Critique* and Hegel's *Philosophy of Spirit*, which he found in his father's library, at the age of fourteen.⁵⁸ What an adolescent of that age, however precocious, could have made of those texts is hard to say, but quite likely little. Philosophy, after all, is the reflection of maturity. In any case, Berdjaev's first publication, written in German and published in Kautsky's journal *Die Neue Zeit* appeared in its 2 May 1900 issue. The article, which dealt with the relationship between "Critical Philosophy" and socialism, also appeared in July of that year in Russian in the journal *Mir Bozhii*. Precisely when

---

55 Struve (2003), 156.
56 Struve (1902), 187.
57 The many entrees on Berdjaev in popular treatments of Russian philosophy make a need for extended discussions of his biography unnecessary. There is also, of course, his autobiography, Berdjaev (1991), of which there is an English translation, Berdyaev (1951).
58 Berdjaev (1991), 92–93. Of course, it is possible that Berdjaev did have in mind the third part of Hegel's *Enzyklopädie*, viz., *Die Philosophie des Geistes*, but this would be an unusual choice taken in isolation from the other two parts. Could it be that he really had in mind Hegel's *Phänomenologie des Geistes*?

Berdjaev composed the piece is unclear, but in a letter to Struve dated 22 November 1899, he wrote, "I think that Marxist philosophy, especially its theory of progress, is badly in need of some positions established in [Kant's] *Critique of Practical Reason* and the *Critique of the Power of Judgment*. This is quite correctly pointed out by Stammler."[59] In another letter to Struve from February 1990, Berdjaev clarified somewhat what he found lacking in Marxism. Namely, he wrote, "It is time for Marxists to recognize finally the significance of the independence of the ethical point of view. I just do not see how to ground this philosophically. I have become more Kantian in this matter than in the theory of cognition and especially insist on the *a priori* character of the moral law, in the sense of it being absolute (despite its varying content)."[60] It, surely, was around this time that Berdjaev sent the German version of his proposed article to Kautsky. For in a letter to him from Kiev, also dated from mid-February 1900, Berdjaev wrote that he had no intention of attacking Engels or Marxism, but merely that the theory needed further development. Obviously referring to the essay he sent Kautsky, he added, "My deepest wish was to critique the philosophical views of Marx and Engels, while at the same time remaining loyal to their spirit. ... It seems to me worthwhile to defend the importance of ethical motives in socialism, while remaining entirely on the basis of social monism."[61] The editor of *Die Neue Zeit*, presumably Kautsky, added a footnote at the start of Berdjaev's German-language version of the article, stating that it had been accepted for publication already "a long time ago," but owing to space constraints it could not appear sooner.[62] With that, let us proceed to the article itself.

Berdjaev both praised and, in effect, condemned Kant's philosophy. It was in his estimation the deepest in human history, but it was half-hearted and inconsistent. The contradictions discernable in it and in that of the neo-Kantians is a reflection in abstract terms of those found in our social life. Unlike the contradictions in Hegel's philosophy, which intend to say something about the nature of being, those in Kant's philosophy stem from how he viewed cognition. The inconsistencies arise from the disharmonious conditions of social life and not simply from the individual's psyche. Some have confused the theory of cognition with a psychology of cognition. The relation between the two is debatable, and Kant himself contributed to this confusion. At this point, Berdjaev agreed with Kant – as the former understood him – that space, time, number, and cau-

---

59 Berdjaev (1993a), 124.
60 Berdjaev (1993a), 128.
61 Berdjaev (1993b), 181.
62 Berdiajew (1900), 132–133 f.

sality are *a priori* forms of cognition. They are not, however, innate in the psychological sense. Were they innate, they could be subject to the process of psychic development that humanity is constantly undergoing. But whereas the individual's mentality is subject to development, the transcendental-logical consciousness, of which Kant wrote, is absolute and inalterable.[63]

The concept of the "thing in itself" was, for Berdjaev, the weakest in Kant's thought. On the one hand, he saw these "things in themselves" as realities, following what we today call the "two-worlds" interpretation. On the other hand, he saw Kant as arguing for the realm of the in-itself as a limiting concept. Berdjaev himself professed a phenomenalism, but one that affirms the vital connection between the object of cognition and the subject of cognition. Thus, what "really" exists is exhausted by appearances. "We categorically assert that to ask whether the lamp in itself is red, disregarding the cognizing subject, is just as senseless as to ask how far Moscow is without specifying another point."[64] This "monistic theory of cognition" is one, which, in its essentials, we have seen in the earliest works of Bulgakov and Struve. That cognition has absolute theoretical limits, beyond which lies the "uncognizable," is a purely fictional concept and epistemologically untenable. Kant approached this view without completely attaining it. Berdjaev wished to stress that although he valued Kantian Criticism and the general theory of cognition, there are specific theories for each individual science. He singled out, in particular, that an epistemological analysis of sociology as a discipline would be of the greatest value. Philosophy, in general, is more than epistemology; its task is to provide an integral worldview, bringing harmony to the various and seemingly chaotic individual truths of the specific sciences. Berdjaev found that Kant's introduction of the primacy of practical reason over theoretical reason presented a great *psychological* truth. However, whereas Kant saw the postulates of practical reason as absolute and eternal, Berdjaev saw them as a time-bound reflection of those within a particular socio-psychological group. Here entered his Marxism, for he goes on, saying that morality within a class society will always bear a class character. The morality of the progressive class in each historical era is the result of the demands posed by the world-historical process. This, of course, leaves much unsaid. Berdjaev would have it that in the era of rising capital, the bourgeoisie represented a progressive class, and, thus, its morality too was progressive. But can we only determine progress in long retrospect? Berdjaev left many questions unanswered.

---

[63] Berdjaev (1900), 233. References to this essay will throughout be to the Russian-language text.
[64] Berdjaev (1900), 240.

As with the others discussed in this chapter, Berdjaev's thought was evolving rather quickly at the time. Berdjaev's next major publication – his first book – entitled *Subjectivism and Individualism in Social Philosophy* bears the date 1901, but if we accept as accurate the date appended to Struve's Preface of "September-October 1900," Berdjaev's work was written earlier than that.[65] Berdjaev, in this book, contrasted transcendental consciousness to psychological consciousness. What is universally valid in cognition is founded in the former. Without clarifying precisely the nature of the relationship between transcendental consciousness and cognition, Berdjaev wrote that cognition has logical conditions, among which, he repeated again, are space, time, causality, and the law of identity. Were we to deny transcendental consciousness, we would have merely the psychological consciousness with its fluctuations and a chaotic world. There would be no objectivity.[66] The objectivity of cognition is guaranteed by the *a priori* elements found in cognition and introduced into it by the cognizing subject. Much the same can be said of ethics. The objectivity of morality is based ultimately on the universal, transcendental consciousness, and just as objective cognition requires logical, *a priori* conditions, so too does objective morality require an ethical *a priori*, an *a priori* moral law. Epistemological reflection shows

---

65 Naturally, this conjecture involves assumptions. We have to accept that there were no typographical errors in the date given. We also have to assume that the given date represents Struve's final draft and not that of an extensive draft to which he could conceivably have added comments after reading Berdjaev's text. In his thoroughly researched study, Kolerov writes, "The year 1901 appeared on the cover, but in fact the book was published in October-November 1900, when the fortieth anniversary of Mikhajlovskij's literary activity was publicly celebrated." Kolerov (2002), 55f. Kolerov's claim is definitively supported by Berdjaev's letter to his father dated 22 November 1900, in which he wrote, "I am very pleased that you read my book and expressed your opinion about it. To clarify its meaning, I can say that in my philosophical views I stand closest to Kant and Fichte, but my social views are closest to Marx and Lassalle, although I think that I do not slavishly follow these thinkers. ... I especially valued the opinion of P. B. Struve, the chief representative of the direction to which I belong, and he has already expressed his opinion in the preface to my book." Berdjaev (1981), 216.

66 Berdjaev made no mention of how objectivity itself arises, though. Is it conceptually impossible for a chaotic world to be objective? What introduces the *sense* of objectivity into the sense manifold? Based on what he has said, would we not conclude that the chaos is how the world "really" is? He wrote that "objectivism" is "a product of norms that are *common to every consciousness*." Berdjaev (1999), 102. His use of the term "objectivism" [ob"ektivizm] is confusing, especially since he otherwise contrasted it with subjectivism. However, if subjectivism is the position that truth lies entirely in the subject, objectivism should be the position that truth lies entirely in the object. Such a stance is barely intelligible. Moreover, is this notion of "objectivism" a *product* of norms, or is the product of norms the definition of objectivism? Berdjaev should have written "objectivity" instead of "objectivism."

that causality is an *a priori* category for objective cognition. Likewise, ethical reflection shows that justice is an *a priori* category for objective morality.[67] It makes moral experience and moral life possible and is given to "our" transcendental consciousness.

Berdjaev provided what he termed an interpretation of Kant's formulation of the moral law, i.e., the categorical imperative. We need not repeat Kant here. For Berdjaev, though, "only those actions which are obligatory for every conscious creature, for every will, merit being called moral."[68] He correctly understood Kant's formulation as elevating humanity, taken as individual human beings, into ends in themselves, although Berdjaev's own interpretation provided no basis for referring to human individuals as ends in themselves. If, Berdjaev continued, the individual is of the highest value, then there is no basis for postulating, as Kant did, something higher, viz., the existence of God and an after-life. There should be only one postulate of practical reason, viz., the moral world-order. However, introducing his lingering Marxism, Berdjaev posited this world-order not in uncognizable noumena, but in the realization of human progress in pursuit of "the kingdom of ends." Such is the culmination of the historical process. While to some degree we can see this as an interpretation of Kant, it is even more strikingly similar to Solov'ëv's vision. The kingdom of humanity is not to be realized in either an after-life or in a noumenal world, as Berdjaev believed Kant and the bourgeois German neo-Kantians held, but in the real phenomenal world. Arguably more explicitly than either Kant or Solov'ëv, Berdjaev believed that although the formal foundations of morality are absolute, its content changes in response to human historical progress. "Moral concepts not only radically change with the change of historical epochs, but in one and the same we encounter several typical moralities, one hostile to another."[69] Berdjaev had in mind here the moralities of the ruling class, of the aristocracy, and of the working class, all three of which can conflict. But only in the "progressive"

---

67 Berdjaev (1999), 138.
68 Berdjaev (1999), 139. Berdjaev placed the entire quotation in italics. We hardly need to point out how in shifting Kant's concern with maxims to his concern for actions, Berdjaev seriously altered and misconstrued the overall thrust of Kant's moral philosophy. Already within Russian philosophy, Solov'ëv recognized the consequences of doing so in his *Justification of the Moral Good*. And what are we to make of Berdjaev's talk of conscious creatures [*soznatel'nye sushchestva*] instead of rational ones? Animals are certainly conscious. Are we to ascribe morality to them as well on the basis of their consciousness? Solov'ëv wished to extend the moral law simply to cover our treatment of them.
69 Berdjaev (1999), 142.

class can we discern a harmony between both the psychological and the transcendental consciousness, between subjective and objective morality.

There is certainly much in Berdjaev's philosophical legerdemain that we can admire. Particularly remarkable is its deft handling of elements from Kantianism and Marxism, all the while remaining highly skeptical of the cogency of the latter's enormous conceptual leap. He also faulted Marxism for its lack of supportive detail. Berdjaev assumed a certain philosophy of history, much as Solov'ëv did, that has roots in Christian eschatology. Whereas traditional Marxism saw the end of history in a classless society, Berdjaev's Critical Marxism saw history as ending with the achievement of objective morality, the attainment of the absolute moral good, independent of social conditions and partisan conflict.[70] To be fair, though, he recognized that this is a postulate, an idea that cannot be proved. But without it we humans are morally blind, just as without the epistemic categories our cognition would be chaotic.

Finally, we turn to Berdjaev's treatment of freedom and necessity. As with Bulgakov and Struve, he too held that the phenomenal world is thoroughly deterministic in theory. Whereas Kant wrote of the antinomy of freedom and determinism, Berdjaev admonished him for not realizing that there is no contradiction involved, since the two are entirely different categories. Berdjaev elaborated that the latter is a cognitive category, whereas the former is psychological.[71] If freedom is taken as indeterminism, as meaning absolutely uncaused, it is an illusion. The human will is indeed caused, but internally caused, for otherwise, morality would be inconceivable. Engels proclaimed that freedom is the recognition of necessity, but that view is deficient by not taking history into account. In each historical epoch, the progressive class represents the unity of freedom and necessity. People are free, feel themselves to be free when their will matches the demands of social progress. Our analysis must also, however, include nature. There is also an antagonism between freedom and necessity rooted in the forces, natural forces, hostile to our mastery over nature. With our ever-increasing recognition of our conscious freedom over nature, that antagonism itself decreases.

---

[70] Berdjaev, writing in 1935, remarked that he believed in 1901 that truth and justice "are rooted in transcendental consciousness. They are not of social origin. But there are psychological and social conditions – pleasant or not – that allow for the human cognition of truth and for the realization of human justice. ... I attempted to construct a theory, according to which the psychological and social consciousness of the proletariat as a class is free of the sin of exploitation, coincides as much as possible with transcendental consciousness, with the norms of absolute truth and justice." Berdjaev (1935), 7. Berdjaev's words in 1935 are a quite important elucidation and amplification of his position in 1901.

[71] Berdjaev (1999), 168.

The end is indeed, as Engels remarked, a kingdom of freedom. The historical liberation of humanity is not merely a psychologically subjective goal, but also the culmination of the struggle with the elemental forces of nature.[72] Berdjaev's first book, *Subjectivism and Individualism*, marked the apogee, the fullest statement, of neo-Kantian Marxism in Imperial Russian philosophy.[73]

In an essay written shortly after the book's publication, Berdjaev openly proclaimed a fight for idealism. Philosophers, he wrote, now understand that positivism, which Berdjaev regrettably did not define, is unsatisfactory. But the advancements of positive science in the nineteenth century were an eternal contribution of the bourgeoisie of the time to the treasury of the human spirit. Such progress, however, did not entail the adoption of metaphysical materialism. Only too often the opponents of progress hide behind the label of "idealism," charging others with materialism. The true idealists are those who serve as the instruments of historical progress, but the bourgeoisie interprets science and all human achievements as an affirmation of metaphysical positivism in order to diminish the spiritual life of its enemies. Berdjaev, in short, still depicted history in recognizably Marxist terms. The greatest merit of Marxism is in having established that only the material social organization of humanity can serve as the ground for the ideal development of life, that human goals can be realized only with economic dominance over nature.[74] The great deficiency of Marxism is its paucity of spiritual and cultural content. It added nothing to the bourgeois ideology that had emerged in the earlier struggle against medieval concepts.

In a sense, humanity aspires to be happy, but happiness is the result of a morally well-conducted life, not the goal itself. Berdjaev along with Kant recognized the absolute value of human life as an end in itself. Only through the recognition of this value now can we successfully strive for individual moral perfection and universal moral progress. These ideas led Berdjaev explicitly to Fichte, just as they did Struve – perhaps under the latter's influence. Indeed, Berdjaev agreed with Struve in "returning" to Lassalle, who understood the need for an idealistic spirit far more than Marx. Contrasting Lassalle's scheme to that of Marx, Berdjaev wrote, "In this respect, we stand closer to Lassalle. The rise of

---

72 Berdjaev (1999), 186.
73 Kiejzik, a contemporary scholar, remarks that Berdjaev's book "became a catalyst for his transition from 'orthodox Marxism' to 'critical Marxism,' because it was a criticism of Marxism in the spirit of Kant." Kiejzik (2018), 180. First, contrary to Kiejzik's statement, Berdjaev was never an "orthodox Marxist." Second, Berdjaev's book did not serve the purpose of being a catalyst, but Struve's preface to it may have done so. It was the following article by Berdjaev's own admission that served as the "transition" from Marxism to idealism. Berdjaev (1935K), 7.
74 Berdjaev (1901), 8.

the new society is not a dialectically occurring cataclysm, but the appearance of a new world-historical epoch which brings with it a great new idea that embodies general human progress. Lassalle pointed to the 'idea' of the modern epoch and to its bearer in a specific social class."[75] With this proclamation, we can see Berdjaev's decisive break with his past, with his neo-Kantian Marxism, into a neo-Fichtean Lassalleanism.

Naturally, the influence of Kant and of Marxism on Berdjaev persisted at least for a time. His next major publication was his contribution to the celebrated collection *Problems of Idealism*, which appeared in November 1902.[76] Berdjaev broke little new ground in his essay "The Ethical Problem in Light of Philosophical Idealism." Although he referred to and quoted Kant, his discussion introduced nothing substantial to what he had mentioned previously. The ethical, "as a principle given *a priori* to our consciousness," is independent of the factual.[77] There is a greater stress in this essay on the human person or individual, but the details too can be found in his earlier work. What is different now in late 1902 and placed Berdjaev outside the neo-Kantian sphere is his belief in the possibility of meaningfully speaking of metaphysics as traditionally understood. He wrote that he now recognized his 1900 book "reflected the shortcomings" of a transitional state of mind from positivism to metaphysical idealism and spiritualism," to which he had arrived.[78] He referred to Kant's construction of metaphysics by means exclusively of moral postulates as a skepticism and affirmed the "possibility of constructing metaphysics by various paths."[79] It hardly need be said at this point that any remaining vestiges of Marxism in Berdjaev's thought would hardly be recognizable by others as quintessential elements of Marxist doctrine.

Berdjaev in 1902 still referred, albeit sparingly, to German neo-Kantians with respect and restraint. By 1904, he became more sharply critical. In his essay "On the New Russian Idealism," which appeared in *Voprosy filosofii*, he claimed that idealism in epistemology largely leads to phenomenalism. Not surprisingly, the Kantian transcendental idealists, whether of the Marburg or Baden School,

---

75 Berdjaev (1901), 24.
76 For more information on the "history" of this work, see, in particular, Kudrinskij (1993). The final set of essays was considerably different from the original intention of the promoters, Struve and Pavel Novgorodcev. In any case, the book was intended for a wider audience than professional philosophers with little similarity to the technical topics being discussed elsewhere in Europe at the time.
77 Berdiaev (2003), 163.
78 Berdiaev (2003), 192.
79 Berdiaev (2003), 194.

come ultimately to a "constipated positivism."[80] Berdjaev's earlier endorsement of Kantian idealism was long spent. He now had high praise for a distinctly Russian intellectual tradition going back to the Slavophiles and Khomjakov, in particular, who had outlined a "concrete idealism" as against German "abstract idealism, which was later developed into an entire philosophical system by Solov'ëv."[81] Berdjaev saw the traditional ontologism of Russian thought set sharply and vividly against the transcendentalism of German philosophy. Whereas many at this time had hardly noticed the Marburg philosophers, Berdjaev singled out Cohen as "a resolute enemy of metaphysics and a genuine positivist." And Windelband of the Baden School was unable "to escape categories and concepts, ideas and values, to arrive at the expanse of being, of what is, its final instantiation being superindividual reason," in short the Deity.[82] By 1910, Berdjaev was even more sharply critical. He faulted the German neo-Kantians for their elevation of epistemology over what he took to be spiritual life, for their narrow focus on combatting psychologism, and for, above all, their attachment to science. He saw Kantianism and neo-Kantianism as a product of Protestant Christianity's devotion to individualism, thereby forgetting Cohen's Judaism and Riehl's Catholic upbringing.[83] We could, of course, continue, but Berdjaev had by this time surely passed beyond the scope of this study.

4.4 A Note on S. L. Frank

Semion L. Frank (1877–1950), a notable Russian philosopher, was a long-time associate of Struve's. He himself remarked in reflection later in life that he came to Marxism while still young, but, though running at times afoul of the law for participation in demonstrations, he never seriously considered leading the life of a committed underground revolutionary.[84] He also felt little attraction in these years to philosophy as a technical discipline.[85] While studying law at Moscow

---

80 Berdjaev (1904), 154.
81 Berdjaev (1904), 163.
82 Berdjaev (1904), 154.
83 Berdjaev (1910), 286. He also wrote, "Kantian epistemology, oriented to the fact of science, is a secondary, a second-rate epistemology." Berdjaev (1910), 289. There can be no lingering doubt; Berdjaev was not a neo-Kantian in 1910.
84 Boobbyer (1995), 15. Boobbyer's biography is not only an outstanding piece of scholarship, but an interesting read. Those interested in more particulars of Frank's life than what is presented here are urged to consult it.
85 In reflections decades later, Frank provided contradictory testimony, writing "In essence already in my secondary school years, I was interested in pure philosophy." Frank (1996a), 53.

University, he attended lectures by the neo-Kantian Novgorodcev, whom we shall see in considerably more detail in the next chapter. But, as he later wrote, it was chiefly under Struve's influence that he combined his understanding of Marxism with neo-Kantianism.[86] Make no mistake, though, he, unlike the others we have discussed in this chapter, produced no philosophical works that could meaningfully lie within the sphere of neo-Kantian Marxism. By the time Frank turned seriously to philosophy, he had abandoned Marxism. Yet owing to that very connection with Struve and his relatively brief flirtation with Marxism, he deserves some mention in this section. Whether he even was a neo-Kantian at all – even if only for a short time – depends on how broadly we understand that designation.[87]

Frank's first publication appeared in 1898, the same year that he met Struve and reflects the impression the latter must have made on Frank. His "A Psychological Direction in the Theory of Value" and subsequent book *Marx's Theory of Value and Its Significance* from 1900 deal exclusively with economics and are outside the scope of the present study – and this author's competence to evaluate.[88] He did call in his 1900 book for a revision of Marxist economics on the basis of new data and saw that work as "an attempt at such a critical evaluation of Marx's theory of value," showing what we must acknowledge as incorrect in it.[89] A second goal of the book was to show that the true significance of the labor theory of value can be revealed only in its organic connection with the theory of marginal utility. However interesting this might be, we find little evidence in the book of Frank's later philosophical interests.

---

[86] Frank (1996a), 54. What Frank meant by this is unclear.

[87] Frank, decades later, wrote, "But my soul did not lie in Kantianism. It was for me only an intellectual construction which never satisfied me inwardly." Frank 1996: 54. The present author certainly does not deny Swoboda's claim that "The influence of Frank's neo-Kantian reading is manifest everywhere in his early articles on philosophy." Swoboda 1995: 259. This "influence," however, did not make Frank a neo-Kantian of either the Marburg or the Baden School. Another view is that "in 1902–1904 Struve, Berdjaev, Bulgakov, and Frank considered themselves to be adherents of 'critical idealism'." Rezvykh (2018), 73–74.

[88] In recollections from years later, Frank stated that his article "A Psychological Direction" was "written in the spirit of Marx's theory." Frank (1996a), 55. We can observe that Frank erroneously gave the title of the article as "A Psychological Theory of Value" and has it appearing in the journal *Russkoe Bogatstvo*, whereas it actually appeared in *Mir Bozhij*.

[89] Frank (1900), v.

Frank's contribution to the 1902 collection *Problems of Idealism* displays little that we could meaningfully call "neo-Kantianism."[90] Frank's essay, entitled "Friedrich Nietzsche and the Ethics of 'Love of the Distant'," concerned not theory of cognition, but Nietzsche and morality. Kant and the German neo-Kantians did not feature at all in it. Moreover, he wrote to Struve in December 1901 that he "could not subscribe to metaphysical idealism or at least to recognizing that it is a compulsory position."[91] We must be cautious here in that Frank's notion of "metaphysics" may be significantly different from ours today or even from, say, Kant's. He also wrote in 1935 that at the time of the *Problems of Idealism*, "I became an 'idealist,' not in the Kantian sense, but as an idealist-metaphysical carrier of a certain spiritual experience, which opened the way to the invisible, inner reality of being."[92] What is not open to dispute is Frank's record of publications. He translated into Russian Windelband's *Präludien* in 1904[93], Kuno Fischer's volume on Spinoza from his series *Geschichte der neueren Philosophie* in 1906, and in 1908 Külpe's *Einleitung in die Philosophie*. These translations show at least Frank's interest in then-current German philosophical scholarship.

Frank's most significant philosophical publication from these years and the one showing his closest approximation to neo-Kantianism was his 1904 essay "On Critical Idealism." It reveals a debt to certain figures within the movement's German propagators, but none to either the Russian figures discussed in the previous chapters or to figures within the Marburg School. This may not be surprising in that Frank's thought was already moving rapidly beyond any neo-Kantianism, and he demonstrated, in any case, little interest in the philosophy of (natural) science. Frank would have us think that he is true to the spirit of Kant's "Critical Philosophy," even though he found Kant's presentation to be rudimentary and relatively incomplete. Still, the seeds were planted for a new philosophical synthesis first sown by Fichte.

True, Frank held Kant to be the founder of a "critical" theory of cognition, but above all Kant was the first to analyze the concept of truth. The German philosopher saw his concern to be with describing and explaining not the object of cognition, but human cognition itself. Kant's detractors often faulted him for

---

90 Swoboda writes but does not show that "neo-Kantian themes curiously intermingle" in Frank's essay. That Frank espoused the existence of absolute values is hardly a neo-Kantian conception. See Swoboda (2010), 211.
91 Quoted in Boobbyer (1995), 28.
92 Also quoted in Boobbyer (1995), 28. I am at a loss to understand what "metaphysics" is if this is not an example.
93 Berdjaev too in a letter from late September 1901 to Struve expressed an interest in translating at least a portion of Windelband's work. See Kejdan (2019), 72.

dogmatically assuming the veracity of judgments both in natural science and in everyday life, judgments the validity of which Hume had cast into doubt. However, "this 'pseudo-dogmatism' represented nothing other than a deeply original and masterful philosophical technique and necessarily followed from the completely new epistemological point of view that Kant revealed and on which he stood."[94] Had Frank stopped with this statement, we could easily see him as a neo-Kantian, provided, of course, that he also shared the position he ascribed to Kant.

Frank held little regard for the correspondence theory of truth. It assumes cognition can fully grasp its object, that reality lies, to use, if I may, a Heideggerian expression present-at-hand (*Vorhandenheit*). Kant, however, for Frank, recognized there to be an impassable chasm separating cognition from reality. Therefore, we cannot place the criterion of truth in some correspondence of our cognitions to externality. Rather, it must lie within cognition itself, in the immediate experience of cognition, in some relationship between the contents of consciousness, lest we lapse into an absolute skepticism. There is, however, no basis for skepticism; there are many judgments that we do recognize as true and must recognize as being so. Again, we must find the criteria of truth within the bounds of consciousness.

Frank omitted entirely the technicalities of Kant's metaphysical deduction of the categories – let alone the transcendental deduction! – but did conclude that for Kant cognition does not copy, but, rather, *constructs* reality. We need not belabor our discussion, but we should note the highly skewed presentation of Kant's epistemology that Frank – and indeed a number of others – presented. Owing to the functioning of our reason, consciousness draws a sharp distinction between our representations of "reality" and our "subjective" representations. The difference between the two is that "we cannot arbitrarily eliminate the representations that we call reality."[95] I cannot wish the objects before me to disappear, but I can wish away the centaur in my imagination. In other words, reality, through a function of reason, takes on the character of an aggregate of things existing independently of our perceptions. Frank held in high regard this allegedly Kantian idea. "One of the chief merits of Kant's theory of cognition is the clarification that objectivity or substantiality is merely a category of consciousness, i.e., a function of our reason."[96] This quotation shows that Frank emphasized the Kantian category of substance, apart from its schema, viz., the persis-

---

94 Frank (1904), 228.
95 Frank (1904), 232.
96 Frank (1904), 233.

tence of the real in time, to account for objectivity.[97] Frank mentioned just the two categories of substance and causality by means of which consciousness connects and integrates the chaotic sense manifold into the substantial beings that populate reality.

However misconstrued we may view Frank's understanding of Kant, one might conceivably view his position as neo-Kantian were it not that he believed others correctly took the next step in eliminating the concept of a transcendent reality. This next step, in his estimation, belonged to Fichte, but the final step was completed only in Frank's day by "immanent philosophy." What Fichte and especially immanent philosophy overcame is the idea that reality and the cognizing subject are utterly distinct and separate from each other. Kant's thing in itself is a glaring and explicit example of such a conception of reality as inaccessible. However, from the proper vantage point, it also reveals that cognized reality is part of an all-encompassing consciousness. What we take to be reality, to be substantial, is not all that is accessible to reason. There is more that does not fall under the category of substance. Other groups of representations are possible. There is the entire sphere of moral, aesthetic, and religious life.[98] These other spheres do not fall under the category of substance; the objects within it, therefore, are not substantial, but, in a sense, spiritual. He, however, did not regard any of this at this time as a lapse into metaphysics.

Frank had much more to say in 1904, particularly concerning ethics, which, as we know, he had sharply separated from the world of substantial being, of what is. A moral "ought" is not a thing, but it does belong to a sphere that reason can grasp. We can see already from what we have said that Frank had by this time left neo-Kantianism behind, though he saw its historical usefulness as a rung on a ladder toward a philosophical or, better, a metaphysical conception that he would elaborate in more detail in the coming years. That he saw matters in essentially these terms is also clear from another essay, this one from 1908 entitled "Person and Thing" that was originally published in the journal *Put'*. He reaffirmed there that Kantian "Critical Philosophy" "was not the end of philosophy, but the beginning. The contradictions between the complex interwoven motifs of Kantian philosophy demanded a new resolution, and the most sensitive Kantians recognized long ago that 'to understand Kant meant to go beyond him' (Windelband)."[99] Of course, we can see today that what Frank meant by "going beyond" was a transition or departure into un-Kantian metaphysics. Un-

---

[97] Frank's total neglect of the *a priori* intuition of time is evident here. Owing to this, he construed Kant along Cartesian, if not Berkeleyan, lines.
[98] Frank (1904), 238.
[99] Frank (1910), 164–165.

like in 1904, Frank in 1908 demonstrated considerably less fear in characterizing his thoughts as dealing with metaphysics. He recognized that although the general intellectual atmosphere at the time considered metaphysics to be dangerous, the task for those with creativity was to justify it and prove its legitimacy and necessity. Frank had definitely sundered any discernable remnant of neo-Kantianism.

4.5 An Obscure Neo-Kantian Marxist

Unlike the others we have discussed in this chapter, all of whom have received ample attention in the West and recently in Russia itself, Davydov is virtually unknown. Biographical information is scanty, and, apart from encyclopedic entries, secondary literature is limited to mentioning his name. Yet of all the neo-Kantian Marxists, Davydov upheld the synthesis the longest, even after the others had abandoned both Marxism and neo-Kantianism for metaphysics and Orthodox Christianity. Iosif Aleksandrovich Davydov (1866–1942) was born in Ryazan, where he attended secondary school, after which he studied for four years at the University in Tartu, in present-day Estonia and for two years in Moscow. He was an early member of the Social Democratic movement in those cities. From 1891, he wrote illegal brochures, which led to his arrest and internal exile for three years to the Russian far north. In 1900, he lived in Germany, France, and Switzerland. While in Berlin, he participated in the preparation of Social Democratic literature and in the smuggling of it into Russia. He returned to Russia in 1901 but was again arrested while attending an illegal meeting, this time receiving a sentence of internal exile to Siberia for three years.

During the early years of the twentieth century, Davydov wrote a number of works, as we shall see. But he was also involved in translating and editing. He participated, for example, in the 1907 translation of Stammler's *Wirtschaft und Recht*, mentioned earlier, and translated Pfänder's *Einführung in die Psychologie* in 1909, and Sigwart's *Logik* in 1908–1909.[100]

Davydov joined the Bolsheviks in 1920 and in September of that year began work at Petrograd University. He taught a course on historical materialism at the University for a short time but was dismissed owing allegedly to teaching in a

---

[100] Gustav Shpet reviewed the first volume of Davydov's translation of Sigwart in 1908. He stated that the translation's "merits and demerits were explained by one thing: its literalness. This leads to precision but sometimes to incomprehensibility in a phrase's construction." Shpet (2010), 86.

Kantian spirit. Nevertheless, he lectured on geography on an irregular basis until at least 1931. Unfortunately, he died in the Siege of Leningrad in 1942.

Davydov's principal works advocating and defending a neo-Kantian Marxism consist of a number of articles published between 1903 and 1905 and collected in a volume entitled *Historical Materialism and Critical Philosophy*. However, in the "Preface" to his translation of Stammler, he, referring to Kant's transcendental idealism, affirmed that the "synthesis of idealism and Marxism is by no means a simple mechanical mixture of disparate elements. On the contrary, it is something integral, organically intertwined that equally honors both great thinkers."[101] Unlike most examiners of Marxism and Kantianism, Davydov saw no opposition between the two philosophical doctrines. Thus, unlike those neo-Kantians whose slogan was "Back to Kant" and Struve who hailed, in effect, "Back to Fichte and Lassalle," Davydov exclaimed, "Forward with Kant, Fichte, and Lassalle but behind them."[102]

Despite the promising "Preface" to his work, though, Davydov is particularly concerned throughout neither with technical philosophical issues nor even with Kant and/or neo-Kantianism. His address is directed much more to his domestic Marxists, who had and would have little patience with any measure of (bourgeois) idealism. He, on the other hand, urged Marxists to be more receptive to metaphysical issues and, in particular, to ethics. Davydov saw both camps, the "orthodox" Marxists and the one-time Marxists who abandoned the creed for liberalism, as harboring a simplistic view of Marxism, as preaching a thorough dependence of thought on one's socio-economic class. One misunderstanding both groups had concerns the use made by the originators of Marxism of the term "materialism" as though it were a matter of philosophical materialism. "The expression '*economic materialism*' as the name of the doctrine is far from successful, or, in any case, it is not precise. For the 'materialism' with which we are dealing here has very little in common with philosophical materialism – only a single word."[103] Davydov claimed that Marx had been misunderstood in writing: "It is not the consciousness of men that determines their existence, but their social existence that determines their consciousness."[104] Marx did not intend it as an affirmation of either philosophical materialism or a reduction to materialism on the part of natural science. What he did mean is that in human socio-historical development there is no single directing

---

101 Davydov (1907), lxxi-lxxii.
102 Davydov (1905), vii.
103 Davydov (1905), 29.
104 Davydov quoted Marx without complete attribution. Davydov (1905), 107. See Marx's statement in the "Preface" to *A Contribution to the Critique of Political Economy* in Marx (1994), 211.

will. Social life "lacks a clear and distinct representation of the goal and of the paths leading to it."[105] Kant provided the philosophical basis for correcting this omission by the postulation of such a goal.

One might well see Marx's statement above that "social existence" determines consciousness as an expression of fatalism, as implying that human actions are necessarily determined by one's socio-economic class, but Davydov held that "nothing could be more erroneous than such a view."[106] Whereas it certainly is the case that within experience all phenomena are necessarily caused, as Kant already showed, we must remember that this causality is not something external to them, not some external force that dictates behavior as if on command. Determinism does not mean or lead to fatalism. Human freedom, i.e., moral and teleological freedom, is located on a different plane than the law of causality. Regrettably, Davydov failed to elaborate and clarify his position significantly. However, he gave every indication that he accepted to some ill-specified degree a Kantian understanding of practical freedom. For he also wrote that Kantian moral duty assumes a mental bifurcation between our sensibility and inclinations, on the one hand, and the moral, categorical imperative that tells us we must subordinate our will to duty, on the other.

The key idea in Davydov's synthesis of Marxism and Kantianism is that he examined phenomena – social phenomena – from a genetic point of view. His synthesis seeks a clarification of the developmental process of consciousness and its objectifying forms across history and social classes. Marxism is not interested in deducing consciousness from existence, but in understanding and explaining the historical development of the "content" of consciousness. For Marx, such an understanding and explanation is possible only if this "content" is connected to human socio-economic existence in general.

Davydov believed Kant's critical point of view has a quite different task, viz., the elucidation of all the *a priori* elements of cognition. Marxism presupposes the validity of Kantianism in its genetic inquiries as shown by its employment of causality and other categories of pure reason. However, Marxism has always relinquished its genetic point of view when it turned to morality when it appealed to human rights and to the eternal values of the self-conscious human individual.[107] Additionally, neither Marxism nor Kantianism renounces the quest for human happiness. Davydov quoted a long passage from Kant's *Critique of Practical Reason* concerning the position that pure practical reason does not

---

105 Davydov (1905), 112.
106 Davydov (1905), 107.
107 Davydov (1905), 128–129.

renounce claims to happiness but merely specifies that as soon as duty is in question one must disregard those claims. Such, Davydov affirmed, "is what Marxist thought always asserted, despite its sharply negative relation to all morality in general and to Kantian morality in particular."[108] Although we may be bewildered by this and many of Davydov's other statements, he found the idea of a Kantian-Marxist synthesis to be thoroughly consistent. The danger as he saw it was that the genetic method of Marxism had inadvertently underestimated, even ignored, the role and the importance of ideal factors in the historical process. The works of Stammler and Natorp, though, correspond most closely to the spirit and character of original Marxism.[109]

By 1907 the first four figures discussed above had made their peace with Marxist philosophy and moved on. Political tensions were rising, and efforts directed at theoretical syntheses were largely spent. Stammler's work had already appeared in a Russian translation in 1899 – and yet Davydov published another translation, as mentioned, in 1907 accompanied by a lengthy preface. Whereas in 1905, he emphasized Marxism as a socio-historical conception, the emphasis now in 1907 was clearly on the need for a neo-Kantian inspired "practical" philosophy. The goal, however, remained the same, viz., "the idea of a synthesis of idealistic and realistic conceptions, the idea of a synthesis of the doctrines of Kant and Marx."[110] Philosophers hitherto have developed their idealistic systems in complete isolation from the real world. Davydov continued to uphold that there was only one way to correct this deficiency while also recognizing the role of the law of causality in natural events, namely through the "materialistic understanding of history."

At this point, we may be brief. Few, if any, cared to continue and further elaborate on Davydov's project beyond what he had himself expressed.[111] Nevertheless, in 1907 his energy had not yet been entirely spent. Davydov still hailed Kantianism as capable of answering epistemological problems as well as those

---

108 Davydov (1905), 162.
109 Davydov (1905), 150.
110 Davydov (1907), lxx.
111 Whether by coincidence or whether he was intimately involved, there also appeared in 1906, the same year as the Stammler translation, a Russian translation of Franz Staudinger's 1897–98 "Die materialistische Geschichtsauffassung und praktische Idealismus." Writing under the pseudonym Sadi Gunter, Staudinger, himself indebted to the Marburg School, argued in this piece in support of many ideas quite similar to those of Davydov. See Gunter (1906). The name of the translator is not given.

that arise in the moral sphere.[112] What is surprising is Davydov's utter neglect of the entirety of Kant's epistemology except for the discussion of causality, something we saw already on the part of other Russian neo-Kantian Marxists. He continued to hold that causality reigned throughout the phenomenal world. However, he also recognized that Kant, in order to secure his moral philosophy postulated or presupposed a free will. Davydov interpreted this central Kantian thesis as meaning that the human will can desire and actively strive to attain material objects and emotional states. To say that the will is free does not imply that it is uncaused, that it does not belong to the phenomenal world. On the other hand, the conception of freedom here is not some noumenal freedom either. Davydov rejected the notion of a theoretically unknowable *Ding an sich*. What Kant had in mind "lies in neither the world of phenomena nor beyond it. But what is the third option beyond these two spheres?"[113]

Davydov failed to answer his own question clearly and analytically. Yet, he did write that the free will is the morally good will, which can never find its full expression in the given empirical conditions. The truly good will is an ideal, a guiding light in the distance to which human reason must strive. In Davydov's conception, the free will is a teleological category. Thus, the idea of freedom has significance independently of the given conditions, of reality. This idea of the ideal, of the absolute goal, is the driving force in historical materialism. To provide a teleological moral foundation to our social aspirations is the task awaiting a neo-Kantian Marxism.[114] Davydov saw the Marburg neo-Kantians, in particular Natorp and Stammler, as having done much of the needed theoretical work. What remained to be elucidated is how to implement the theory, how specifically to act.

Davydov's problem largely remained unaddressed. He surely thought that Marxism was the key, but he himself neither advanced further after unlocking the door nor explained the mechanism involved in the lock. The quest for a Russian neo-Kantianism by the end of the first decade of the twentieth century was a distant memory, like a "puppy love" that one looks back upon with heart-felt tenderness. Only Frank would continue in the years ahead to pursue something resembling academic philosophy. Bulgakov became a theologian and Orthodox priest; Struve was absorbed more and more in politics; Berdjaev turned into a

---

112 Although someone sympathetic to Davydov's neo-Kantianism might feel some satisfaction with his audacious confidence, Davydov did not address the major issues associated with adopting this standpoint. He appears, in fact, not even to be so much as aware of them.
113 Davydov (1907), xxxiii.
114 Davydov (1907), xvi.

public intellectual and religious existentialist; and Davydov, alone of the neo-Kantian Marxists remaining in Soviet Russia, was largely sidelined.

Chapter 5
Baden versus Marburg on Russian Soil

Two noteworthy and antagonistic philosophers of law in late Imperial Russia, serving as proxies, in effect, for the two most well-known schools of German neo-Kantianism, came to sharp blows in print not in Berlin, Marburg or Heidelberg, but in Moscow. The senior of the two figures published a number of works and has received significant and careful attention in English-language scholarship, arguably arising more, though, out of his political involvement than for his scholarship and juridical theorizing. Our focus here, however, is not on his political activism[1] or even his overall intellectual position, which would require an entire treatise of its own, but, more narrowly, on his stand as a *neo-Kantian philosopher* (of law) promoting natural law theory significantly influenced by German neo-Kantianism. The second figure we look at has received much less recognition in Western scholarship. He died a relatively early death and was not intimately involved in the political tumult of the day. His frank and defiant opposition to his teacher's position hardly endeared him to that teacher who, were it not for this dispute, was typically noted for his philosophical openness, warm personality, and defense of liberal values. In the most direct manner possible, the younger philosopher expressed an allegiance also to neo-Kantianism but of a *competing* German variety. Surprisingly, he expressed neither any apparent gratitude to those closest to him and who could have aided his career prospects nor did he express any appreciation for their overall philosophical stand.

Of the two German schools of neo-Kantianism best-remembered today, the Baden School is, arguably, the less documented and therefore less familiar to English-speaking students. However odd it may appear, though, the Baden School struck a deeper cord in Russian philosophy than did the Marburg School, undoubtedly owing to the former's greater emphasis on ethics and values,[2] whereas the latter at this stage in its development generally viewed philosophy as a second-order reflection on established (natural) science. Although the Mar-

---

[1] For a summary of Novgorodcev's political involvement and much more, see Walicki (1987), 294–298. Owing to his political involvement, he was eventually forced to emigrate. Novgorodcev settled in Prague, where he helped found an émigré school of law attached to the city's Charles University. He allegedly "gradually abandoned Neo-Kantianism in favor of a kind of Eurasian Orthodox mysticism." Dmitrieva (2016a), 384.

[2] The influence of the Baden School in Russia is also comparatively poorly documented. There is as yet no systematic treatment of that influence not just in English, but in any language including Russian.

burgians emphasized mathematics and physics, they did turn to jurisprudence and psychology. The respective approaches of the two individuals discussed in this chapter were largely shaped by these two divergent approaches.

With regard to our first figure, we could say, with only slight exaggeration, that natural-law theory formed the leitmotif of all his writings. This undoubtedly was to a large degree in reaction to what he perceived as the abandonment of natural law in the previous century. Lest we get ahead of ourselves, let us look briefly at some biographical facts concerning our first protagonist and then at how natural law fared in the 1800s in both Germany and Imperial Russia.

5.1 Natural Law in Russian Jurisprudence

Among the first to be impacted by German neo-Kantianism and the general spirit of the Baden School, in particular, was Pavel I. Novgorodcev (1866–1924), who led a distinguished professorial career at Moscow University and who, unlike the essentially apolitical Vvedenskij and Chelpanov, played a prominent role in both university and national politics. As a committed liberal during the tumultuous years in the early twentieth century, Novgorodcev opposed both those who would set the clock back and those who sought to destroy everything in the hope of realizing a future utopian ideal.

After his graduation from secondary school, Novgorodcev in 1884 enrolled initially in the department of natural sciences at Moscow University, but after a mere month requested a transfer to law. It is from the latter faculty that he graduated in 1888. Clearly an outstanding student, Novgorodcev was encouraged to further his studies, and, consistent with the Russian tradition of going abroad for additional training, he spent a considerable period of time abroad, particularly in Germany. Intending to visit the universities of Berlin and Heidelberg, he found a wealth of resources in the history of the philosophy of law at the former and remained there "almost two semesters," apparently during the academic year 1889/1890.[3] At the end of that year, Novgorodcev departed for Heidelberg in order to attend the lectures of Kuno Fischer, a prominent historian of philosophy and a commentator on it.[4] In the immediate years that followed,

---

3 Dmitrieva (2007b), 149.
4 In a short piece "Pamjati Kuno Fishera" ["In Memory of Kuno Fischer"] written shortly after Fischer's death, Novgorodcev wrote that he heard Fischer lecture in 1890 and then in 1894. Novgorodcev (1907), vii.

Novgorodcev would return to Berlin for further studies but also went to Paris and Freiburg.[5]

Novgorodcev defended his *magister*'s thesis *The Historical School of Jurists, Its Genesis and Fate* in 1897, although it had appeared in print already the previous year, a common practice at the time. He remarked in it that he had initially hoped his work would amount to a study of the opposition to natural law in nineteenth-century German literature. The intention was to cover that reaction as manifested in more than a single literary movement. However, as he familiarized himself with the so-called Historical School of Law and the enormous role it played in nineteenth-century German jurisprudence, he decided to devote his thesis solely to that movement including its emergence and evolution.[6]

Whereas many within the German Enlightenment of the late eighteenth century had championed natural law, seeing it as embodying reason and thus as the standard by which to judge a country's positive laws, there certainly were others who questioned the very existence of a body of natural law that transcended earthly temporal institutions of any kind. For these critics, positive laws had to be seen in the context of their particular time and place, i.e., within their respective cultures. These early representatives of historicism, however, refrained from carrying this train of thought to its logical end for fear of lapsing into relativism.[7] Its later German proponents would not be so reluctant to take that final step.

Novgorodcev thought the Enlightenment's championing of natural law was undermined and abandoned with the rise of a sense of history in the nineteenth century and its comparatively subdued political mood. A major figure in this movement away from natural law was Friedrich Karl von Savigny (1779–1861). Whereas the proponents of natural law saw it as a measuring rod for judging the rationality of positive law, Savigny came to reject the reliance on reason in

---

[5] The specific dates and locations of his studies abroad are poorly recorded. Walicki writes that in these early years he was "mainly in Freiburg, where he was strongly influenced by the Baden School of neo-Kantianism." Walicki (1987), 294. Walicki provides no reference. Possibly, he relies on Putnam, who also alleges without reference that Novgorodcev worked "mainly in Freiburg, where he came strongly under the influence of the Neo-Kantians." Putnam (1977), 34. Note the similar wording in the two quotations. In an introductory essay to his recent publication of Novgorodcev's doctoral dissertation, Al'bov, again curtly and without reference writes, "In the period from 1890 to 1899 Novgorodcev was on a study trip to Berlin and Paris." Al'bov (2000), 7. There is no mention of either Freiburg or Heidelberg! Glukhikh writes that Novgorodcev "spent more than four years abroad at German and French universities." Again, the basis for this claim is not provided. Glukhikh (2002), 322.

[6] Novgorodcev (1896), i.

[7] Beiser (2011), 216.

the Kantian and Fichtean moral systems. There is no transcendental grounding of law and no resolute human nature upon which political institutions could be erected. Human beings are molded by their institutions.

Novgorodcev's focus throughout his *magister*'s thesis was on German philosophy of law. This was not just politically expedient, but intellectually as well, since Germany was viewed in Imperial Russia as the most sophisticated nation philosophically. Nevertheless, one can hardly imagine that he was unaware of the current situation within his own country and of the history of natural law theory therein. During the last decades of the eighteenth century and the first decades of the nineteenth, teaching of natural law, with the proviso that it included a defense of the monarchical form of government, was common in the universities of the Russian Empire.[8] Instruction was to prepare an enlightened officialdom, the representatives of which were to serve as functionaries of the government in its necessary interactions with the people. For example, in one of, if not the first, Russian-language works in jurisprudence, Vladimir T. Zolotnickij (1741-?) in 1764 explicitly adopted positions drawn from others primarily in Western Europe. However, his text also contained such deep [sic] insights as that the human mind, being naturally rational, recognizes the difference between good and evil and inclines toward the former while rejecting the latter. The first precept of natural law is that of the many actions that are within our power, we should choose those that are useful and honorable.[9]

It did not take long, however, for someone to appear who understood that an advocacy of natural rights could lead to radical political and social conclusions – and he went on to pay a heavy price for his forthrightness. Aleksandr Radishchev (1749–1802) was the first Russian writer to invoke the notion of natural rights in determining the legitimacy of autocracy. Realizing that positive laws serve to bind society into a reasonably harmonious whole, he cautioned against appealing to natural law except in extreme cases of the abrogation of justice.[10] Articulating his stance under the reign of Catherine the Great, Radishchev saw copies of his chief work *Journey from St. Petersburg to Moscow* confiscated wherever they could be found. He himself was first condemned to death but through

---

8 For a much more detailed and systematic treatment of the teaching of natural law in Russia, see Berest (2011), 105–142.
9 Zolotnickij (1764), 10. Zolotnickij wrote in addition to this text on natural law, *An Abridgment of Natural Law*, a work entitled *Rassuzhdenie o bessmertii chelovecheskoj dushi* [*A Discourse on the Immortality of the Human Soul* (1768)] and yet another related work, *Dokazatel'stva o bessmertii dushi* [*A Proof of the Immortality of the Soul* (1780)]. Unfortunately, the present author was unable to obtain either.
10 Walicki (1987), 21–23.

an abject demonstration of contrition had his sentence commuted to Siberian exile.

Despite the example of Radishchev and the lengths to which it could be employed, natural law continued to be taught in the nation's higher educational establishments. For example, Lev A. Cvetaev (1777–1835), one of the first ethnic Russian professors of law and who was sent abroad for Western training in Göttingen and Paris, taught natural law – and morality – on the basis of Kantian philosophy at the University of Moscow. At the very start of his 1816 work *First Principles of Natural Law*, Cvetaev stated, "Natural law is the science [*nauka*] of external and absolute human duties and laws. ... It is a science gleaned from reason and the properties of its relations."[11] Cvetaev did manage to conclude from his essentially Kantian standpoint that according to natural law other human beings should not be treated as means, but he held back – wisely perhaps – from outrightly condemning any politically sanctioned institution, such as serfdom.[12] He, unlike Radishchev, encountered no career or personal obstacles. The same cannot be said of the next proponent of a Kantian-inspired theory of natural law.

Kunicyn's two-volume text *Natural Law* from 1818–1820 would not in the normal course of things have received any special attention from governmental authorities. Aleksandr P. Kunicyn (1783–1840) was another in a highly select group of young students sent abroad for further academic training at the start of the reign of Tsar Alexander I. Undoubtedly exposed to Kant's thought while in Göttingen, he, unfortunately for us, neither kept a diary nor wrote detailed written impressions of German university life. Kunicyn, upon his return to Russia, received an appointment to teach at the newly opened and highly selective Tsarskoe Selo Lyceum, undoubtedly revealing in 1811 the ministry of education's great esteem for Kunicyn's intellectual and moral qualities. Later in that decade, Kunicyn left the Lyceum to become a professor of law in St. Petersburg at the Main Pedagogical Institute that would shortly later become St. Petersburg University.

Natural Law was intended from the start to be a textbook along the expository lines common at the time. It stated – not argued – that there are universally valid natural laws concerning human rights independent of political institutions,

---

[11] Cvetaev (1816), 1. The Kantianism of Cvetaev's moral teaching is clear from his formulation, "The universal moral law is: Do not treat humanity either in oneself or in others merely as a means, but always as an end." Cvetaev (1816), 6.

[12] Berest (2011), 127. Berest is certainly correct in her assessment, but we have with Cvetaev another historical case of someone preaching one thing but who refrained from advancing one's thought when he realized it ran counter to one's own interest.

that these rights follow from reason itself, that ethics, as a rational discipline, teaches us to hold all those who are also rational as thereby free, and that we should never treat others merely as means to an end. Certainly, we can find these moral injunctions in other Russian texts of the time, albeit perhaps with less force and bluntness than in Kunicyn's. Whatever the case, the mood of officialdom was already slowly turning away from any encouragement, however tepid it may have been, of Enlightenment values that we find in Alexander I's first years. In 1821, Kunicyn's book was found to contain ideas contrary to Christian teaching and harmful to state and family relationships. Kunicyn was fired from the University, and the copies of his book that could be located were seized. Also in that year, a campaign began in earnest to forbid the teaching of natural law and philosophy in the country's educational institutions. The prohibition was not entirely successful, but the teaching of natural law where it continued to be done largely reverted to pre-Kantian models.[13] The Decembrist Revolt or Uprising of late 1825, though summarily quelled, taught the new Tsar Nicholas I that it was best to avoid the codification of law based on abstract conceptions of justice derived from some universal human nature. The adoption of the German Historical School's approach, viewing a nation's laws as expressions of its particular characteristics, was far less likely to lead to criticism and political dissatisfaction.[14]

We should also not forget the contribution to the attack on natural law afforded by the anti-Western Slavophile movement. Although the ideas of its representatives were shaped by what they regarded as the spirit of their Russian Orthodox faith, their leading thinkers were versed in post-Kantian German philosophy. In particular, their "views on law were determined by their opposition to juridical rationalism, which they saw as peculiar to the West."[15] In the words of one of the movement's leading ideologists, "Western jurisprudence infers from each legal case logical conclusions, stating 'the form is itself the law,' and tries to connect all forms into one rational system, where each part by the necessity of the abstract mind is correctly inferred from the whole. ... Law in Russia was not invented in advance by scholars of jurisprudence ... but, as a rule, was written on paper after it was formed in the popular mentality and little by little was forced to become part of the nation's customs and way of life. A logical development of law can occur only in those cases when society is itself

---

13 Berest (2011), 106.
14 Wortman (1976), 43.
15 Walicki (1987), 35.

based on artificial conditions."[16] Thus, whereas the German Historical School outlook could be characterized as politically conservative, the Russian Slavophiles were manifestly anti-Western reactionaries in principle, yearning for a past way of life that largely existed only in their imagination.

In any case, the ground in Russia for the rise of legal positivism against any talk of universal natural rights was, thus, well prepared by the second half of the nineteenth century.[17] Its representatives rejected all metaphysical philosophy, including in their eyes the existence of natural law. These positivists dismissed the dualism promoted by the defenders of natural law, who set it against established law. The positivists saw natural law as an unwarranted doubling of the legal order that recalled Plato's regrettable bifurcation of the factual and the ideal. What has hitherto been called "natural law" was but an assessment of established, positive law from an ethical point of view, i.e., an evaluation of positive law in terms of what the advocates of natural law thought should be positive law.[18] These opponents of natural law saw their position as a "scientific jurisprudence" in that the understanding of law was to be kept within the bounds of empiricism, and the study of law was to be of phenomena of real life, not of metaphysical principles.

5.2 Novgorodcev on Natural Law

We can view Novgorodcev as agreeing with Shershenevich (1863–1912) that the idea of natural law – if there is such – provides the basis for moral evaluations of positive institutions and laws, and such evaluations are possible only by investigating the basis of legal right. Since moral criticism of natural phenomena is impossible – moral judgments apply exclusively to human actions – and if the Historical School of Savigny is correct in holding that positive laws are formed naturally and thus involuntarily, then we would have to exclude the possibility of morally evaluating them. From Novgorodcev's perspective, the direct consequence of the positivist and historicist denial of natural law is a moral and political quiescence, which, given Novgorodcev's personality, was totally unacceptable. Thus, assuming that we must be able in principle to make moral

---

[16] Kireevskij (1911), 207–208. These words, of course, express merely Kireevskij's opinion. He offered no substantiation in the least for them, no evidential argument.
[17] For a general survey of Russian positivism at this time, see Nemeth (2018).
[18] Shershenevich (1911), 35. Shershenevich's work is representative of late legal positivism. Written with full knowledge of Novgorodcev's own views, Shershenevich, nevertheless, clearly summarized the long-held positivist stand on natural law.

evaluations of positive laws, how is that possible? Unless we agree to lapse again into relativism, the conditions for such evaluations cannot be seen as arbitrary and essentially contingent. Such is the train of thought that led Novgorodcev to resuscitate the concept of natural law. In his *magister*'s thesis, he concurred with Windelband in viewing philosophy (of law, in Novogorodcev's case) as a normative discipline.[19]

Novgorodcev's thesis sought to show the untenability of denying natural law. He held that Savigny's original formulation itself already contained substantial limitations and qualifications. As the Historical School of Law developed, the inspiration behind the original rejection was left behind, leaving a resolution to the problems that directly conflicted with Savigny's conclusions.[20] This result eventually led to a revival of natural law. And it is in this sense that we can see a Hegelian historical dialectic operating in Novgorodcev's argument. His neo-Kantianism was only implicit; those without knowledge of his later trajectory after his 1896 thesis could just as well see him here as a defender of a traditional metaphysics. Novgorodcev's neo-Kantianism came to the fore in trying to resolve additional issues involved in his defense of natural rights, although there is no reason to doubt that he was already at this time a neo-Kantian.

In his contribution to the 1902 collection *Problems of Idealism*, Novgorodcev mentioned at the start that his attempted revival of natural law in his *magister*'s thesis had met with "mistrust and doubt" in some corners of the Russian legal profession.[21] This reaction notwithstanding, however, the idea of natural law, which encountered such derision in the nineteenth century, was undergoing a revival. Novgorodcev pointed to developments within German jurisprudence, but he could just as well have mentioned those within his own circle, namely the other contributors to that 1902 collection. He also could have referenced his own work, *The Teachings of Kant and Hegel on Law and the State*, published the year before and which he defended as a doctoral dissertation in 1902. It is hardly surprising, given Novgorodcev's confrontation with natural law, that he would turn to Kant for at least a hint concerning how to answer one of his questions.

---

19 I am not aware of Novgorodcev ever commenting in detail on the methodology of the natural sciences. Since, however, he viewed philosophy to be fundamentally evaluative and we can evaluate only human actions, there can be no *philosophy* of nature. His discussion of the methodology of the natural sciences is always evoked in order to contrast it to philosophy proper, i.e., seen as a normative discipline.
20 Novgorodcev (1896), 217.
21 Novgorodtsev (2003), 274.

Novgorodcev had the highest praise for Kant's ethical philosophy. "The history of philosophy has up to now," he wrote, "not seen a more serious attempt to ground an independent position for the ethical method."[22] Owing to Kant's demonstration of this independence, the ethical method had found its genuine foundation, which eluded other attempts that tried to ground morality on the objectivism of natural science. Unlike scientific cognition, which, according to the Kantian scheme, rests on the sensible manifold and is formed by us, our own reason in the practical sphere creates the law for our will to act upon. In the theoretical sphere, objects are ultimately the cause of representations; in the practical sphere, the will itself creates its object. The objective reality of the moral law is given immediately by and to reason. Like Windelband with his methodological distinction between the normative and the natural, Novgorodcev too saw Kant as arguing that every action can be both free and necessary at the same time depending on the point of view taken. An action "is necessary if we take it as part of a series of temporal causes and effects, and free if, released from this series, we ascend to its intellectual foundation, to the subject as thing in itself."[23] Novgorodcev's approach allowed for a never ending quest for causality in the natural sciences, for the discovery of laws of existence by the natural scientist, while allowing the philosopher to undertake without hindrance the establishment of the norms of what should be. "Those who make moral evaluations of phenomena do not attempt to explain the causes of those phenomena. Evil remains evil, whatever its cause may be."[24] In this way, Novgorodcev following

---

22 Novgorodcev (2000), 31.
23 Novgorodcev (2000), 142. His invocation of the "thing in itself" here is quite odd, and it played hardly any role in his thought. Windelband in his 1910 essay would remark, "In this sense transcendental idealism has no more need of 'another world', as Kant originally would have deemed necessary in the concept of the 'thing in itself'." Windelband (2015b), 323. It appears, then, that Novgorodcev was by no means a slavish follower of Windelband even in epistemology.
24 Novgorodcev (2000), 150. A most disconcerting element in his discussion here is its neglect, unlike previously, of the rationality of ethical evaluations. He wrote, "Moral duty is an idea (*ideju*) immediately given to our consciousness and which cannot be decomposed into any further elements. ... To the question whether we should or why we should, there is no other answer except for the testimony of the moral feeling. This is why for empirical science the moral problem is insoluble." Novgorodcev (2000), 148. The danger here is that if moral duty is a feeling (*chuvstvo*), we lapse into subjectivism, a fear that became increasingly acute later in Russian neo-Kantianism.

Windelband avoided the two-worlds interpretation of Kant's idealism and any possible charge of psychologism.[25]

Novgorodcev saw that despite his high regard for Kant's ethics it had limits and gaps. One problem is that Kant's concern was exclusively with the purely subjective moral will. Kant allowed no measure of experience to enter into the means of determining the principle upon which the will is to be determined. How, then, can an abstract law be unambiguously applied in real life? A related question, one which Novgorodcev found particularly compelling, concerned how to transition from Kant's subjective ethics to an objective ethics. The moral law makes no claim on others.[26] "The chief inadequacy of Kant's formalism is that it is closed to real relationships in the world and its conceptions are restricted to abstract determinations."[27] The absence of an objective ethics in Kant is the best evidence of this deficiency. If normative considerations cannot be conjoined with those in social philosophy, this still does not mean that there is no room in philosophy for the latter. Hegel, in Novgorodcev's eyes, sought to remedy the situation by turning to the objective side of morality. However, Hegel was inclined to go to the other extreme, subordinating the individual to the social and forfeiting the middle ground between the idea of socially organized morality and the principle of the individual's moral autonomy. Novgorodcev saw the correct resolution to lie in accounting for their coexistence and mutual recognition. Such was the task and the challenge he set for himself and one that he saw as pressing in his day.[28] Novgorodcev's inspiration in practical philosophy remained Kant, even though he understood that Hegel had offered adjustments.

In his essay "Morals and Cognition" from 1902, the same year as his dissertation defense, Novgorodcev largely reiterated and thereby reaffirmed the positions he had articulated previously. His concern as before was with normative

---

25 Windelband devoted his 1882 essay "Normen und Naturgesetze" to this matter. See the reprint of this essay in the collection Windelband (1911b), 59–98. Interestingly, Novgorodcev mentioned in a footnote Windelband's article as well as two others, but he did so for their "interesting and also thorough clarifications of the *a priori* nature of moral principles." Novgorodcev (2000), 148.
26 Closer to our own day, Stuart Brown trenchantly framed the quandary. Kant attempted "to apply the principles of his moral philosophy to positive law in order to distinguish between what is just or dutiful or right and what may be merely lawful, a legal right or a legal duty. But Kant fails in this attempt. He fails, not because he was careless or inept in stating his principles and applying them to positive law, but because his principles have no application to positive law. For this reason, Kant has no philosophy of law." Brown (1962), 33. Kant never shows how the categorical imperative can be applied to test the morality of positive law.
27 Novgorodcev (2000), 236.
28 Novgorodcev (2000), 351–352.

determinations of what should be, not with what factually is. Practical reason is not subject to the conditions or limitations of theoretical reason. If it were, practical reason would merely be an extension of the theoretical and lose its autonomy. Whereas a check on the validity of theoretical reason is empirical evidence, the only evidence in support of the validity of practical reason is the voice of our immediate conscience.[29] In this essay, Novgorodcev found the "deficiencies" in Kant's ethical philosophy to lie in its inability to explain how to realize unambiguously the moral law, which stems from his "hopeless dualism between the inner and the outer worlds."[30] There must be a means to reconcile the antithesis between what is and what should be. If we are to maintain that the two worlds can be reconciled, that the non-factual "ought" can be realized in the "is," we must hold to a higher synthesis of the two, a synthesis that can take place only in the transcendent sphere. But this synthesis cannot be at the expense of either the factually given or moral demands. This third sphere of metaphysical presuppositions or convictions reveals the need for a final harmony. A recognition of the connection between these convictions and our moral awareness affords a new firmness to that awareness, but it offers no proof that we *must* be moral.

Referring to his discussion in his dissertation, Novgorodcev mentioned there that Hegel had attempted a synthesis of what is and what should be in the concept of the absolute existent. However, since such a reconciliation is achieved in the sphere of the transcendent and only therein, Hegel's effort in this respect represented no advance over Kant. The reconciliation can be manifested in human consciousness only as an *a priori* imperative and not otherwise in the empirical sphere.[31] Novgorodcev's own thinly disguised invocation of belief in a deity technically differs from Kant's postulation of the existence of God, but in both cases, the result is the same: we must believe in order to insure the effectiveness of our moral convictions in the real world. For both Novgorodcev and Kant, the desire is to see moral progress in successive generations.[32] For the former, the fulfillment of this progress requires an objective ethics as found in a complete ethical system, which Novgorodcev found lacking in Kant. Despite realizing such a need, Novgorodcev himself did not present one.

---

29 We can hardly give much credence to such a view. Unlike a conclusion of theoretical reason, which can be falsified by empirical evidence, there is nothing that can falsify practical reason's "voice." If another "voice" in conscience should proclaim that the first voice is wrong, there is no standard by which to adjudicate the dispute.
30 Novgorodcev (1902a), 831.
31 Novgorodcev (2000), 266.
32 Novgorodcev (1902a), 834–835.

In his already mentioned 1902 essay "Ethical Idealism in the Philosophy of Law", arguably his best-known work, Novgorodcev provided a highly erudite, though quite verbose, assault on historicist philosophies of law, and particularly on any deprecation of natural law and a reduction of morality to what is or has been. Although he conceded with Kant that the supreme moral criterion has a purely formal character, he agreed with Stammler concerning the idea of natural law with changing content and that this directly followed from the basic concepts of ethical idealism. However, Novgorodcev explicitly added here an idea at least adumbrated earlier that "the formal moral principle is the recognition of the idea of eternal development and improvement."[33] The moral principle excludes only stopping at some concrete goal, viewing it as atemporal.

Novgorodcev faulted Hegel for treating social development as a process separate from the development of the individual person and "as reflecting the manifestation of absolute spirit."[34] Instead, he believed objective ethics is a combination of the rational moral principle with the circumstances of social development. Without social development, there can be no improvement in the objective ethical condition of society. However, the objective ethical condition is a means, not the end to individual human development, which remains, as Novgorodcev, put it "the foundation and goal of morality."[35] That development is not guaranteed, but must be actively sought with the goal being continuing improvement without end. He emphasized this infinite striving for development in his later *On the Social Ideal*, published in book form in July 1917.[36]

Novgorodcev in 1917 reaffirmed, as if necessary, the unbreakable connection between the moral principle and the value of the human individual. It, he remarked, is Kant's greatest legacy. However, it is impossible to derive from it the concept of society. The transition from the individual to the social in his works remains unclear. The supplementation of a social philosophy onto his

---

33 Novgorodtsev (2003), 309. Let us assume from the context that Novgorodcev had in mind moral development and not necessarily physical improvement. Still, we must ask what is the source, the basis, of this notion of development and improvement, which implies temporal duration as well as content. If the form is purely rational, how can it undergo development and improvement? Is it possible for pure practical reason to become purer practical reason? Novgorodcev wished to resolve the issue of reconciling the normative and the empirical spheres, but how is that possible? This was his question from the start but to which he supplied no rigorous answer.
34 Novgorodtsev (2003), 312.
35 Novgorodtsev (2003), 312.
36 The book had been serialized over the course of ten issues of the journal *Voprosy filosofii* from 1911–1917. The "Preface" to the first edition of the compiled book is dated 16 July 1916, and a second edition quickly appeared the following year.

commitment to Kantian ethical philosophy is what made Novgorodcev a neo-Kantian. The social philosophy he envisaged is at least Kantian in spirit. The goal of society is not the accumulation of power or wealth; it is not even the greatest happiness for the greatest number. The point, rather, is the realization of equality and freedom for all. Happiness will be a *consequence* of this realization, but not the goal as such. Novgorodcev sincerely believed that among the conditions for the achievement of freedom and equality were communication and mutual recognition.[37] Despite the Kantian spirit of this train of thought, it harbored many assumptions that Novgorodcev never addressed. His legal and social philosophy remained philosophically inspiring but technically superficial. We saw when discussing his book *Kant and Hegel* that he found Kant's formalistic ethics to be replete with abstract determinations at the expense of real world relationships. A year later in his "Ethical Idealism in the Philosophy of Law" he, in effect, retracted his position, writing "it is necessary to turn to the *a priori* instructions (*ukazanie*) of moral consciousness, which in its essence, independent of any experience, contains the bases (*dannye*) for evaluating any empirical material."[38] Thus, apart from reversing his earlier position and despite saying moral consciousness has the ability to evaluate the *a posteriori*, Novgorodcev has not shown (a) that that is the case, nor (b) what those "instructions" are. He simply declared the problem was solved.

5.3 A Marburgian Philosophy of Law

Whereas Hermann Cohen, the founder of the Marburg School, is chiefly known, insofar as he is known, in English-language philosophical circles for his interpretation of Kant's epistemology as an inquiry into the possibility of Newtonian physics, he did not enter Russian philosophy out of that concern. The first monograph in Russian devoted to the exposition of Marburg School neo-Kantianism was the *magister*'s thesis of Novgorodcev's student, Vasilij A. Saval'skij (1873–1915),[39] who defended at the expense of virtually all else, the ethical and legal philosophy of Cohen. Regrettably, we have little information concerning Saval'skij the man. He left no memoirs, no diary. From several sources, however, we gather that he was born into the rather modest family of a local Russsian Orthodox priest in a Caucasian village, that he studied first at a theological seminary,

---

[37] Novgorodcev (1991), 107–111.
[38] Novgorodcev (1902b), 255. Cf. Novgorodtsev (2003), 287.
[39] Belov (2019), 154f.

then at a classical secondary school, which he finished in 1895. He, then, enrolled in the law faculty at Moscow University, where he showed promise particularly with a work on Spinoza. After his university graduation in 1899, he remained at the University for further preparation and study with Novgorodcev. In the first years of the new century, he appears to have gone to Heidelberg and on returning to Moscow taught philosophy at the Higher Women's Courses, and in 1904 he became a *privat-docent* at Moscow University.[40] Saval'skij was awarded a two-year scholarship to study in Germany in January 1905. He spent the first year in Berlin and Freiburg and for some indeterminate time, but at least the summer semester of 1906, in Marburg and then Halle. He published his thesis *Foundations of the Philosophy of Law in Scientific Idealism* in 1908 and defended it in May 1909 in Moscow.[41]

Saval'skij in his first publication "Critique of the Concept of Solidarity in the Sociology of A. Comte" published in 1905 displayed a debt to Kant but not to any one school of neo-Kantianism. Although the references to works by Russian neo-Kantians (Novgorodcev, Lappo-Danilevskij) are few, they are all respectful. The central theme of the piece is that Comte used the term "solidarity" in two discordant senses. He employed it in connection with causation in the physical world and also in a teleological sense when writing of development in the biological sphere. In the one case, it is constitutive and in the other regulative.[42] Critical Philosophy, however, has shown that the constitutive and the regulative must not be confused. Saval'skij ended his essay, writing "Critical Philosophy teaches that the phenomena of natural science and of history in the broad sense, i.e., the phenomena of nature and the phenomena of the human spirit, differ not quantitatively, but qualitatively and, consequently, fundamentally."[43] One could hardly anticipate from this the polemical tone that would emerge from Saval'skij's pen in a couple of years.

Although scholarly attention has understandably focused on his much more detailed thesis, Saval'skij already in 1907 published a relatively short article in which his commitment to the Marburg School was clear enough and largely at the expense of his mentor in Moscow. Saval'skij shared along with Novgorodcev the view that natural law forms the foremost problem in philosophy of law. The

---

[40] Belov (2019), 152.
[41] The copy in this author's possession clearly gives the date of publication as 1909. Dmitrieva in the bibliography for her excellent secondary study of Marburg neo-Kantianism also gives 1909. Dmitrieva (2007b), 440. Yet, she and Belov in their respective texts give the date as 1908. Dmitrieva (2007b), 193; Belov (2019), 152.
[42] Saval'skij (1905), 105.
[43] Saval'skij (1905), 106.

former added, however, that whereas there had been brilliant and fundamental critiques of positivistic theories of law by Russian authors such as Boris Chicherin and Vladimir Solov'ëv, the thrust of the "back to Kant" movement was to free philosophy from mysticism and metaphysics. Saval'skij acknowledged that with its concern for the inscrutable Kantian concept of the thing in itself, the movement broke into hostile camps. One, which centered around Windelband and Rickert, saw Fichte as a step forward and thereby revealed an inclination to metaphysical constructions. These German philosophers ultimately could not rid themselves of psychologism. In contrast, the Marburg philosophers rejected the alleged advancement made by Fichte and presented a quite different conception of Critical Philosophy.[44]

Had Saval'skij ended with charging the Baden philosophers with neo-Fichteanism, he may have encountered measured rebuke but little more. He did not. Saval'skij pointedly claimed that Novgorodcev and Kistjakovskij, a figure we shall see in the next chapter, belonged to the Baden School in significant aspects of their Kant-interpretation. However, he added that their commitment to Baden neo-Kantianism had its roots to a significant degree in the fact that that School's conceptions were related to those of Solov'ëv. "Novgorodcev presents the concepts of his critical philosophy together with an understanding of Kant that is chiefly given by Solov'ëv and Windelband."[45] Saval'skij continued by criticizing Novgorodcev for not providing elaborations of the positions he advocated that clashed with the Marburg reading of Kant. "In none of his claims does the author defend his critical philosophy from the directly contrary assertions of the Marburg School's critical philosophy. The mentioned aspect of Novgorodcev's critical philosophy obviously needs replenishment."[46] Statements such as this were hardly likely to endear Saval'skij to any professor and were, to be frank, an imprudent claim on his part toward one who could at least influence his career prospects.

Saval'skij's piece, despite its brevity, was not without criticism of the Marburg philosophers, particularly as represented by Stammler, who, the former charged, lacked a concept of culture. Properly speaking, Saval'skij wrote, nature is not opposed to the "social," but to culture, in which the "social" sphere belongs. However, should someone ask for an introduction to the philosophy of law, Saval'skij incongruously opined that he would answer the *Critique of Pure*

---

44 Saval'skij (1907), 10.
45 Saval'skij (1907), 11.
46 Saval'skij (1907), 11.

Reason and that for a deeper understanding of that text one should be guided by Cohen's *Ethik des reinen Willens*.[47]

Saval'skij's thesis was purportedly an intervention into and resolution of the dispute between what he took to be the neo-Fichtean and Marburg Schools. There was no mistaking at the very start how Saval'skij viewed the situation. "In this work, the author wishes to show that if in general a 'return to Kant' is necessary, this sought-for-Kant is provided not by the neo-Fichteans, but by the Marburg School and only by it."[48] The post-Kantian German idealists, not understanding the Critical method, moved toward the construction of metaphysical systems, sundering a vital connection with science. Saval'skij considered this connection with natural science to be paramount including for the understanding of the Critical or transcendental method.[49] The representatives of the Baden School, however, were unfamiliar with science; their philosophy could properly be termed normative-critical in distinction from the Marburg's transcendental-critical philosophy. Whereas the former elaborated Critical Philosophy as a historical fact and in doing so returned to Fichte, the Marburgians were concerned with the sense of Kant's philosophy, viz., the logical structure of the concepts used in natural science, in particular mathematical physics, and thus the legitimate bounds of their application. Cohen took the natural science of the day as established fact and inquired into the theory grounding it. The Baden School, on the other hand, worked with invented cognitions. The object of cognition for its representatives, such as Rickert, was not what exists, but what should exist. In short, they treated natural science as if it cognized what *should* exist, not what *does* exist.

Saval'skij's thesis, understandably, is largely an exposition of Cohen's reconstruction of Kantian philosophy, which in Saval'skij's eyes was necessary owing to its many vacillations.[50] We need not repeat here the many details, for they can be found directly in Cohen's texts. Nevertheless, in order to get a clearer conception of Saval'skij's own position, some specifics are needed. Cohen distinguished theoretical ethics from applied ethics. Theoretical ethics, i.e., the exposition of the principles of ethics as a "science," is the task of Kant's second *Critique*. The scientific facts of ethics, analogous to those in natural science, are offered not by religion, history, or psychology, but by jurisprudence. But these facts

---

47 Saval'skij wrote, "it is impossible to enter into contemporary philosophy of law without entering into this [the *Critique*] book." Saval'skij (1907), 13.
48 Saval'skij (1909), iii.
49 Saval'skij uses the two terms interchangeably. See Saval'skij (1909), 12.
50 Saval'skij (1909), 48 f.

are governed by a form, a regularity that is of a different sort than that found in mathematical physics but that can be formulated in terms of a principle.[51] Thus, the best known element of Kant's ethics, the categorical imperative, must be completely eliminated from the list of concepts that ground ethics. Just as the concepts or categories involved in scientific regularity need a deduction, so too do those that ground ethical regularity and therefore make a pure ethics possible. Cohen – and Saval'skij – distinguished three sorts of deduction in general: psychological, metaphysical, and transcendental. The metaphysical deduction specifies the *a priori* concepts and shows that the given list of *a priori* concepts is a complete tabulation. In this way, ethical regularity is shown to be logically possible and elementary, i.e., irresolvable. The psychological deduction shows the psychological possibility of ethical regularity by revealing its origin in the empirical mind. The transcendental deduction shows not only that such regularity is logically possible, but that the concept is a being of a special sort.[52] The transcendental deduction does this by demonstrating that there are *a priori* concepts involved in the sphere under investigation, e.g., physics or jurisprudence, that are necessary logical conditions of the facticity of present knowledge in that sphere. The distinction between these forms of deduction is important, for Saval'skij held that whereas Cohen avowed the logical priority of the transcendental deduction, the Baden School in pursuing the metaphysical deduction verged on accepting some elements of psychologism.

More interesting to us than Saval'skij's summary of Cohen's positions is the opposition he saw of Cohen to the Baden School and, in particular, to Novgorodcev. In Saval'skij's eyes, Novgorodcev, like Windelband, was inclined to make what he saw as necessary additions to Kant's theory from Fichte and Hegel, specifically by adding tenets to the spheres of objective ethics and social, legal, and political theory. Novgorodcev's positions contrasted directly with those of the Marburg School in holding that we need not subject the moral principle to critique and grounding, that the genetic and systematic methods move along parallel lines, and that the thing in itself is something real. Saval'skij, presumably, rejected all of these positions as well as a number of others, although he also stated that his teacher (Novgorodcev) departed from Windelband at times owing to the powerful influence of Solov'ëv. Saval'skij contended, however, that, on the whole, Novgorodcev remained in the neo-Fichtean camp of metaphysical idealism. Once one exits the "theoretical" sphere for a separate "prac-

---

51 Cohen wrote, "Jurisprudence serves as the analogy to mathematics. It may be called the mathematics of the social sciences [*Geisteswissenschaften*], and above all it serves for ethics as its mathematics." Cohen (1904), 63.
52 Saval'skij (1909), 37.

tical" sphere, one abandons the possibility of theoretically grounding, analyzing, and evaluating a position.

Saval'skij elaborated how he understood Critical Philosophy's "grounding" within the bounds of experience. In short, such "grounding" consisted of the three deductions we saw above. To be sure, in dealing with ethics the psychological and the metaphysical deductions establish the originality of its principle. However, its role in cognition is determined by the transcendental deduction, and this is of decisive significance.[53] The transcendental deduction relates the moral principle to the fact of moral experience. A failure to recognize the centrality of this deduction is what Saval'skij saw as the fundamental deficiency of Novgorodcev's treatment of ethics. Novgorodcev concerned himself with only the other two deductions, which have merely a secondary and provisional significance, and completely omitted the transcendental deduction, which is the condition *sine qua non* of Critical Philosophy. He could find no way to ground morality within the bounds of natural-scientific experience. As a result, Novgorodcev grounded morality – insofar as he was able to do so – in the metaphysical reality of things in themselves.

Saval'skij conceded that the grounding of ethics is among the most difficult problems of philosophy. He turned to the issue in an extended footnote, wherein he again assailed his teacher for misunderstanding the genuinely philosophical question. First, in confronting the problem of grounding ethics, we must distinguish how ethics, as a science, is possible from how freedom is possible. The genuinely "Critical" approach of Cohen asks for the transcendental ground of ethics and is directed not on showing the cause of freedom, but on proving the possibility of ethics. It proceeds "entirely and exclusively in the sphere of the logical relations between the concepts of pure cognition."[54] Determining the conditions of the possibility of ethics as a science no more depends on providing answers to why freedom exists or what is its cause than the possibility of mathematical physics depends on answering why space, time, and causality are its *a priori* conditions.

We see from the preceding that Saval'skij recognized his adoption of Critical Philosophy stood in stark opposition to metaphysics and psychologism. Indeed, he himself remarked, "The liberation of concepts from psychologism and meta-

---

[53] To support this claim, Saval'skij alluded to a passage in Cohen. In it, after referring to Kant, who saw that the empirical derivation of concepts by Locke and Hume cannot be reconciled with the reality of the *a priori* in mathematics and general natural science, Cohen wrote, "Thus, the transcendental deduction is clearly and concisely grounded and oriented on the fact of science." Cohen (1907), 53. For the reference to Kant, see Kant (1997), 225–226 (B127–128).
[54] Saval'skij (1909), 95 f.

physics forms the chief task of Critical Philosophy."⁵⁵ To this end, he saw the Marburg understanding of natural law as avoiding the traditional theories that are ultimately religious through an appeal to God-given rights or are psychological through an appeal to the constancy of human nature. Whichever route tradition takes to support its view, it is unprovable and groundless. Natural law in Marburg neo-Kantianism is neither a metaphysical reality nor is it to be simply dismissed as fallacious as the positivists do. "Natural law must be understood as the idea of law, as the transcendental idea of juridical experience."⁵⁶ An idea expressed as a concept within systematic philosophy lacks being. Rather, it is a thing in itself. The idea of natural law within the sphere of jurisprudence serves, on the one hand, as a thing in itself, as a meta-juridical concept, but, on the other hand, it is also a method, a standard by which to examine positive law. It is a method for a critique and evaluation of positive laws. Although not in total agreement with Stammler, Saval'skij sympathized with the former's formula of natural law with changing content, a conception that had as much sense as to speak of natural science with changing content. "Just as science is recognized not by its content, but by its form, i.e., its method, so natural law is recognized not by its content, but by its form, i.e., its method."⁵⁷ Moreover, just as the content of natural science today is part and parcel of science, positive science, so too the content of natural law forms present, positive law. Just as the results of physical experiments may alter the "content" of physical science without ceasing to be science, so too may our positive laws change in accordance with our understanding of natural law. Natural law, then, is a *Grenzbegriff*, in the Marburg understanding of the thing in itself.

Although Saval'skij was rather dismissive of the Baden School, he did at least acknowledge in the name of Critical Philosophy its opposition to positivism in general and in particular to its recognition of the dichotomy between what is and what should be, with the latter being irreducible to the former. Saval'skij lauded Novgorodcev for his principled stand against historicism and for his defense of the independence of ethical principle.⁵⁸ However, the former rejected as

---

55 Saval'skij (1909), 214.
56 Saval'skij (1909), 209.
57 Saval'skij (1909), 214.
58 However odd, Saval'skij paid only negligible attention to Emil Lask (1875–1915), generally associated with the Baden School despite the similarity of their concerns. Lask, like Saval'skij, sought to find a way to speak meaningfully about natural law that avoided the traditional metaphysical approach as well as the historicist view that led to relativism. Saval'skij recognized that much. See Saval'skij (1909), 209f. Lask wrote, "Natural law and historicism are the two cliffs against which legal philosophy must protect itself." Lask (1905), 13.

false the line that led from Fichte to the Baden philosophers, which included his teacher, alleging that the independence of ethics was at the expense of the "logical principle," i.e., epistemology. Just as we cannot obtain the "ought" from the "is," so too we cannot obtain the "is" from the "ought." "The opposition and yet a connection between what is and what should be, logic and ethics, nature and culture, is achieved not in the normative philosophy of Fichte-Rickert-Windelband-Novgorodcev, but in the Critical Philosophy of Cohen."[59]

Certainly, there is a unity of consciousness, but we can distinguish its different "moments" or aspects by its different concerns: epistemological, ethical, and aesthetic. Consciousness in each sphere of concern is an end in itself. This does not mean that when concerned with ethical issues, consciousness is unconcerned with logic. But its use of logic is to provide principles, as norms, for ethical aims. In this way, the logical principles employed in moral analyses are ethical demands. Just as natural scientists use the law of causality in their inquiries as to "what is," so the investigator in ethical issues uses the categorical imperative. The scientist determines the factual by looking for the cause; the ethical investigator determines what should be by looking whether the categorical imperative was heeded. Rickert and the other "neo-Fichteans" do not uphold the distinctions consistently. For them, something is true not because it is factually true, but because it must be true. Whereas the positivists reduce the "ought" to the "is," the Baden School reduces the "is" to the "ought." "It is as if Newton's laws or Mendeleev's periodic table were not true, but should be true."[60] Such is how Saval'skij viewed Cohen's neo-Kantianism, which, with its principles, continues the Platonic-Kantian idealistic tradition.

5.4 A Russian Dispute between Marburg and Baden

Saval'skij's highly opinionated book quickly aroused a number of restrained though largely positive reviews by a number of young figures, who were associated with the same ideas as those propagated by Saval'skij. Despite not being entirely free of criticism, one reviewer found particular value in the work's presentation of a "whole number of quite important philosophical problems in the form in which they are posed and resolved in the works of representatives of various directions of the neo-Kantian movement."[61] Another young scholar associ-

---

59 Saval'skij (1909), 355.
60 Saval'skij (1909): 357.
61 Fokht (1909), 74. This itself is quite an exaggeration. Saval'skij hardly presented the views of a number of neo-Kantian schools.

ated with the movement remarked in a footnote in an article dealing with Cohen, "Savalsky's voluminous work on Cohen is imbued with a naive spirit of worship, and in this respect quite rightly aroused the kind of response that was given by Prof. Novgorodcev."[62] Of more interest is the reaction offered by a newly appointed (1906) professor of the history of the philosophy of law at Moscow University Evgenij N. Trubeckoj. In a lengthy article devoted primarily to Cohen's *Ethik des reinen Willens*, Trubeckoj managed to incorporate a few remarks on Saval'skij's thesis. Trubeckoj could not believe that Saval'skij seriously claimed jurisprudence to be a science. "For, in the first place, jurisprudence as a science does not exist apart from the theories that contradict each other."[63] In the second place, to speak of law as factual is hardly possible when that fact, as a "motley mixture," consists of good and evil. Saval'skij was fully aware of this possible line of attack already when composing his thesis. He recognized that some regarded "the orientation of logic in mathematics and natural science as acceptable, since we are dealing with sciences. However, the orientation of ethics in jurisprudence is unacceptable, since jurisprudence is not a science."[64] Unfortunately, Saval'skij, rather than engaging in an exposition of what characterizes science, continued his thought saying that the argument proceeds from an arbitrary concept of science. To the critic who says that jurisprudence with its contradictory theories is an incomplete science and, thus, lacks the dignity of a science, Saval'skij vaguely responded that it is not a matter of theories about a fact, but the fact itself, the conditions that make that fact logically possible. We can and must presume that Trubeckoj was familiar with Saval'skij's words but found them wanting and unconvincing.[65]

Trubeckoj's criticism was brief and pointed, but it was nestled in an article of much broader scope and lacked the features of an *ad hominem* attack. It also stemmed from a philosopher of law with a known personal relationship to Vladimir Solov'ëv and an intellectual relationship to the latter's religious metaphysics. In short, then, Trubeckoj's line of attack on Saval'skij's work was neither glaring nor unexpected. Saval'skij himself did not pay particular attention to the differences between Marburg conceptions and those of either positivism or the Russian metaphysical tradition. His scattered comments on Novgorodcev and the Baden School struck a nerve that Novgorodcev, who had invested so

---

[62] Jakovenko (1910a), 240. It should be kept in mind, however, that this was *all* Jakovenko had to say concerning Saval'skij's book. Jakovenko, in short, did not dwell on Saval'skij.
[63] Trubeckoj (1909), 141.
[64] Saval'skij (1909), 198.
[65] Expressing a more determined position, Dmitrieva writes, "This explanation does not satisfy Trubeckoj." Dmitieva (2007b), 215. She does not provide documentation for this factual claim.

much in his academic and political activities in support of moderate liberal aims, could not let pass. Saval'skij's treatise *could* be read as an endorsement of conservativism and the status quo, even if that was not its intention. How does one fight in the political arena for upholding human rights in the present, if they are, so to speak, an ephemeral concept, a limit-concept? Where is the human individual in Saval'skij's grand scheme? Novgorodcev, as, or perhaps despite being, Saval'skij's teacher at Moscow University, could not and would not remain silent.

With Novgorodcev's attack on Saval'skij, we have what we can surely see in retrospect as a sharp break in the philosophical opposition, such as it was, within Russia to religious metaphysics.[66] This public dispute between the two would be the only instance when the competing visions for updating Kant would directly confront each other in Imperial Russia. Saval'skij was so deeply absorbed in his commitment to Cohen's idealism that he could not see potential allies around himself. Anyone who did not adhere to Cohen's understanding of Kant was an outcast.[67] Given the substance and the tone of Novgorodcev's review-article from early 1909 simply entitled "A Russian Disciple of Hermann Cohen" and directed against Saval'skij, Novgorodcev must have felt deeply offended by the latter's criticisms and thought too much was at stake. Novgorodcev launched at the start into a damning assault on the thesis and particularly on the person behind it. "Those who take on themselves to read through this by no means easy work are in for a painful disappointment. The reader will find that the effort of the author does not correspond to the difficulty of the task and that the path selected by him is not the correct path leading to the goal."[68] Instead of simply stating and interpreting Cohen's work, Saval'skij had, in Novgorodcev's words, the "excessive pretension" to resolve the dispute between Critical Philosophy, positivism, and mysticism. Saval'skij's partiality was to the detriment of his scholarly analysis and gave his work the appearance of being incomplete and overly subjective. Moreover – and this was Novgorodcev's chief complaint – Saval'skij's

---

66 Belov has reasons, though they be exaggerated, for claiming, that previously "Russian neo-Kantianism appeared on the whole as a united front, because there was a common worthy opponent dominating the Russian philosophical scene, namely Russian religious philosophy." Belov (2019), 156. Insofar as there was such a front, it was that offered by the Marxists, with whom the small number of secular philosophers in Russia had even less in common than with the metaphysicians.
67 To be sure, Novgorodcev was certainly not a-religious, let alone anti-religious, but Saval'skij only glancingly looked for a common thread with the Baden School. Novgorodcev, after all, upheld Stammler's notion of natural law with changing content, with which Saval'skij could have emphasized but did not.
68 Novgorodcev (1909), 636.

veneration for Cohen was evident throughout his work and explained all of its essential defects. Saval'skij's sheer enthusiasm for Cohen made it impossible for him to present his material correctly. He proceeded in a dilettantish manner, careless with details, and with a slovenly elaboration of his theme. Novgorodcev adduced a number of what he took to be inaccuracies in Saval'skij's depictions of historical figures and ideas, after which he remarked that these were but some of the characteristic shortcomings of Saval'skij's book. Even in his analysis of Cohen's works, Saval'skij was by no means clear, which showed that he did not fully understand Cohen. His exposition of Cohen was replete with the grossest errors, blunders, and defects. Saval'skij "wrote much but said little, and this is because his book is completely unpolished."[69] Whereas he attempted to throw light on the relationship between ethics and jurisprudence, his explanation, in Novgorodcev's eyes, was completely unsuccessful. Indeed, Saval'skij provided an entire series of clarifications, but they contradicted each other and were essentially wrong.[70]

Given all the deficiencies that Novgorodcev found in Saval'skij's work, it is surprising that Novgorodcev devoted little attention in this review-article to his student's criticism of his own work. Novgorodcev devoted only a single long footnote to commenting on it, and for the most part, the criticism stemmed from Saval'skij's alleged misquoting or misrepresentation of his mentor's ideas. Novgorodcev charged that Saval'skij "depicted my views not as they are set forth in my book, but as he needed them for the purposes of his criticism."[71] However, Novgorodcev oddly did not proceed even to attempt any clarification of his own position vis-à-vis Saval'skij's supposedly intentionally misleading or false construal of it.

Novgorodcev's final appraisal of Saval'skij's book is harsh in the extreme, though perhaps unsurprising given the criticisms that he presented.

> My goal was to show to what extent Saval'skij's presentation of Cohen is correct. After analyzing the most basic points of Saval'skij's presentation of the *Ethik des reinen Willens*, we can be sure that he has not cleared up *a single one of these points*. ... This journal has neither favored nor suppressed any direction. But a special organ of Russian philosophical thought must require equally of all directions consistency and completeness. Analyzing Saval'skij's work from this viewpoint, we must say that it deserves the full measure of censure that can take place in such a case.[72]

---

69 Novgorodcev (1909), 653.
70 Novgorodcev (1909), 654.
71 Novgorodcev (1909), 652f.
72 Novgorodcev (1909), 660.

With such a damning indictment of Saval'skij's book, it is not hard to understand that young scholars would in the immediate future think twice about writing directly on the Marburg School. Novgorodcev's tone also revealed a side to his personality that had hitherto largely kept hidden. He sought not to understand Saval'skij, to wrestle with his philosophical ideas on a philosophical plan, or even to elaborate the foundations of his own position. One cannot but conclude that Novgorodcev sought, above all, to destroy Saval'skij and his prospects whatever others might think.[73] Novgorodcev succeeded in neither. Saval'skij would not go quietly into the night.

Saval'skij replied to Novgorodcev at the end of the following year fortunately in the very journal from which Novgorodcev would have a writing such as his proscribed. Saval'skij accepted none of his former teacher's criticisms. Frankly, they were talking past each other. Saval'skij wrote that he would address only the issues that concerned the "scientific merit" of his work, which Novgorodcev had called into question. We need not go through these point by point, but only insofar as they illuminate the neo-Kantianism of Novgorodcev's opponent, as it were, in the dispute. In fact, Saval'skij took this opportunity to detail his neo-Kantianism as clearly, if not more so than in his thesis – and with less parroting of Cohen.

Saval'skij saw the *a priori* of judicial experience as analogous to that of (natural) scientific experience. The aim of an investigation into both types of experience is to determine the *a priori* concepts involved, i.e., to answer how judicial and natural scientific experience is possible. Such is the transcendental method. Just as any natural phenomenon is a concrete result of the law of causality – and through it we obtain special instances of that law such as the law of gravity – so too any phenomenon of the moral order is subject to moral reason and through it we have special instances within the "kingdom of freedom."[74] A now standard criticism of Kant's ethical theory is that being general, i.e., purely formal, the categorical imperative lacks content and thus cannot serve as a practical guide. This objection, Saval'skij contended, makes no sense, for it simply shows that the objector has no idea of what the transcendental method is. Just as it is impossible to deduce from the law of causality this or that particular physical law, such as the law of universal gravitation, so too is it impossible to deduce from the categorical imperative a maxim with specific content. Without the observance of causality, we would not have physical experience; without the observance of the cat-

---

73 Such is how Belov characterizes Novgorodcev's attempt. He is likely correct, given the evidence. However, Belov is wrong in asserting that Novgorodcev succeeded. Of course, it is a matter of what one considers to be "successful." See Belov (2019), 155.
74 Saval'skij (1910), 357.

egorical imperative, we have no moral experience. Each is a condition for experience in the respective sphere. Just as one cannot accuse natural science of formalism, so one cannot accuse scientific ethics of formalism. The categorical imperative is the principle of morality. As such, it is the principle by which evaluations are made of the various elements in ethical experience.

Saval'skij held that a recognition of the role of ethics in jurisprudence is one of the central elements of the transcendental method. A rejection of that role is tantamount to a rejection of the role of epistemology in the natural sciences and mathematics. Thus, to deny the role of the latter is to deny the transcendental method in its entirety. It is precisely a recognition of the role of ethics in jurisprudence, the pursuit of which yields a scientific ethics, that insures it being free of metaphysics, mysticism, psychologism, and naturalism. Saval'skij certainly saw the Baden School philosophers as flirting with psychologism owing to their indifference toward concrete science, be it current natural or legal science.

> If scientific logic is possible, it must not be a theory of cognition in general and not a theory of ideal cognition, but ... of the *a priori* conditions of mathematical and descriptive natural science. If scientific ethics is possible, it is only as a theory of cultural values, seeing culture as an entire system of means and ends in which the dignity of the human person is seen as an end in itself and all other elements of culture as the means for attaining this finite end.[75]

Saval'skij believed his fears concerning the Baden philosophers were being confirmed with the appearance of Rickert's lengthy article in 1909 "Zwei Wege der Erkenntnistheorie," wherein its author, fearing an irresolvable duality between being and sense, value and reality, appealed to a transcendental psychology to help solve how values could serve as norms in actual quests for scientific knowledge. Saval'skij viewed Rickert as hoping to avoid a two-world Platonic idealism by appealing to a psychology, albeit of a "transcendental" sort, but in doing so Rickert could not resolutely avoid a lapse into psychologism. In adopting the Baden approach to philosophy, Novgorodcev's analysis was by no means a renewal of Kantian philosophy. Such, at least, was Saval'skij's counter-attack.

Novgorodcev replied to Saval'skij at once. Indeed, it appeared in *Voprosy filosofii* immediately following Saval'skij article. Regrettably, it was little more than a series of *ad hominem* attacks. Apart from pointing out specific errors in Saval'skij's statements and attributions to others, Novgorodcev failed to take the opportunity either to engage philosophically with Saval'skij or to elaborate his own stance with regard to Kant or to the Baden School philosophers as

---

75 Saval'skij (1910), 369.

against the Marburg School. Novgorodcev himself, however, exaggerated in saying that Saval'skij's book "encountered strong condemnation from those who it would seem would come to his defense."[76] As though aggrieved for some slight, Novgorodcev wondered why Saval'skij did not attempt to reply directly or refute his particular objections; he could not understand why Saval'skij chose not to play by Novgorodcev's rules. The senior combatant reiterated again that many of Saval'skij's claims were incomplete and inaccurate and that his former student's reply went awry. However, the former added that he did not wish to pursue the matter further. "I already did my work once before on his thesis, and I now have neither the time nor the desire to repeat the same for his article."[77] As remarked with regard to Novgorodcev's previous review-article, he also never addressed Saval'skij's criticism of his views. "I recognize them as based on a completely false representation of the most important points of my exposition of Kant's philosophy." In what way it was false, Novgorodcev did not say.[78] It was always Saval'skij who had to reply to charges, not Novgorodcev. The polemic ended with the Novgorodcev's counter-reply, minimal though it was. Although Novgorodcev did pen a number of writings subsequent to this exchange with and about Saval'skij, he never addressed his philosophical relationship to the Baden School – or, for that matter, the Marburg philosophers.

Saval'skij could not reasonably expect any of the established professors in Moscow to help him secure a professorship, but he did receive an appointment in 1910 in the law faculty at the University of Warsaw. He worked there until the outbreak of hostilities marking World War I, at which time he was evacuated and lived with the family of a friend in Moscow. Whether those around him knew of his physical condition is unknown. He died unexpectedly from cancer in 1915.

Saval'skij's time in Warsaw is unrecorded apart from a manual for students that he wrote in conjunction with a course for the academic year 1911/1912.[79] The tone of this introductory text is subdued and relaxed, generally free of polemics, and its contents explanatory. His debt in all matters, he wrote, was, above all, to Kant, and he emphasized at the start that he would "try to maintain the view of

---

76 Novgorodcev (1910), 383.
77 Novgorodcev (1910), 386.
78 Novgorodcev (1910), 387 f.
79 Belov reports that at the time of his death Saval'skij was preparing an article "Gosudarstvo kak predmet vozmozhnogo opyta" ["The State as an Object of Possible Experience"], which he had begun while in Warsaw. Although the writing was almost complete when Saval'skij died, it was not published. Belov (2019), 152. If this proposed title is any indication, we would have to conclude that Saval'skij remained committed to Marburg-style neo-Kantianism to the end.

Kantian Criticism" in his course.[80] Whereas there are a few mentions of Cohen, these are largely in conjunction with other names including Rickert, Windelband, and Lipps – with one exception. He stated, as he did in his thesis, that each sphere of study works within its own framework, has its own paradigm. Saval'skij, in this regard, differentiated the viewpoint of the natural sciences, which look on the human being as a physical thing not essentially different from other physical things, from the cultural viewpoint, in which every phenomenon can be seen as a means to the attainment of an end. Whereas from the viewpoint of the natural sciences all things, including human actions, must have a cause and an effect, human beings from a cultural viewpoint, viz., that of idealism, are free to initiate actions seemingly independent of external causes. Such is "one of the central concepts of Kant's entire system and also the apex of his system."[81] The best development and elaboration of the problem of freedom in Kantian idealism is, Saval'skij contended again, that given by Hermann Cohen. Although Saval'skij presentation is clear, in his own mind it did not depart from that of his Marburg master. Saval'skij remained to the end Cohen's disciple. With his premature death, Russian neo-Kantianism lost its most vociferous and outspoken advocate.

---

[80] Saval'skij (1912), 5.
[81] Saval'skij (1912), 119.

Chapter 6
Baden Makes Inroads

In Russia, as in Germany, in the early twentieth century, social scientists and philosophers influenced by Baden neo-Kantianism wrestled with a dual and possibly conflicting commitment to ideal Kantian values and ontological realism. They uniformly shunned the idealism of Cohen's *Logik* which seemingly eviscerated the real content of everyday experience. These social scientists and philosophers more immediately than their Marburg-inspired colleagues sought involvement, albeit to varying degrees, in their country's socio-political life. With their realism – understood both in the philosophical and in the popular sense – they knew the struggle for a stable, liberal democracy would face formidable obstacles. In light of their personal fates, they arguably may have underestimated just how formidable those obstacles would prove to be.

6.1 A Ukrainian Activist Finds the Baden School

Another figure often mentioned along with Novgorodcev as a supporter/disciple of Baden School neo-Kantianism is Bogdan Kistjakovskij[1] (1868–1920), who was only two years younger than the former.[2] Like Novgorodcev, Kistjakovskij was politically involved, arguably even more so and at the expense of a scholarly career. As a result of his political activism, he too has attracted a measure of attention in English-language scholarship, albeit focused on his political involvement and decidedly less so on his narrowly construed philosophical efforts. For that reason, we can quickly summarize the major events in his life, as we did with Novgorodcev. Bogdan's father, who taught law at the University of Kiev, was an advocate of progressive causes, and a moderate champion of Ukrainian ethnic identity. The young Bogdan inherited all of his father's intellectual gifts, zeal, and then some. His advocacy of Ukrainian identity caused his expulsion from secondary school a number of times. The same happened at the universities of

---

[1] He himself gave his first name as "Theodor" in his German-language study in 1899.
[2] In her valuable book devoted to Kistjakovskij, Heuman writes that with his 1902 article he "established his position as the leading spokesman for neo-Kantianism in the Russian Empire." Heuman (1998), 25. This is a wildly exaggerated, indeed downright incorrect, claim, as the previous chapters in the present work show. Heuman refers for support to Gurvitch's encyclopedic entry, but it simply reads, "Kistyakovsky was one of the most able representatives of the neo-Kantian movement in Russia." Gurvitch (1932), 575.

Kiev and Kharkov. Although he then attended the University of Dorpat, his political work during the summer vacation of 1892 back in Ukraine resulted in his arrest. This was not the only time he was incarcerated on one charge or another, but after being expelled from Dorpat he had exhausted his options for studying within the Russian Empire. He spent a year in a town on the Baltic coast to improve his German, and when granted a passport, which had previously been denied, he went to Berlin in January 1895. At the university there, Kistjakovskij attended lectures by Georg Simmel, who clearly impressed him. In addition to venturing afterward to Heidelberg and Paris, he heeded the advice of Simmel to go to Strasbourg University, where Windelband was teaching at the time.[3] Kistjakovskij there completed and defended a doctoral thesis entitled *Gesellschaft und Einzelwesen*, which he dedicated to Windelband and Simmel, in 1899.

Kistjakovskij's thesis, written in German and submitted to a German university, lies on the periphery of our concern here.[4] But it does reveal many of the issues he would later develop. Although some within the Russian Empire were acquainted with it, particularly those whom its author knew personally, its recognition there was a result of the attention it had received in Germany.[5] The fact remains, though, that a Russian-language translation of the book did not appear until recently (2002). As for the work itself, a direct influence from the Baden School is clear enough, as well as from Simmel, who, though deeply interested in the philosophy of sociology and the social sciences, was an idiosyncratic neo-Kantian but by no means a representative of the Baden School. One cannot help but also notice that although Kistjakovskij's work is replete with references to contemporary authors of the day, including many who are now regarded in retrospect as neo-Kantian, there is little reference back to Kant and to distinctly Kantian themes. For example, although he approvingly quoted Windelband and developed the idea that there are many laws of nature, the only common feature in these various laws is that they speak of a causal connection between the phenomena that they study. Whereas the Marburgians found these laws to be particular manifestations of the Kantian category of causality, Kistjakovskij made no

---

[3] Simmel's position as a private lecturer and then *Extraordinarius* at the university in Berlin did not legally allow him to supervise theses. Windelband, with Simmel's agreement, formally supervised the theses of several students of Simmel's.

[4] Although the focus in this study of Russian neo-Kantianism is the Russian-language works of the various philosophers, Kistjakovskij's 1899 German-language thesis is an exception in that so many of its themes would be elaborated later in his Russian-language writings.

[5] Walicki comments that Kistjakovskij's early recognition in Germany was "a big asset to an academic career in Russia." Walicki (1987), 346. The reader should decide for oneself based on the facts to what extent that was the case.

such claim.⁶ More typical of the Baden understanding of Kantian idealism was the contention, which Kistjakovskij endorsed, that "this common feature of all laws of nature is itself not a law, but a norm of our thinking."⁷ Regrettably, he did not further elaborate the point, which would allow some assessment of the depth of his neo-Kantianism.

Kistjakovskij's first concern in his 1899 thesis was to follow the already well-trodden path of critiquing positivism, although he would have it be a still vibrant sociological option. Kistjakovskij's approach was to distinguish the different understandings and use made of the concept of law in the various sciences. An error often occurs at the start, particularly among advocates of positivism, in assuming that the term "law" has an unequivocal meaning. A law in physics, however, is understood differently than, say, a law in biology. Kistjakovskij took note of this ambiguity particularly as it was manifested in the "organic theory of society" promoted by Pavel von Lilienfeld.⁸ Kistjakovskij lauded Lilienfeld for seeing society as a complex living being. This, however, was the only merit of Lilienfeld's "organic conception," which was, in any case, self-evident, since society consists of people.⁹ The organic theory provides no answer to questions, such as what are the rules that govern society and what are the determining features of social life. Kistjakovskij felt that whatever we can glean from Lilienfeld regarding these and other matters is not just inadequate, but contradictory.

Kistjakovskij rejected Aristotle's assertion that the human being is social by nature. The Aristotelian claim, in fact, was simply assumed henceforth to be correct, and this dogmatism delayed the development of the social sciences. The former countered, on the contrary, with the reverse, that humans *become* social by being in society. "The social qualities of the human being have gradually formed through living together with others and first come to light in society."¹⁰ Kistjakov-

---

6 Kistiakowski (1899), 33. Kistjakovskij's omission is all the more surprising in that it features prominently in, for example, the Transcendental Deduction in the first *Critique*, where Kant writes, "But all empirical laws are only particular determinations of the pure laws of the understanding, under which and in accordance with whose norm they are first possible." Kant (1997), 242–243 (A128).
7 Kistiakowski (1899), 33. This claim, which teeters on psychologism, faces the obvious objection that causal laws in physics are not norms of human thought, but natural laws.
8 Lilienfeld was a Baltic German bureaucrat and social scientist in Imperial Russia. His views attracted a surprising amount of international attention.
9 Kistiakowski (1899), 21.
10 Kistiakowski (1899), 140. He referred in this context to Kant – which he rarely does – who wrote in his "Idea for a universal history": "The human being has an inclination to become *socialized*, since in such a condition he feels himself as more a human being." Kant (2007), 111 (Ak 8, 20).

skij utilized Kant's analogy of society and individuals to a forest and trees. In a forest trees grow much straighter and taller owing to a constriction of free development.[11] Similarly, the talents of human individuals are spurred on by social antagonisms. *Through* society, *through* social life, the human being develops from a material being into a social being, and in society has a completely altered character.[12]

Of more interest to us here, however, is Kistjakovskij's thought, embraced by the Baden School, that different methodological points of view are required in different investigations of social phenomena. If we are dealing with society as a collective whole, we adopt a different viewpoint than if the cognitive object is an aspect of a segment of society. We can observe social movements, for example, as causally conditioned, but we can also independently view the results of these movements with an interest in assessing them legally, ethically, or even aesthetically.[13] Kistjakovskij believed that this methodological standpoint stemmed from Kant's ethical theory and provided the philosophical grounding for treating sociological problems completely and separately from all ethical and political considerations. This stood, he believed, in sharp contrast to Hegel's position that could not justify a formal jurisprudence or ethics.[14]

Kistjakovskij recognized that Kant held the cognitive synthesis to be grounded in human consciousness itself as its spontaneous function, not in empirically-given relations between the contents or sensible manifold. In doing so, Kant made this synthesizing process the focus of the epistemological endeavor. Kistjakovskij saw modern psychology, in contrast, as justifiably and correctly investigating the functions of consciousness apart from any talk of an ego or I. Social science can and must follow in these footsteps "investigating social connections without the state and without any externally binding forces."[15] Kistjakovskij admonished us not to forget that such a sociology without a unifying subject is the result of a purely methodological approach, and that, though this procedure is of indubitable utility, there is a consciousness, on the one hand, and a political state, on the other. Social unity cannot and does not arise merely from the psychic interactions of individuals and social groups. As we see already from his indisputably well-researched thesis, Kistjakovskij's approach with its blend of Baden neo-Kantianism with other contemporary German trends differed consid-

---

[11] Kistijakovskij surely had in mind here Kant's metaphor in "Idea for a universal history with a cosmopolitan aim."
[12] See Kant (2007), 113 (Ak 8, 22).
[13] Kistiakowski (1899), 156.
[14] Kistjakowski, somewhat surprisingly, cites Hegel far more than Kant throughout his work.
[15] Kistiakowski (1899), 198.

erably from Saval'skij's adamant Marburg-School approach. Indeed, its open advocacy of pluralistic methodologies in sociology, while largely indebted to the Baden-School approach, was hardly an exemplar of that approach and even inclined toward empiricism rather than the transcendentalism often associated with neo-Kantianism.

6.2 Russian-Language Works

Writing years later in the introduction to a collection of essays, many of which had been published previously, Kistjakovskij stated that the studies originally were carried out under the influence of the idea that there are many valid methods of social-scientific investigation. These methods thereby allowed for an understanding of the "irreconcilable contradictions and infinite diversity" of social life.[16] We can easily see his enduring plea for the acceptance of pluralistic methodologies in the social sciences beyond his 1899 thesis and into an article published the following year. Kistjakovskij, there at the very start, held that the growth of our social-scientific knowledge has led us to inquire to what extent the search for causality so prominent in the natural sciences is also warranted in investigating social phenomena. As the work of Windelband and Rickert has shown, our recently acquired positive knowledge has provided us with no "guiding thread" to aid in penetrating social phenomena and what occasions them.[17] To be sure, everything in nature takes place in accordance with natural laws. Strictly from the point of view of physics, even social actions must have a cause.[18] Thus, from this abstract standpoint, there is no difference between the social and the natural worlds. However, the physicist's point of view when study-

---

16 Kistjakovskij (1916), iii.
17 Kistjakovskij (1916), 120. The essay, "Kategorii neobkhodimosti i opravedlivosti pri izsledovanii social'nykh javlenij" ["The Categories of Necessity and Justice in Investigations of Social Phenomena"], originally appeared in the May and June 1900 issues of the journal *Zhizn'*. The present author regrets that he was unable to compare the essay as published in the 1916 collection *Social Sciences and Law* with the original from 1900.
18 Kistjakovskij, in this regard, made some disconcerting remarks. Although acknowledging the universality of physical laws, he wrote, "Even in the sphere of the most common phenomena obeying the all-encompassing law of gravity, each single phenomenon is completely by chance. Why, for example, does the Earth occupy the third and not the second or fourth place in the solar system?" Kistjakovskij (1916), 122. He gives here no indication of recognizing the distinction between absolute and relative contingency, i.e., that something merely appears to be by chance owing to our ignorance of the causes. This distinction was pointed out by Windelband in his 1870 dissertation, *Die Lehren vom Zufall*. Windelband (1870), 68.

ing a natural phenomenon is completely different from that of the social scientist looking at, for example, a historical event. The scientific interest in each case is not the same. The physicist is looking for or has in mind the *general* laws accounting for the object under study, whereas the historian – the paradigmatic social scientist for Kistjakovskij – seeks the *individual* features and details.[19]

The differences between history and natural science notwithstanding, the practitioners of both insist that a relationship exists between the disciplines. History is interested in establishing facts, but it is also interested in investigating causal connections. Therefore, causality, as a category, logically unites them, and their techniques and investigative methods are the same. Such, at least, is a common argument, but one that Kistjakovskij believed was misleading. It rests on an ambiguity in the nature of the causality that both see as a common object of investigation. Whereas the historian, to be sure, seeks causality, it is of a very different sort than the causality sought by the physicist.

> The causes of an event that historians seek and the causal correlation between phenomena that natural scientists establish have, from an epistemological and logical point of view, nothing in common. The historian investigates causes that are as individual and singular as the historical events themselves.[20]

Implicit in this description of the historian's methodology is Windelband's famed contrast between the idiographic and the nomological approaches to cognition. Kistjakovskij had no quarrel with the natural sciences or their perspective as such. Their concern is with causally-connected and necessary correlations. The unconditional necessity of the correlations, as he saw it, being a necessity, is identical with aspatiality and atemporality. All such higher scientific truths are fully consistent with Kant's theory of the apriority of space and time and do not contradict it despite the fact that the universe is filled with matter. Kistjakovskij added, but does not further explain, that the apriority of space, which he obviously upheld, has nothing in common with the psychological apriority of our spatial and temporal representations.

---

19 Kistjakovskij (1916), 126–127. Undoubtedly, one may have considerable reservations here concerning Kistjakovskij's representations. However, Windelband in his 1894 "Geschichte und Naturwissenschaft" wrote, "We have here now a purely methodological division of the empirical sciences based on reliable logical concepts. The principle of this division is the formal character of their cognitive goals. Some seek universal laws; others seek particular historical facts. To put this into the language of formal logic, the goal of the one group is the general, apodictic judgment, that of the other is the singular, assertoric proposition." Windelband (1911b), 144. For an alternative translation of this passage, see Windelband (2015a), 291.
20 Kistjakovskij (1916), 130.

Turning to an investigation of social laws, the category of necessity undoubtedly plays a role in procuring the truth. It is an *a priori*, and thus unavoidable, category of thought. However, we can look on natural phenomena alternatively from the viewpoint of contingency.[21] Kistjakovskij mentioned, though, that natural science does not employ this viewpoint owing to its futility. Its use produces nothing of value. The situation need not be the same in the social sciences. In them, the investigator need not look for causal connections, but for something else. From the point of view that accepts chance, the scientist looks for justice and, connected with it, the idea of duty. When examining social relations, we constantly judge – at least to ourselves – whether they are just. "In opposition to the category of necessity, which is applicable equally to natural as to social phenomena, the category of justice is always applicable to judgments about the social world but inapplicable to natural phenomena."[22]

Even though social phenomena from the cognitive or "theoretical" point of view appear to be contingent events, judgments viewed ethically – and, therefore, formed on the basis of the category of justice – have the same character of universal obligation as those based on necessity. Kistjakovskij, in other words, held that the verdict of ethical judgments is just as necessary within its sphere as the verdict of cognitive judgments in the theoretical sphere. Ethical truths are not relative, and cognitive truths are not relative. However, we apply the category of cognitive necessity to understand or explain, but ethical necessity, i.e., justice, is a category of evaluation. With the latter, we determine whether some social phenomenon is morally good or bad, but justice explains nothing.[23]

In his contribution to the much celebrated collection *Problems of Idealism*, published in late 1902, Kistjakovskij reiterated many of his ideas from 1900, albeit in a less technical but, unfortunately, more verbose and polemical form.[24]

---

21 Kistjakovskij called contingency, as well as necessity, a category. Are we, then, to conclude it is as much an *a priori* Kantian category as is necessity? What sense can we ascribe to opposing categories of thought? Was Kistjakovskij contending that the categories of thought can be employed at will by the empirical I? Without further clarification, it seems Kistjakovskij is flirting with a quite un-Kantian conception.
22 Kistjakovskij (1916), 174.
23 Kistjakovskij (1916), 187.
24 The circumstances behind the inclusion of Kistjakovskij's essay in the *Problems* are not entirely clear. In late September 1901, Struve wrote Novgorodcev concerning his proposal for a collection of articles dealing with "freedom of conscience," mentioning contributions he hoped to elicit from thirteen individuals. Kistjakovskij's name was not included. See Kolerov (2002), 120 – 121; Poole (2003), 19. However, Struve and Kistjakovskij were in communication during this time period and would go on to collaborate in editing the journal *Osvobozhdenie* [*Liberation*].

He again wrote that the natural sciences seek unconditionally necessary correlations. In the social sciences, though, the objects of study are linked to the "most vital" of human interests and therefore (*poetomy*) "different points of view" are evoked for their study.[25] Kistjakovskij objected to the position of the Russian Populist Mikhajlovskij, who, in the former's telling, held that the "objective method" of the natural sciences, i.e., the search for necessary causality, as the *exclusively* objective method is not just impossible, but was and is never used by anyone. Kistjakovskij's aim in this essay, though, was not to contribute as such to a neo-Kantian understanding of sociology. Rather, its intention was to combat those sociological theories that from one or another direction would, in his eyes, obstruct both the realization of moral ideals and the development of an overall scientific outlook. This outlook hopes to preserve the methodology of natural science within its own domain "while also recognizing the equal rights of all other types of scientific and extra-scientific (e.g., metaphysical) thinking and creativity."[26]

Kistjakovskij's 1907 essay "In Defense of Scientific-Philosophical Idealism" was his lengthiest excursion into philosophy, and at least from our perspective today we can see it as the fundamental articulation of the Baden School message in Imperial Russia.[27] In it, he threw his full weight against, above all, positivism but also against other currents at the time, including the religiously inspired ethics of Vladimir Solov'ëv that was gaining ground among Russian philosophers.

---

[25] Poole (2003), 328; Kistjakovskij (1916), 34. The *Problems* in its Russian original exists in many editions. I refer to Kistjakovskij's essay as reprinted in his 1916 collection solely for the sake of convenience.

[26] Poole (2003), 148; Kistjakovskij (1916), 108. Kistjakovskij discussed at some length the concept of probability and, though acknowledging the indisputable service of Johannes von Kries to its study, the former held that "the entire theoretical construction of Kries for explaining the epistemological significance of objective possibility is quite erroneous, since it is based on an incorrect definition of the concept of cause." Kistjakovskij (1916), 105. It is interesting in this regard that Max Weber, in the third of his "Critical Studies in the Logic of the Cultural Sciences" entitled "Objective Possibility and Adequate Causation in Historical Explanation," not only discussed Kries but even offered qualified criticism of Kistjakovskij's criticism. See Weber (1949), 167 f.

[27] The essay as reprinted in Kistjakovskij's 1916 collection contains many additions not found in the 1907 version without indications that they were newly incorporated. We must assume, then, that what is common to the two versions represents his enduring position, but we cannot conclude that he would not have defended in 1907 ideas found only in the 1916 version. To complicate matters, Kistjakovskij noted in the 1907 version that the essay "was written already in the winter of 1903/04 and in December of 1904 was promised to readers of the journal *Voprosy Zhizni*. He added, however, that owing to some accidental circumstances concerning Kistjakovskij alone it could not be published in 1905. Kistjakovskij (1907), 57 f.

There can be no mistake, though, that Kistjakovskij favored the recent rise of idealism as opposed to the views of those such as Comte and Mill. The reaction to positivism stemmed from a recognition that ethical questions could not simply be neglected or dismissed as unscientific. Idealism took these questions and their associated problems seriously. The virtually palpable sense of moral obligation and demand for justice in the face of social inequities could not be reduced to mere facts connected by causal necessity. However, the emergent idealism was represented by two directions of unequal strength and depth. One of them held to a metaphysical, even mystical, interpretation, whereas the other was sketched by supporters of a "scientific philosophy" that had little to do with the positivistic understanding of philosophy and ethics in particular.

Although both directions embraced the urgency of dealing with ethical issues and the need for doing so independently of the natural sciences, the numerically larger camp rushed to conclude that the problems could be solved only by an appeal to metaphysics. Its representatives in a short time declared that ethics had to be grounded on faith, particularly on faith in a moral world order and a supreme being. Kistjakovskij recognized from the start that such an outlook could be refuted neither by scientific means – since it was explicitly metaphysical – nor by logic – since it was internally consistent. However, for these very reasons this direction or stream also quickly ran dry. There was no way to return from it to scientific knowledge or to real-world solutions to real-world problems.[28]

Kistjakovskij held that the recent appearance of idealism was due to a recognition of the urgency of "the ethical problem" as an independent problem outside the bounds of natural science.[29] Its solution was idealism's fundamental task. Although it could not be solved by natural science, it could and must be solved by scientific means, by what he called "scientific philosophy." Both streams or directions of idealism have the same starting point, the same data. Both start with the fact of evaluation; both set out with judgments of true and false, of good and bad, of beauty and ugliness. That we establish norms and rules of evaluating the facts we gather already testifies to our autonomy as human beings and indicates our freedom from purely natural confines in general. Autonomy and freedom ground the ideals embodied in our judgmental norms. Both idealist directions agree to these principles, but their relationship

---

28 Vucinich would have it that Kistjakovskij had Berdjaev specifically in mind when arguing against "mystical and metaphysical idealism." No doubt Vucinich is likely correct, but Berdjaev's name is not explicitly mentioned in this regard. Thus, to be precise Vucinich's statements are presumptuous. See Vucinich (1976), 129–130; Kistjakovskij (1916), 253–254.
29 Kistjakovskij (1916), 192.

to them is different. For metaphysical idealists, human autonomy and freedom point directly and above all to something that transcends human experience. Their concern is ontological. Instead of investigating and analyzing these principles, they see them as evidence of an underlying metaphysical source. They devote their energy to the disclosure of what they consider to be higher truths obtained by some metaphysical intuition or sheer religious faith.

The scientific-philosophical idealist, on the other hand, rejects as absolutely unacceptable the move from human autonomy and evaluation to an alleged source in some metaphysical entity. Moreover, we have no special organ for cognizing metaphysical beings. An idealist of this direction sets as one's task the purely scientific elaboration of the data in evaluations and of human autonomy together with all the phenomena of the human spirit connected with these evaluations.[30] Kistjakovskij maintained that scientific philosophy itself demonstrates the fecundity of such a formulation.

Just like natural science, which can be subdivided into various fields, scientific philosophy too can be subdivided into separate sciences. Whereas the various natural sciences investigate natural events with an eye to discovering the causal laws connecting these events, i.e., a nomological interest, the various disciplines of scientific philosophy establish and analyze not what is, but what should or ought to be.[31] Thus, scientific philosophy, according to Kistjakovskij, seeks the norms or universally obligatory rules in each philosophical discipline, i.e., in cognitive, or theoretical, thought, in practical activity, and in cultural creation. In each case, the unifying category – unfortunately he provides no definition or elaboration of the term[32] – is a consciousness of this "ought" (*dolzhnoj*). Kistjakovskij regarded the operative significance of this category in all cognitive, ethical, and aesthetic endeavors as indubitable. We refuse falsehoods and the telling of lies not just on ethical, but also on theoretical, grounds. We reject a mistake made in a document written long ago, even though there is no chance of correction today, out of a consciousness of truth-telling. We condemn lying not just on ethical grounds, but also with cognitive truth as the goal. In short,

---

30 Kistjakovskij (1916), 194.
31 Although writing in a journal of professional philosophy, Kistjakovskij largely refrained from posing his own "task" here in terms of the contrast between the nomologic and the idiographic that we have come to associate so closely with the Baden School.
32 Kistjakovskij (1916), 195. He went on to write that a clarification of "category" would lead "into the sphere of metaphysical problems." Kistjakovskij (1916, 197). Presumably, he thought he could speak of *a priori* categories in "scientific philosophy" without thereby venturing into metaphysics. But we can ask of him what philosophical basis he has to use a concept, for which he, in principle, cannot provide a thorough and precise exposition.

we seek to uphold truth, broadly understood, in all matters and in all fields. There is not only a theoretical or cognitive truth but an ethical as well as an aesthetic truth. Moreover, as we see from the example of lying, these "truths" are intertwined and serve as *norms* of behavior.

Kistjakovskij alleged that in the history of human thought the norms of one area were incorrectly identified or derived from those of another, resulting, for example, in various forms of reductionism, one of which is positivism with its dissolution of the "ought" as merely an expression of one's individual approval of some action. Although the norms of ethics and aesthetics are rarely identified, both fields are often seen as derivable from a common source, viz., a person's psychological makeup. One may acknowledge that Kistjakovskij was correct here while still noting its sheer ambiguity. Instead of investigating such a charge – or rather in order to do so – Kistjakovskij addressed its opposite, namely, that there is no "ought" in the process of scientific cognition. Kistjakovskij believed his opponents were those who upheld that the search for truth is a matter of what is, not what should be. That is, these opponents hold that the quest for truth is a matter of discovering the necessity in nature. Kistjakovskij alleged that this is incorrect. There is a confusion here between the object sought and the scientific procedure involved in that search. Nature does operate according to necessity. Even our very own thinking takes place on the basis of psychological laws. As a process within nature, it also proceeds with natural necessity. Whatever the laws be that correlate our thought processes, they are natural laws of the psyche, i.e., psychological laws. However, we continuously face a seemingly endless number of possible choices. The decision which maxim is to serve as the basis of our action is not done strictly on the basis of some natural necessity. The choice is made on the basis of a principle. In other words, our particular decision is based on an evaluation of the situation, be it practical or theoretical, i.e., what is relevant to our individual investigation. The caveat here is that the evaluation must, in turn, be grounded not on contingent, subjective motives, but in general rules and universally binding norms. Kistjakovskij wrote, "We therefore conclude that in order to select and systematize the definite contents of thought into scientific knowledge, certain rules or logical norms must be observed."[33] We can obtain genuinely scientific, i.e., universally obligatory knowledge only in this way. We must be careful here so as not to misunderstand Kistjakovskij (and the Baden theorists in general). The choice is made based on a principle that serves as our rule; in natural science this can be the choice to establish universally binding natural laws.

---

33 Kistjakovskij (1916), 211.

Kistjakovskij then followed up with themes that are familiar to the reader from the "Introduction" to the second edition of the *Critique of Pure Reason*. All logical laws or principles are, what Kant called, "analytic" and will not expand our knowledge in the scientific sense. As such, they cannot precisely be said to be scientifically true. Just as *a priori* analytic propositions cannot expand knowledge, *a posteriori* synthetic judgments do not provide scientific truths. Any talk about a particular tree in the meadow ahead, while it may be true, is not a scientific truth. It becomes of interest to science when we talk not about a given tree, but about trees or plants in general, what Kant called "judgments of experience."[34]

In framing judgments, we can and must recognize that our natural mental faculties that allow us to make judgments have neither set limits for their employment nor sure, guiding principles for making generalizations. The result can just as well lead to incorrect affirmations of scientific truth. What we must recognize is that our mental faculties do not naturally act according to goals, but act causally. In short, their activity, as a part of nature, is subordinate to natural necessity. The rules or norms we set are those to be used in the evaluation, and they are the general formulas that determine what we take to be the appropriate or expedient actions for achieving goals. We, human beings, create our values, those that we wish to serve as the basis of our norms. These values are independent of nature. Scientific philosophy enters the picture as the investigation of this world of values. Such an investigation and, thus, the determination of the values in cognition and the means for their attainment is called "logic in the broad sense." Likewise, the determination of what is valuable in practical activity along with the corresponding norms forms the sphere of ethics. Finally, the determination of what is valuable in the creation of culture and of the forms of its embodiment, i.e., what is beauty, is called "aesthetics." Kistjakovskij refers to each of these three studies as a "science" under the general heading, as mentioned, of scientific philosophy.[35]

Ever mindful of an unintentional lapse into metaphysics, Kistjakovskij asserted that the first step in logic, broadly understood, is to establish similarities and differences on the road to generalization. Windelband had done much fundamental work in this regard, placing distinction as the first category and sim-

---

34 Kistjakovskij framed his exposition here without any indication that he had Kant's distinction in mind between judgments of perception and judgments of experience. He neither cited Kant in this regard nor did he even use Kantian terminology. Nevertheless, it is hard to avoid thinking that he did have Kant's distinction in mind, even if only second-hand, i.e., transmitted to him through the writings or teachings of others.
35 Kistjakovskij (1916), 226.

ilarity as the second. Kistjakovskij, gently dismissed such priority as debatable. Moreover, to speak of similarities and differences in practice, we must have criteria in mind for their correct application and for noting the limits of that application. Kistjakovskij clearly feels here that the very determination of the appropriate criteria involves principles that lie outside the sphere of formal logic. Normative, as contrasted with formal, logic enters at this point. Normative logic is concerned with determining what is essential and inessential for the purposes of concept formation. It is not hard to see that Kistjakovskij owed a substantial debt not only to Windelband but also to Rickert, who had devoted considerable attention to just this train of thought.[36] Normative logic recognizes that the theory of concept formation must be based on methodological premises, not on those of formal logic or psychology.

> In order to solve this problem correctly in each individual case, we must know the special goal for which a concept is created for use as an instrument of cognition. Methodology, not formal logic, provides instructions for the special goals of cognition.[37]

Kistjakovskij gives as an example the concept of a human person, which can vary depending on whether the concern is physiology, psychology, physics, etc.

As touched on above, Kistjakovskij saw the work of the natural scientist as totally consistent with the message stemming from his scientific philosophy. Such a scientist values the discovery of a law of nature even though the procedure requires a complex combination of rules. However, in the context of this discussion Kistjakovskij provided a puzzling remark inconsistent with his portrayal thus far of the "logic" of cognition, at least if it is made from the viewpoint of his "scientific philosophy." Abandoning an understanding of the evaluative norm as a rule to be followed in judging, he wrote,

> the most important point, however, is that the very concept of a law or causal connection rests on a higher transcendental norm, namely on the category of necessity. Kant's analysis in the *Critique of Pure Reason* of the thought process that leads to the establishment of natural laws serves as the ultimate clarification of the meaning of formal norms in general and transcendental norms or categories in particular. The norms or categories are peculiar to our thought as obligatory forms, the meaning and value of which are clarified in the creative process of science itself.[38]

---

36 See, of course, Rickert's 1896 *Die Grenzen der naturwissenschaftlichen Begriffsbildung*, which appeared in a Russian translation already in 1903.
37 Kistjakovskij (1916), 228.
38 Kistjakovskij (1916), 229–230.

Kistjakovskij here distinguishes between "formal norms" and "transcendental norms" without providing an explanation of what the latter could possibly be.[39] Given Kistjakovskij's exposition as is, it conflicts with Windelband's in his 1882 essay "Normen und Naturgesetze," in which we find,

> psychology with its laws explains how we actually think, actually feel, actually wish and act. The "laws," on the other hand, which we find in our logical, ethical, and aesthetic conscience have nothing to do with the theoretical explanation of the facts, to which they refer. They express only what these facts should be so that they can be universally acknowledged as true, as good, as beautiful. Thus, they are not laws by which the event must objectively take place or can be subjectively conceived, but ideal norms by which the value of what naturally happens necessarily is judged. These norms are therefore the rules of judgment.[40]

Windelband explicitly rejected viewing norms as a theoretical explanation of facts, which is something a transcendental account would provide.

Apparently unaware of the ambiguity of his key operative terms such as "norm" and "logic," Kistjakovskij asserted that it is impossible to refute the normative character of logic – we should note that he does not qualify here the word "logic" – by referring to "irrefutable facts" of immediate psychological experience. The concern of logic and of the theory of cognition is to establish the means by which we proceed to acquire scientific truth, and by doing so provide its justification. Kistjakovskij's own example here, interestingly, incorporated an appeal to history. He stated that the meaning of a scientific truth attributes something given with more power and significance than has a fact given to our immediate perception. "With the passage of time, our psychic mechanism adapts so that we gradually begin to recognize an originally rejected truth as though it were given with a psychological force comparable to immediately perceived facts."[41] For example, we now after so many centuries naturally, as it were, think that the Earth revolves around the Sun, rather than vice versa, as though this were an irrefutable fact of immediate perception, though nothing has changed in those immediately perceived facts over the centuries. Kistjakovskij's example is, from a strict Kantian standpoint, puzzling. How did this appeal to the historical development of a conception enter the picture? Even more fundamentally – and alarmingly – Kistjakovskij's conception flirts with historicism. Do

---

39 This is not to say that Kistjakovskij could not possibly provide some account of what a "transcendental norm" is – elaborated conceivably on the basis of Kant's discussion of "categories of freedom." Kistjakovskij, however, does not do so. In any case, it would depart from Windelband's conception of philosophy. For Kant, see Kant (1996), 192–194 (Ak 5, 65–66).
40 Windelband (1911b), 67.
41 Kistjakovskij (1916), 231.

our scientific propositions receive scientific justification historically through the gradual alteration of our mental conceptions? Or is it, rather, that our representations asymptotically approach an ideal norm or value, as a more orthodox Baden-School approach would have it? If we opt for the latter, what serves as a check that history does not err?[42] Kistjakovskij did not grapple with these matters, largely referring the reader looking for additional details and elaboration to the works of Windelband and Rickert.

It is not clear that Kistjakovskij recognized the fundamental importance and potential dangers lurking in his "turn" to history, but he expressed full confidence that the direction taken by his former teachers and colleagues in Germany was the proper one. He had no doubt that the paths by which the methodologies of each scientific discipline are established run parallel to each other despite their differences. These paths, he affirmed, must be laid down following the particular tasks or problems the respective discipline pursues. There is an inherent complexity to following a cognitive "ought" in the human sciences compared with investigations in the natural sciences. In the latter, only the cognitive process itself is subordinate to obligation while the object of cognition falls within the sphere of necessity. However, "the cognitive concern in the human sciences is with a world in which conscious human activity is manifested. The human sciences have to investigate not only elemental social processes, which arise with natural necessity, but also the effects of any sort of norm on socio-cultural life."[43] Thus, not only must the investigative procedure in the human sciences be subordinate to the principle of obligation, but also the very object of cognition is subordinate both to natural necessity as well as to obligation.

Kistjakovskij managed to pay tribute to Kant's efforts to provide a theoretical foundation of pure ethics and in doing so showed that pure ethics, contrary to the position of mystics and metaphysicians, is independent of religion. Its autonomy follows from the legislative faculty of the human will, whereas a prescription for action stemming from another source, even if it be a Supreme Being, is not an ethical prescription, but a heteronomous command. The ethical "ought" does not remain in the subjective consciousness, but calls for action. The effective implementation of these "oughts" leads to a world of values and culture. It was to Hegel's credit to recognize this and pose an understanding of this social dimension of ethics as a philosophical task. But Kistjakovskij con-

---

42 Moreover even if we accept this depiction of the historical change from the geocentric model of the solar system to the heliocentric, what has this to do with the Kantian quest for the *a priori* synthetic conditions that make science possible? Has Kistjakovskij abandoned the Kantian transcendental method?

43 Kistjakovskij (1916), 238.

demned Cohen for creating a "special sort of being" from ethical actions instead of seeing ethics as concerned with volitional decisions based on the Kantian categorical imperative. The consequence of this is that Cohen's ethics bears a consistently deontological character and from it drew "quite erroneous conclusions concerning cognition in the social sciences and in particular concerning cognition of the essence of law."[44]

There can be no doubt that Kistjakovskij presented the fullest exposition and defense of a Baden-inspired neo-Kantianism in the era of Imperial Russia. He appreciated the fact that all the directions of neo-Kantianism recognized the need to develop further the principles laid down by Kant especially in order to apply them to the needs of the emerging human and social sciences. But he held that the various schools of neo-Kantianism saw this task in different ways and, consequently, came to different results. In his eyes, "the most fruitful direction for the development of scientific knowledge in general and of the human sciences in particular is that neo-Kantian direction represented by Windelband, Rickert, Lask, and others"[45] More than many others, Kistjakovskij remained committed over the years to Baden neo-Kantianism. How he would have reacted to its demise within Germany and its seemingly spent force within the confines of the Soviet political space after its consolidation is unclear. He died in the capital of his beloved Ukraine in 1920 at the rather youthful age of fifty-two. From our vantage point today, his personal passing also marked the passing of the Baden School's most passionate and skillful defender within the borders of what had been Imperial Russia.

6.3 A Historian's Use of Baden Tenets

Another Russian scholar indisputably indebted to Baden neo-Kantianism but who was also indebted to Russia's homegrown variety was one we have already seen when dealing with the problem of other minds, namely the historian

---

44 Kistjakovskij (1916), 251. This attack on Cohen also provided Kistjakovskij with the opportunity to note Saval'skij's "unsuccessful" attempt to attack the Baden School's direction. The basis of the dispute, according to Kistjakovskij, was Saval'skij's mistake in not recognizing the differences between the ideas Cohen promoted in his commentaries on Kant and Cohen's own systematic writings. It is particularly in those later writings that Cohen veered astray. See Kistjakovskij (1916), 252f. For a lengthier discussion of Cohen by Kistjakovskij, see his essay "The Methodological Nature of the Science of Law" in Kistjakovskij (1916), 384–389; 394–404.
45 Kistjakovskij (1916), 388.

Lappo-Danilevskij.[46] It would not be a fanciful exaggeration to compare the Russian academician's efforts in history to those of Max Weber in economic sociology. Much as the latter sought to test the usefulness of Rickert's thoughts for the methodology of his discipline,[47] so too Lappo-Danilevskij sought a methodology for historical investigation adopting to a significant degree the ideas of the Baden philosophers supplemented importantly by the neo-Kantian reflections of his colleague Aleksandr Vvedenskij. As we shall see, this did not mean that Lappo-Danilevskij slavishly followed them. Instead, he took from them what he deemed useful in light of and for his own historical investigations.

Lappo-Danilevskij enjoyed a spectacular academic career in St. Petersburg that began with the defense of his *magister*'s thesis in May 1890. We have little direct information from his own hand by which we could categorize his philosophy of history, assuming he had one, at this early date, but he did assert in the *de rigueur* speech at his thesis defense that universal history has a meaningful sense, that it "studies the norms of social development common to all humanity, or at least the civilized part of it."[48] This universal history does not exclude that of individual nationalities but abstracts general norms from them. Whether on the basis of such meager statements or through personal conversations, those who knew Lappo-Danilevskij saw his general view of world history at the time to run along the lines of Comtean positivism. Presnjakov wrote, "During his

---

46 Lappo-Danilevskij's allegiance to Baden neo-Kantianism, and even to neo-Kantianism in general, is challenged by some Russian scholars. Whereas some, such as Marina Rumjanceva, focus on the Russian sources of Lappo-Danilevskij's ideas at the expense of an influence from the Baden philosophers, at least one other focuses on Lappo-Danilevskij's disagreements with the Baden School without mentioning Vvedenskij. In this way, Jurij Vasil'ev avoids characterizing Lappo-Danilevskij as a neo-Kantian at all. Vasil'ev contends that the portrayal of Lappo-Danilevskij as a neo-Kantian was created by Soviet scholars in the 1940s. See Vasil'ev (2017), 187, and for Rumjanceva's presentation, see Rumjanceva (2014). Contra Vasil'ev, N. I. Kareev (1850 – 1931), who knew Lappo-Danilevskij well, wrote that the latter saw epistemology "as the necessary grounding of any methodology. He sided philosophically with Kantian Criticism, and its representatives, Windelband and Rickert, were the greatest influence on him. ... I am not saying that Lappo-Danilevskij submitted passively to their views. He criticized them, but there can be no doubt of their general influence on him. In any case, Lappo-Danilevskij's philosophical orientation was neo-Kantian." Kareev (1920), 121. Another distinguished historian, A. E. Presnjakov, wrote that Lappo-Danilevskij "in his neo-Kantianism, started with a quest for a synthesis of 'pure' and 'practical' reason, constructing his views with a desire for the ideal of such cognition in which the cognitive values of critical thought and the 'categorical imperative' would be realized." Presnjakov (1920), 89.
47 Weber (2012), 6.
48 Lappo-Danilevskij (1890), 284. This article – actually a report – provides a significant portion of Lappo-Danilevskij's speech.

youth Lappo-Danilevskij's investigative and philosophical thought developed under the banner of positivism. ... He studied Comte's philosophy thoroughly and with sympathy." [49] However, within a few short years his outlook changed, making him an attractive possible contributor to the Novgorodcev-Struve collection *Problems of Idealism*.[50] Novgorodcev wrote to Lappo-Danilevskij on 1 June 1902 with the invitation – or, rather, plea – to help in the effort.

> We need an article about Comte for our collection that would reveal the unsatisfactory nature of his point of view. ... If, in addition to all this, you have the chance to mention *albeit briefly* the impossibility for positivism to pose and solve the moral problem, we would be infinitely grateful for your support of the *Leitmotiv* of our edition. ... Your reputation and your name in scholarship free you from any suspicion of harboring metaphysical phantasies and passions and make a blow against positivism by your hand particularly attractive.[51]

Since *Problems of Idealism* was published in November of that year and Lappo-Danilevskij's essay was rather lengthy, we can assume that he not only accepted Novgorodcev's invitation but began to work on it expeditiously.

To be sure, Lappo-Danilevskij's essay is hardly a place in which we would or could expect to find an elaboration of neo-Kantian historical methodology. Nevertheless, we can find in it themes and expressions of ideas that we associate with the German – and Russian – neo-Kantian movement. For example, although Lappo-Danilevskij noted "certain points of contact" between Comte and transcendental idealism, the former noted that the French "sociologist" never firmly grasped the Kantian conception of the unity of apperception.[52] Comte's careless comments about self-consciousness, which he saw as nothing more than a "sense of harmony" among our various mental functions, prevented him from understanding the unity and continuity of the individual consciousness.[53] Comte's total disregard for psychology as a social science, in fact, excluded from the outset any possibility of grounding the social sciences on consciousness. Lappo-Danilevskij drew from these criticisms of positivism the conclusion, as a warning to others, that sociology, a science of society, must first explain its foundations in order to be a science. It is with such an analogous lesson that he also turned to history as a discipline.

---

49 Presnjakov (1922), 53.
50 Quoting Presnjakov again, Lappo-Danilevskij's "philosophical development went ... from dogmatism to Criticism." Presnjakov (1922), 53.
51 Quoted in Kolerov (2002), 145–146; cf. Grekhova (1976), 264–265.
52 Lappo-Danilevskij (2013), 363.
53 Lappo-Danilevskij (2013), 374.

6.3 A Historian's Use of Baden Tenets — 173

Lappo-Danilevskij from the very start of his scholarly career was interested in historical methodology with the hope of turning the academic discipline of history into a rigorous discipline, a "science." Already in the 1899/1900 academic year, he conducted seminars on the sixth book of Mill's *A System of Logic* entitled "On the Logic of the Moral Sciences." But most importantly for us, he gave a general course specifically on the methodology of history as a scholarly discipline already in 1906, a course that he continuously reworked during the ensuing years. The resulting text serves as the basis and fundamental argument in support of his allegiance to neo-Kantianism.

Lappo-Danilevskij remarked at the very start of his *Methodology of History* that any elaboration of scientific methodology must be grounded on a theory of cognition. Such a methodology needs the support of principles, both general ones grounding all the sciences and particular ones applicable to the specific discipline.[54] A theory of cognition will provide the *a priori* elements of the framework for our empirical knowledge. As such, then, it gives the bases by which we regard cognition as valid and universally obligatory. Although the genesis of knowledge, regardless of whether that genesis be psychological or sociological, can be useful in understanding the foundations of our knowledge, it cannot validate that knowledge. Even erroneous knowledge-claims have a genesis. Given the close connection between the theory of cognition and scientific methodology, it can hardly come as a surprise that the latter can aid in the development, correction, and supplementation of the former. Lappo-Danilevskij provided his own concerns, not surprisingly, as an example, viz., those associated with the methodology of the social sciences. Surely with the Marburg philosophers in mind, he wrote that for too long the theory of cognition had been elaborated one-sidedly taking natural science as the paradigmatic scientific endeavor. Lappo-Danilevskij, nevertheless, was encouraged by the recent attention paid to the logical structure of historical knowledge and its bearing on the theory of cognition. Whereas historical methodology aims to ground history as a rigor-

---

[54] Lappo-Danilevskij delivered a lengthy report paralleling and summarizing much of the first pages of his academic text at a January 1918 meeting of the history and philology department of the Russian Academy of Sciences. He opened his report, writing that "it is impossible not to recognize the most intimate connection between the theory of cognition and the methodology of science. Only from a certain epistemological point of view can we construct an integral doctrine that would fundamentally establish its task and consistently outline the path leading to its completion." Lappo-Danilevskij (1918), 239. He continued his report at a subsequent meeting of the Academy in March of that year. This continuation provided a detailed intellectual history of the nomothetic point of view and demonstrated Lappo-Danilevskij's superb scholarship and mastery of the original texts.

ous, i.e., "scientific," discipline, the fundamental principles it reveals concern not just historical knowledge, but the very possibility of any knowledge.

Based on the above considerations, Lappo-Danilevskij wrote that the elaboration of a methodology of history requires first a general theory of cognition and additional concepts or principles specific to the scientific discipline that the theory is to ground. Thus, the main task is to establish the foundations that provide science with a meaning (*znachenie*), i.e., to clarify the meaning of the scientific principles. The second or derivative task is the systematic theory of the methods employed in the discipline.[55] This second task is of little concern to us here, although Lappo-Danilevskij devoted considerable attention to it in his book. As for the first task, he turned abruptly, one may say, to the theory of historical knowledge. The theorist is concerned with such questions as the role of causality and expediency in historical reconstructions of the facts and what criteria the historian is to choose in evaluating the subject-matter. In retrospect, such issues have been resolved differently by different schools of historical thought. The historian can look at the facts of historical situations from either the point of view of whether there is a common thread, some connecting link between them that explains one on the basis of the other, or the historian can look on them as unique and individual facts, events or occurrences. If the historian, as a social scientist, studies the subject matter looking for the involvement of a causal relationship throughout, this historian adopts the nomothetic point of view. If, however, this historian is concerned with the uniqueness and individuality of the concern, the idiographic viewpoint is taken. In either case, though, which point of view the historian adopts is based on what the historian considers efficacious for the chosen goal(s).

Lappo-Danilevskij was particularly concerned with demonstrating the defects of the nomothetic approach if taken as the exclusively appropriate method in all of the sciences. Generally speaking, its proponents typically employ already prepared concepts from the natural sciences in their own investigations instead of independently developing a system of concepts more applicable to historical material. Drawing upon the well-known critique of causality, viz., that regardless of how many instances we observe of a phenomenon we cannot logically infer from those finite number of observations a universal and necessary law. We can at most appeal to an ever-increasing likelihood of some correlation with an ever-increasing number of observed occurrences. Admittedly, advocates of the exclusivity of the nomothetic direction in social science buttress their stand by invoking an alleged uniformity of human psycho-physical nature. But

---

[55] Lappo-Danilevskij (2006), 18; Lappo-Danilevskij (1918), 243.

by doing so and having in mind an interest only in generalizations, such a social scientist, in effect, neglects or dismisses the variety and originality of reality. The exclusive application of the nomothetic point of view in the social sciences cannot embrace individuality of any sort except as an anonymous member without unique features. "From the nomothetic point of view, a historian easily overlooks or arbitrarily excludes from one's observations facts (such as persons, events, etc.) that history cannot neglect."[56] The human being, whether we have in mind an individual or a constituent element in some group, organization, or society, is not like an elementary particle in physics where one is indistinguishable from another.

Lappo-Danilevskij also criticized the nomothetic "construction" of history from what he called the practical point of view. The nomothetic construction cannot "provide an idea of the totality of the given conditions of space and time in which our activity takes place."[57] Without such knowledge, we cannot be sure a historical person has acted in accordance with some supposed law. In other words, we cannot be certain that the lessons we learn from history are correct since we lack a complete idea of the past situation. Other difficulties present themselves in a consistent application of physical causality in historical explanation, to which nomothetic theorists so assuredly turn. Their invocation of causality, albeit in a psychological sense, when writing of, for example, the motivations of actions, amounts to a claim of the qualitative equivalence between the cause and effect, whereas, of course, in natural science the equivalence is quantitative. Conservation laws in the latter are mathematical expressions. The nomothetic theorists do not even so much as attempt to provide quantitative formulations linking historical events. Even if we should be able to formulate general historical laws, the historian's interest lies in determining under what conditions those laws were followed. Only an ideographical point of view can venture such a determination of where and when an event or phenomenon followed a general law. In other words, the historian's concern is not simply with the laws as such, but also in their implementation.[58]

Of course, the proponents of the exclusivity of the nomothetic point of view reject the opposing idiographic point of view, charging it with being unscientific. According to the former viewpoint, genuine science consists of the construction of general concepts. An individual and unique something, such as a feature, a characteristic, or a phenomenon, is not a general concept and therefore cannot

---

56 Lappo-Danilevskij (2006), 131.
57 Lappo-Danilevskij (2006), 131.
58 Lappo-Danilevskij (2006), 177.

be the object of scientific cognition. Lappo-Danilevskij believed that this conception of science is inaccurate and the portrait painted of the historian's work is incorrect. The historian finds the elements needed for his/her work by assembling concepts of the individual drawing from the infinite variety of elements found in reality. Although refraining from any generalization, this does not mean the historian does not abstract the needed elements. History as a discipline can claim scientific significance despite not being a generalizing science. In short, the idiographic point of view applied to historical material can have a scientific character.

We see, then, that Lappo-Danilevskij opposed the exclusive employment of the nomothetic point of view in the study of history and by extension all methodological reductionisms, such as "economic materialism," i.e., Marxism.[59] His numerous criticisms, in his opinion, legitimized the idiographic point of view, which he found already present in Kant's works despite the underdevelopment it received in them. "Kant established the general foundations by virtue of which it is possible to discuss the significance of the individual, but he paid too little attention to clarifying the theory of historical knowledge proper and to the logic of history as a discipline."[60] Kant recognized the value of a unique event or phenomenon in history but only to the extent that it contained something general and rational. Consequently, he was unable to establish a fundamental distinction between the knowledge given in the natural sciences and in historical studies. Nonetheless, his "critical idealism" ushered in a new period in the development of the idiographic point of view, and, as we know, it formed the basis of subsequent neo-Kantian theories of historical methodology.[61] Kant realized that a depiction of the story of humanity needed a guiding thread, a measuring stick, which consistent with his Enlightenment outlook, was our rationality. He tied this, in turn, to morality, which resulted in viewing history from a moral point of view, with history being the expedient realization of the moral ideal.

Lappo-Danilevskij hoped he had shown with his critique of the nomothetic point of view that we must also make use of the idiographic viewpoint. The con-

---

[59] Lappo-Danilevskij (2006), 70–71. We should mention that he did not dwell on Marxism, apparently viewing it as of no more "scientific" importance than any other reductionism.

[60] Lappo-Danilevskij (2006), 152.

[61] Lappo-Danilevskij was clearly more gracious toward Kant than was Windelband. The latter wrote originally in 1904: "Although historically understandable, Kant's concept of 'science' is limited to the methodological character of theoretical natural science given by Newton's *Principia*." Windelband (1911b), 153.

cern of the latter is not with a particular object as an object. It is with an understanding of its particularity or individuality. He wrote:

> One and the same object can be considered not just from the nomothetic point of view, but also from the idiographic. Thus, if our interest is directed to a given object not because it has properties in common with other objects, but because its individuality is of interest, it is natural to appeal to a special kind of construction, viz., the idiographic.[62]

If the object's individuality is paramount, then the idiographic viewpoint predetermines the nature of the historian's investigation. The historian, however, is undoubtedly interested in real historical events and seeks to utilize the methodology of the nomothetic viewpoint as much as possible to reach an understanding of those events. We should remember, though, that the understanding of the individual events, etc., is the goal and the use of general concepts is with that goal in mind.

The historian, Lappo-Danilevskij contended, needs a basis, a criterion, by which he/she chooses what has historical significance from the enormous number of possibilities reality offers. What has such significance in the entirety of experience is a value. They need not be positive. Falsehood, evil, and ugliness – all have negative significance. Such values, whether they be negative or positive, are concepts based in an agreement or violation of certain norms or demands. In this Lappo-Danilevskij, defending the rationality of values, appears as more "Kantian" than Rickert. Lappo-Danilevskij's overall scheme will better preserve the Kantian priority of the practical over the theoretical while yet viewing reason as operative in both spheres.[63]

Lappo-Danilevskij was certainly not without criticism of the Baden-School philosophers, whom he thought had not engaged in concrete historical research

---

[62] Lappo-Danilevskij (2006), 179. In his discussion, Lappo-Danilevskij referred to Rickert's work on the formation of concepts. Thus, it is instructive to compare the two and see the former's philosophical debt to the latter. The Russian historian often made use of Rickert's terminology and conceptual scheme. We should also recognize the role of Windelband who recognized years earlier that the natural sciences were interested in individual events insofar as they served to validate – or falsify – proposed laws just as the social scientist was interested in determining general laws if they could aid in the understanding of individual events. For Windelband, "the same objects can be made the object of a nomothetic and also of an idiographic investigation." Windelband (1911a), 145.

[63] Lappo-Danilevskij must yet account for these norms without an appeal to psychology lest he too should ultimately lapse into a psychologism or a sociologism. Part of the problem is that he believed we "easily transfer concepts, i.e., values, to the objects themselves, which have in our eyes the corresponding significance." Lappo-Danileveskij (2006), 190. Clearly this transference must be elaborated in more detail.

and therefore were not aware of the practical problems associated with it. First of all, though, on a theoretical level they had not concerned themselves sufficiently with the *a priori* conditions of consciousness in general, seeing them and the values involved in historical investigations as purely ethical values. Another important disagreement with Baden, as Lappo-Danilevskij understood those philosophers, was that the historian had to explain how the particular, whether an individual person or event, arose from the general conditions. Thus, the Baden philosophers did not take into account the concept of historical development with its irreversible sequence of events. But his principal criticism of Windelband and the other "founders of the idiographic construction" concerned their sharp distinction between natural science and the historical discipline. The philosophers forgot that some natural sciences use principles commonly used by historians. In any case, the terminological distinction between natural and historical sciences "seems artificial in many respects."[64] Although setting the universal against the particular may seem logical, its practical implementation in research work is difficult. Each science is concerned in part with both generalization and individualization. One can actually speak only of a *predominance* of one of the two points of view – the nomothetic and the idiographic – over the other in each science. History, similar to natural science, concerns itself with relative generalizations. The historian, for example, lacking the needed general concepts, produces them in conformity with the objects studied and the cognitive goals pursued.

In conclusion, history as a discipline, elaborated as a "rigorous science" – though Lappo-Danilevskij did not use that expression – can combine the nomothetic and the idiographic points of view. Nevertheless, the two should not be confused, but consciously distinguished. Indeed, the historian does in fact use both viewpoints in processing the raw material of the researcher's investigations. Lappo-Danilevskij appreciated the efforts of the Baden philosophers, but their work too needed further elaboration and emendation derived from the work of the practicing historian. An invocation of the idiographic point of view seeks the individual, a viewpoint concerned most often with the actions of an individual human being. Any attempt to understand another human must be by way of external signs. Hence, our historical inquiry via the idiographic point of view requires entering into the philosophical problem of other minds, lest our overall philosophy of historical methodology not be from the outset a rigorous science. For this reason, Lappo-Danilevskij entered into a discussion, as we saw in Chapter 2, of the issue of other minds and believed he had found in Vvedenskij's ap-

---

64 Lappo-Danilevskij (2006), 229.

proach some insights that coupled with the Baden dichotomy of scientific methodologies could advance the science of history.

While not active in politics – in fact, he avoided such engagement – he was a member of the liberal Constitutional Democratic Party (Kadets) in the last years of Imperial Russia. After the February 1917 Revolution, he served on the commission charged with writing the electoral law for the convening of a Constituent Assembly. The October Revolution apparently sapped Lappo-Danilevskij's creative energy, although he devoted subsequent months to scholarship and archival work. He was hospitalized in early 1919 in bad health as a result of exhaustion and malnutrition, but he continued as well as he could, including making corrections for a new edition of his *Methodology of History*. A surgical operation was not entirely successful in that an infection set in and spread rapidly. On his death bed, he was found reading Hegel's *Phenomenology*. He whispered to a friend the day before he died, "There was never time to study it carefully. I will start now."[65]

6.4 A Forgotten Neo-Kantian

As we saw previously, Venjamin Khvostov has largely gone unrecognized – and thus unappreciated – in the slim secondary literature on Russian neo-Kantianism. Yet, he offered some of its clearest statements. Unlike other Russian neo-Kantians who were influenced by the Baden philosophers, Khvostov directly expressed his opinion of Kant's ethical system, finding it to be the deepest of all moral theories presented thus far: "Kant's theory has quite a lot of true ideas, and I consider its fundamental conception of morality to be correct."[66] This, nonetheless, did not prevent him from finding faults or deficiencies in it. The first charge Khvostov leveled against Kant's basic formulation of the moral law is one known to all students of it today, namely the ambiguity in its practical application, an ambiguity that results from its universality. More specifically, the categorical imperative, owing precisely to its pure formality, can be used to justify any behavior. Another problem with Kant's ethics, indeed one of its weakest points in Khvostov's eyes, concerns the link between the noumenal and the phenomenal worlds. If our actions are free in the noumenal sphere, how can they act upon the determined phenomenal human will? Is it not the case that natural causality is the only causality in the phenomenal world? But, then, a noumenal

---

65 Quoted in Malinov (2001), 57.
66 Khvostov (1913), 126.

causality is – or at least should be conceived as – operative only in a noumenal world, not in the phenomenal.

Khvostov saw himself as offering a different interpretation of Kant's conception of freedom than that presented by Hermann Cohen in his *Kants Begründung der Ethik*. Cohen there affirmed that physical causality can explain all phenomenal action. In light of this fact, freedom cannot serve as a constitutive condition of the possibility of experience, the determination of which is the concern of Kantian transcendental philosophy. Freedom, not being a constitutive condition, is a regulative principle. In other words, according to Cohen, Kant's conception of freedom means a person should fulfill one's moral duty *as if* one were free. Our phenomenal determinism does not prevent us from conceiving ourselves as outside the limiting conditions of the phenomenal world. That we have a free will serves as a guiding principle of our behavior. It is a necessary ideal in the practical sphere. Khvostov, however, found Cohen's standpoint unacceptable. He understood Cohen as arguing for the realization of freedom as an asymptotic ideal, one that can never be reached. But for Khvostov, "all ideals must remain within the bounds of the possible in order to have a practical significance and not be transformed into the purest phantasy. I do not want to say that free will, as a regulative idea, must be excluded from the ethical sphere. But I insist that free behavior must be placed on some foundation in the empirical human will."[67]

Khvostov insisted that Kant's sharp dichotomy between sensible human nature and the ethical was unclear. He sided with Kant that the attainment of happiness could not serve as the fundamental ethical principle, and instead he offered the feeling of human dignity as the basic *psychological* foundation of morality. A person filled with a sense of dignity chooses to pursue the moral ideal without reservation or in pursuit of some contingent goal, and not in order to achieve happiness. For the pursuit of the moral ideal is part and parcel of one's dignity. "From the point of view of dignity, the moral imperative acquires a *categorical character*."[68] Khvostov asserted that in posing morality to be grounded on human dignity it joins with the fundamental categorical aspiration of Kantian ethics.[69] While he recognized that the concept of human dignity plays a significant role in Kant's theory, Khvostov recognized that, unlike Kant, he wished to introduce an empirical and psychological element, of which Kant

---

[67] Khvostov (1913), 135.
[68] Khvostov (1913), 273.
[69] We hardly need to point out that Khvostov's notion of "categorical" is different from Kant's.

would sharply disapprove. However, he wrote that he, unlike Kant, did not fear doing so.[70]

Khvostov elaborated his exceedingly sketchy moral philosophy in university lectures on the history of ethical theories. We can, given the context, be somewhat forgiving toward the lack of specificity in the statement of his own position. However, in an address to the Moscow Psychological Society on 20 December 1908 and published in *Voprosy filosofii* in early 1909, Khvostov added some clarification. He stated there that although science rests on a recognition of the universality of the principle of causality there are spheres, particularly the social sciences, in which the concept of freedom is routinely employed without wishing to introduce metaphysics. Khvostov wrote, "My goal is to show in what sense the concept of 'freedom' not only can, but also should, be taken as included in empirical science, leaving aside all metaphysics."[71] Khvostov, thus, wished to retain the legitimacy of freedom in the empirically given world. He did this, he believed, by invoking a psychic – and not, as in Kant, a transcendental – causality, by which, he stated, a reconciliation can be had between freedom and necessity that is adequate for the purposes of empirical knowledge.

Khvostov's interpretation of this psychic causality may be different from Kant's causality from freedom or freedom in the practical sense, but his understanding of psychic causality closely approaches that of Kant if we limit ourselves to the empirical sphere. For Kant, freedom, practically understood, is "the independence of the power of choice from necessitation by impulses of sensibility."[72] Thus, we humans have the ability to determine our own actions independently of any coercion from sensible impulses. Whereas Kant initially here framed freedom negatively as an absence from constraint, Khvostov sought to frame it positively. If we do not act from constraint, what, then, is the basis of our actions? When speaking of moral actions, this basis is moral duty. "A person is free who is able to subordinate all of one's actions to the dictates of moral duty."[73] We see someone as free who despite temptations to do otherwise performs one's ethical obligations. There is no reference here to a noumenal causal-

---

70 Khostov went on to confuse matters further by writing that the ultimate basis of moral duty lies "not in theoretical, but in practical reason, i.e., not in knowledge, but in faith." Khvostov (1913), 273. Is he saying that theoretical reason is associated with knowledge, whereas practical reason is associated with faith? In any case, where did Khvostov find a connection between practical reason, psychology, and now faith?
71 Khvostov (1909), 34.
72 Kant (1997), 533 (A534/B562).
73 Khvostov (1909), 42.

ity. Kant's metaethical theory erred in ascribing the free will to the noumenal world and thereby expelling it entirely from the phenomenal world. Kant – and Schopenhauer – understood causality exclusively in terms of mechanical causality. Khvostov believed that with this we have a "new" concept of free will bereft of even a hint of metaphysics.[74] He, in effect, recognized the "conflict" between the claim of universal natural causality and that of the necessity for a causality through freedom, but unlike Kant, he did not wish to see freedom as involving "a transcendent or non-empirical component, which requires the resources of transcendental idealism in order to be reconciled with the 'causality of nature'."[75]

Instead of turning to Kant's treatment of freedom in the "Third Antinomy" in the *Critique of Pure Reason*, Khvostov invoked Lopatin's notion of a creative causality as the most important feature of our power of free will. Whereas mechanical causality operates according to physical laws by which we can with assurance predict future events, we cannot predict with any degree of accuracy what a person, who is in possession of a free, creative will, will decide. Unlike Lopatin or Kant, Khvostov explicitly refrained from concerning himself "with the metaphysical question of the essential relationship between the physical and the psychic worlds."[76] He stated that he found no reason to limit our understanding of causality solely to that found in physics, but on the other hand, he did not look very hard for one. His philosophical myopia prevented him from seeing the transcendental.

Far more indicative of Khvostov's neo-Kantianism was his 1914 *Theory of the Historical Process*.[77] Khvostov, without equivocation, associated philosophy with Critical Philosophy. Its concern is not with the creation of a set of abstract speculations independent of scientific findings but, rather, to help science understand its own concerns. In order to do so, philosophy, properly understood as a critical enterprise, starts with scientific facts and seeks to clarify the first principles involved in scientific knowledge. Thus, the fundamental problem of philosophy is epistemological: what cognition is, what its proper object is, how cognition is formed, and the determination of its limits. "Philosophy aims to investigate the general principles that underlie already acquired knowledge, at-

---

74 It is certainly disputable whether this conception of the free will from a purely empirical viewpoint is "new."
75 Allison (2006), 381.
76 Khvostov (1909), 48.
77 Nonetheless, Khvostov sought to place some distance between himself and Kantianism and neo-Kantianism, saying that when he used the term "criticism" he was employing the term in a broader sense than they do. Khvostov (1914), 38 f.

tempting to render them into a systematic form, and if need be offer hypotheses for filling the gaps left by science but remaining aware of the nature of the suppositional character of what it offers."[78] Philosophy, in this understanding, is ancillary to scientific investigation, indeed a necessary prerequisite for it to proceed successfully toward ever greater knowledge of reality. This conception of philosophy and of its tasks originated with Kant, who in his three *Critiques* clearly posed and formulated it.

Some would certainly charge that Khvostov's conception of philosophy as virtually a handmaiden to the natural sciences is totally consistent with characterizing him as a neo-Kantian. The reader acquainted with "mature" German neo-Kantianism will surely then be surprised to come upon Khvostov's apology for a measured dose of psychologism. Khvostov affirmed that epistemological problems must take into account psychological data. The only cognition and cognitive faculties that we can investigate are our own, which are quite finite. Any attempt to dispel psychology from epistemology is doomed to utter failure. We possess no other knowledge except that obtained by ultimately psychological means.[79] To think otherwise is to think that we have some connection to the supernatural, a connection that obviates our mental limitations. That philosophy must take into account human psychology does not mean that epistemology is simply a psychological investigation of our cognitive processes. Epistemology is not psychology, and psychology is not epistemology. Whereas epistemology at the very start must recognize the relevance of psychological data for its concerns, psychology's concerns are exclusively with the factual structure and origin of our cognitive processes. It is a *descriptive* science. Epistemology, on the other hand, is a reflection on that data as well as on that of the other sciences. It seeks the significance of the cognitive processes in terms of an understanding of the natural world in the broad sense. "Epistemology poses as its task to answer the question of what value is our cognition as a means for penetrating into the real being of the world of which we are members and what conditions must the cognitive process satisfy in order to lead us to that goal."[80] In contrast to psychology, epistemology is a normative discipline.

Khvostov in an article in 1910, largely aimed at Rickert, expounded on his own measured psychologism in the ethical sphere. Even Rickert recognized that universally obligatory values serve as a psychological factor in determining

---

78 Khvostov (1914), 7.
79 Khvostov (1914), 31.
80 Khvostov (1914), 32. Khvostov here refers for support to Harald Höffding's *The Philosophy of Religion* in a Russian translation. We take it in any case as Khvostov's own position and not necessarily as Höffding's.

the actions of people. However, Rickert did not appear to realize that those values cannot be simply sundered from the factual psyche in which they were realized. Those, such as Rickert, who strive to divorce being and value accuse each other of lapsing into psychologism. However, "the psychologism that they kick out the door, they inevitably let back in through the window."[81] This, in Khvostov's eyes, was quite understandable. An ethics that would be free of psychologism would be an ethics free of anything human.

We can infer from the discussion above concerning free will and the function of philosophy that Khvostov departed sharply from those who advance an idealist reading of both Kant and philosophy in general. Indeed, Khvostov saw himself as firmly positioned among representatives of "critical realism," who

> believe that the forms of perception and thought that are inherent in people are in conformity with the nature of things in themselves. This is the reason they assert that the more complete our knowledge based on a study of phenomena becomes, the closer it brings us to an understanding of the world as it is in itself, independently of our consciousness.[82]

Khvostov explicitly then remarked that he shared the viewpoint of critical realism, one of whose representatives was Wilhelm Wundt.[83]

Khvostov's neo-Kantianism certainly did not mean that he refrained from criticism of Kant or even of other neo-Kantians. He rejected Kant's position that space and time were entirely *a priori* intuitions, although as he saw them they certainly have *a priori* elements. For example, we seldom encounter perfectly straight lines and perfect right angles in our representation of space. However, there are spatial properties that can be explained only by deducing them from what is given to us empirically. "We represent space as three-dimensional and cannot intuitively represent another space of a larger number of dimensions. ... Since space intuitively exists for us as three-dimensional, this is clearly due to the fact that such space is most consistent with the data of experience."[84] Khvostov drew from this the general conclusion that our representations of space and time are a result of the interaction between our cognitive faculties

---

81 Khvostov (1910), 366.
82 Khvostov (1914), 40.
83 Wundt is among the most often cited individuals in Khvostov's text. It is surprising how little Khvostov made use of Alois Riehl's own "critical realism" – so similar to his own in spirit – even though a Russian translation of parts of Riehl's *Der philosophische Kriticismus* already existed since 1887 and thus would be accessible to Khvostov's readership.
84 Khvostov (1914), 46.

and the manifold, or content, of experience. Thus, the properties of space and time that we represent as having are indicative not only of contributions from our cognition but also from what is independently given in experience. Khvostov generalized this conclusion, making it consistent with his overall metaphysical realism, saying "the logical unity of experience corresponds to the objective order of nature."[85]

In light of recent Kant-scholarship that directly and intimately ties transcendental idealism to "the thesis that things in themselves, whatever else they may be, are not spatial and temporal,"[86] we should note that Khvostov as a critical realist did not entirely disavow this thesis. Our representations of space and time, with their particular properties and features, may not reflect space and time as they are from a hypothetical theocentric point of view. Naturally, we cannot say what "objectively" corresponds to what we represent as space and time. What objectively is, is a matter for science, not philosophy even though the scientific quest must continue indefinitely. As long as the scientific enterprise continues, it will continue in search of a more complete and perfect determination of objectivity. But unlike the Baden and Marburg neo-Kantians, Khvostov did not reject the thing in itself. Kant accepted it but saw the thing in itself as the logically necessary basis of phenomena. The thing in itself, for Kant in Khvostov's interpretation, does not temporally precede phenomena; they are not the cause in the sense of one billiard ball impacting another, causing this second ball to move. The thing in itself is a logical foundation, but it is also transcendent in that it is not given in experience. That we must logically think it shows that it is conceptually necessary, i.e., is conceptually *a priori*.

Khvostov certainly expressed strong objections to Rickert's neo-Kantianism from his own critical realist viewpoint. He particularly took exception with Rickert's use of the Kantian expression "consciousness in general" [*Bewusstsein überhaupt*].[87] Rickert understood such a consciousness to be merely a concept, neither a transcendent nor an immanent reality. How, then, Khvostov asked, can it help to ground reality? He further remarked, "I do not understand how a non-real epistemological subject can actively cognize the content of consciousness. How can consciousness be both active and not real?"[88] Rickert first rejected

---

[85] Khvostov (1914), 50. Someone, of course, could reply to Khvostov that his charges are not applicable to Kant's position, that his criticisms show he misunderstands Kant.
[86] Guyer (1987), 333.
[87] Khvostov devoted much more attention to the Baden philosophers than to those in Marburg. This, undoubtedly, was due to the Baden interest in the *Geisteswissenschaften* compared to the Marburg concern with the *Naturwissenschaften*.
[88] Khvostov (1914), 117.

this consciousness as having any hint of real existence and then hypostatized it into the epistemological subject. If this "consciousness in general" has a real significance, as grounding knowledge of reality, it can be understood only as the content of consciousness, which we all have in common and attained as the ideal or asymptote of scientific knowledge.[89]

Khvostov saw his critical realism, additionally, as in opposition to those philosophies that attempt to expunge even the slightest hint of psychologism, whether it be in ethics or epistemology. He singled out Rickert, in particular, for attack. Rickert had agreed with his mentor Windelband that the laws of logic were fundamentally different from the laws of nature. The latter act with necessity, whereas the former are only norms, or guides, of thinking. It makes no sense to ask why we should obey the laws of nature, but we can ask why we should obey the laws of logic. Khvostov found Rickert's answer quite disappointing: what logic commands is an ideal, a value, and as a value it is transcendent. How can logic, supposedly being a value, serve as the super-logical basis of logic?

> Instead of seeking this basis in a truth of the first order, in the immediate perception of an extra-sensible reality contained in Kant's moral knowledge, in the mystical or religious experience of James, or in the immediate intuition to which Bergson turns the attention of philosophers, Rickert resorts to the concept of transcendent values in order to get the basis he lacks. It is impossible to accept such a technique. How can a transcendent value create a basis for logic, a value that he says is not a reality (*bytie*), but is not nothing, but is some "something"?[90]

Khvostov immediately generalized this criticism of Rickert's invocation of transcendent values. Khvostov found Rickert's appeal to irreal transcendent values and a consciousness in general specious. How can something that is not objective guarantee objective reality; how could it possibly refute subjectivism and solipsism? In short, it cannot.

As already mentioned, Khvostov's critique of Baden neo-Kantianism was conducted from a critical realist viewpoint largely indebted to Wundt. Admitted-

---

89 Khvostov (1914), 65. Whether Khvostov understood Rickert correctly here is open to debate. After remarking that the "consciousness in general" is a concept, Rickert then wrote, "We form this concept not without the content that belongs to it, and the content alone is what belongs to reality, namely the reality of immanent being. The epistemological consciousness is at least for the time being, therefore, nothing else but what is common to all immanent objects and that which cannot be further described." Rickert (1904), 29. Khvostov and Rickert were not as far apart in this matter as Khvostov thought.
90 Khvostov (1914), 116.

ly, Khvostov was not satisfied with the ideas of any of the historical figures he considered. What we have seen here represents his final venture into philosophy. Although he was not trained as a professional philosopher and despite its many lacunae, his 1914 work is an outstanding example of the realist direction in neo-Kantianism.

The three figures examined in this chapter are among the best social-scientific minds in late Imperial Russia. All three in their respective ways sought inspiration from neo-Kantianism of one form or another in order to grapple with issues in their own discipline, be it law, sociology or history. Arguably, Kistjakovskij presented the most carefully executed and detailed philosophical elaborations. Curiously – but then on second thought perhaps not – all three suffered from the political turbulence in which they lived and all three died within a short period of time of each other, Lappo-Danilevskij in 1919 and the other two in 1920. They were only in their fifties when they perished. Had they lived in a calmer era, they could have, indeed surely would have, contributed much more to their professions and to the philosophical underpinnings of their respective disciplines.

Chapter 7
Baden School Philosophers Who Scattered

Whereas Germany had a number of universities, each competing, as it were, for students and for broad recognition, the Russian Empire had far fewer, a reflection not only of governmental policies but also of its belated entrance into modernity. The administrators of the Russian Empire, recognizing that their domain still had a predominantly agricultural economy, believed the Empire could not afford the luxury of too many educated youth with insufficient career prospects. If we ascribe the veritable abundance of German neo-Kantian "schools," – the Marburg and Baden Schools being only the two most prominent – to be rooted in or at least influenced by competition between domestic universities, then one might, given the paucity of Russian universities, expect that there should have been scant divergence within Russian neo-Kantianism. This, however, did not turn out to be the case.

7.1 The Critical-Realist Objection to Baden

The young neo-Kantians who returned to Russia from a period of study at German universities more often than not returned with an allegiance to the particular "school" associated with where they had been. We see this happened time and again but perhaps most clearly of all in the clash between Novgorodcev, who went to Freiburg in the German state of Baden, and his own student Saval'skij, who imbibed neo-Kantianism in Marburg. They returned from their studies abroad with sharply differing understandings of neo-Kantianism.[1] However, we can easily find differing understandings even among its proponents of the older generation. Vvedenskij's neo-Kantian stance bore little resemblance to that of Chelpanov, and the latter, in turn, was quite unlike that of Novgorodcev's or Struve's.[2]

---

[1] Even in light of what we have seen thus far, we must look askance at Belov's claim that "Russian Neo-Kantianism was a rather uniform movement and did not attach significant importance to the differences between the approaches of both [the] Marburg and Baden schools." Belov (2016a), 396. He merely considers the Novgorodcev-Saval'skij "episode" to be an exception.

[2] Although likely of little philosophical significance, Vvedenskij wrote a scathing review of Chelpanov's 1907 *Uchebnik logiki* [*Logic Textbook*], which was meant for secondary school students, finding it to be replete with "major flaws that, despite all of its mentioned values, cannot be considered satisfactory." Vvedenskij (1908), 223. We should note, though, that Vvedenskij devoted

Certainly, the dispute between Novgorodcev and Saval'skij can hardly serve as the basis for any generalization, but we do not have to look far and wide for others. Indeed, the disputes took place along lines not dissimilar to those in Germany at the time and with an eye toward the German scene. Struve in his 1900 "Preface" to Berdjaev's first book devoted significant attention to the Baden School's positions, a critique generally overlooked in the secondary literature. In the context of that earlier discussion, we saw that Struve rejected the Baden claim that logic is normative. "Logic describes the forms in which human thought necessarily moves. It does not prescribe; its norms are natural laws of thought."[3] The position Struve explicitly rejected and which he called "normative formalism" had, in his eyes, a deep flaw that was reflected in its attempt to combine the true, the ought, and the beautiful in the single concept of the universally obligatory norm. He saw this single, general concept of an objective norm present in Berdjaev and found it to be a major mistake. Berdjaev "erases the profound and fundamental difference between logical laws, between the laws of cognizing being and the norms of obligation."[4] Struve had no issue with its use in metaphysics – presumably to be understood in terms of *religious* metaphysics – but objected to its usage in the theory of cognition, where it leads to the conception of a pseudo-formal unity of the differentiated experiences, a conception that is futile and perverse. Struve particularly singled out Windelband as this position's classical exponent and saw it as advocating an "unnatural parallelism (or analogy) between cognition and morality."[5] Windelband would have, according to Struve, the cognitive process be understood in terms of ethical definitions and would have the categories of one discipline subordinate to that of another, thereby distorting the nature of both cognition and morality. Whereas we can find objective reality in our empirical consciousness, there is absolutely no objective obligation to be found in everyday consciousness.

Rickert did not escape Struve's realist neo-Kantian barbs. Struve, in effect, charged Rickert, of course, with making all of Windelband's mistakes and then some. All judgments, be they logical, ethical, or aesthetic, assert a connection between representations, a connection that is evaluative. For both Windelband and Rickert, "What I assert, I necessarily deem correct; what I deny, necessarily arouses my dissatisfaction. Cognition, consequently, is a process,

---

most of his attention to the failures of Chelpanov's work as a textbook, not as a piece of original philosophy, which it definitely neither was nor was it intended to be.
3 Struve (1999), 28.
4 Struve (1999), 26.
5 Struve (1999), 15.

determined by feelings, i.e., by satisfaction or dissatisfaction."[6] Struve found that in this way Rickert has a subjective – indeed a purely hedonistic – understanding of truth. Struve, to be sure, recognized that Rickert's philosophy involves more twists and turns, but it ultimately returned to its original standpoint that all cognitive judgments are valuations. Rickert's arguments reproduced Fichte's untenable epistemology with its deep metaphysical sense. Struve held that there is no sharper opposition in epistemology than that between being and obligation, between what is and what should be.

Semion Frank, whom we saw in a previous chapter as a close associate of Struve's, published in 1904 a translation of Windelband's *Präludien*. He accompanied his translation with a relatively brief but informative preface dated December 1903. He remarked there that apart from historical works Windelband's chief significance for philosophy was due to the unique grounding he gave to Kant's philosophy, to its thoughtful interpretation of Kant's system, to that system's adaptation to the latest needs and conditions, but also to its eloquent literary style that freed the essence of Kant's thought from its cumbersome and lifeless litany of concepts. The result was that Windelband had omitted many details, the omission of which Frank apparently agreed were unnecessary for the conveyance of Kant's main idea. Still, the general outline of the Kantian system that Windelband drew thereby stood out all that more starkly, albeit in a subjective interpretation as a result of Windelband's own understanding of it.[7] Windelband's scattered judgments on social issues were in Frank's estimation the weak side of the former and showed little philosophical depth, but they were few in number and stood quite apart logically from Windelband's scholarly views. Frank remarked that in any case, he was recommending Windelband as a philosopher, not as a sociologist.

7.2 From Neo-Kantianism to a Realist Phenomenology

As we saw in Chapter 5, those associated with a neo-Kantian Marxism in Imperial Russia comparatively quickly moved on first from Marxism and then from neo-Kantianism in general. The same, however, cannot be said without significant qualification of all the Russian adherents of Baden School neo-Kantianism. Indeed, it attracted new disciples in the early years of the twentieth century. Unlike the Baden-School philosophers we saw in the previous two chapters who regret-

---

6 Struve (1999), 41.
7 Frank (2007), 5.

tably died prematurely, there were others who endured, albeit in emigration, well into the century. One such was Nikolaj N. Alekseev (1879–1964), who studied in the law faculty of Moscow University and who did, unlike Saval'skij, come under the decisive influence of Novgorodcev. Awarded a diploma with honors in 1906, he stayed on to prepare for the *magister*'s degree examination the following academic year. His success made him eligible to serve as a *privat-docent*, and in 1908 he went to Germany for two years. In Berlin, he heard Riehl and Simmel; in Heidelberg, Windelband and Jellinek; and in Marburg Cohen and Natorp. Alekseev, while still in Marburg in February 1910, reported to the Ministry of National Education on his first year abroad and wrote concerning his choice of a topic for his thesis:

> The fundamental interest of my scholarly work has revolved for several years around the question of the historical origin and theoretical value of the latest naturalistic theories in the sphere of the social sciences. … I have decided to focus on the possibly most philosophically sound, the possibly most universal, and possibly most consistent theory. My choice rested on the system of the German sociologist Karl Marx.[8]

Alekseev defended his thesis in 1912 at Moscow University, but before turning to it let us look at a piece he published in 1909 and thus likely written while its author was still in Germany.

If we should need additional evidence that Rudolf Stammler's socio-legal philosophy received significant attention in Imperial Russia, we have it in Alekseev's piece "Rudolf Stammler's Social Philosophy" that appeared in *Voprosy filosofii*.[9] Alekseev stated at the outset that his aim was merely to determine the significance of Stammler's ideas for social philosophy. This, he claimed, is best done by comparing Stammler's conceptions in social philosophy with those of the two "fundamental" neo-Kantian schools – he presumably meant Baden and Marburg.[10] Alekseev characterized the Baden direction, particularly as expounded by Kistjakovskij, as methodological pluralism, meaning that the investigator can conduct his/her investigation of any object from quite different, even incompatible viewpoints. We can view, for example, the general concept of society juridically but, if we choose, also sociologically – an investigation of a single object but conducted with two different interests or from two different angles. However, to be precise, Stammler was not an adherent of such pluralism. He saw as his aim the determination of the essence or absolutely common ele-

---

8 Quoted in Dmitrieva (2007b), 198.
9 Alekseev quotes liberally from the 1907 Russian translation by Davydov of Stammler's work.
10 Alekseev (1909), 6.

ment in every social being, an aim that Kistjakovskij would go on to consider logically senseless. The concern of Stammler was with the social object, whereas the concern of Kistjakovskij and the Baden philosophers was with methodology. The goal of Stammler's work was to find the underlying unity in social life, whereas the concern of the Baden School was with how best to approach social life.

Alekseev, generally speaking, supported Kistjakovskij's position, which the former saw as an empirical realism, set against what he took to be the position of Stammler. However, for Alekseev, Kistjakovskij's social methodology employed naturalistic concepts, i.e., concepts drawn from natural science. Kistjakovskij, thereby, saw society as an ordinary spatio-temporal unity. Just as we see a forest as a collection of trees that can be described and studied without departing from naturalism, so too society in this conception is a collection of human individuals and so forth. Alekseev admitted that Kistjakovskij was willing to concede that the study of society may require invoking talk of mental interaction between the individuals, but the latter interpreted even this qualification naturalistically. Alekseev objected, saying that naturalistic concepts alone cannot suffice to describe society. "In order to constitute society, we must transgress the concepts of natural science. We must step beyond naturalism."[11] Alekseev held that Kistjakovskij did not make this move, but Stammler did see the possibility – albeit by speculative means – of finding the fundamental unity in the concept of human social life, a unity that cannot be found by the methods and concepts of natural science. In this lies the entire significance of Stammler's social philosophy. Although we often today associate the Baden neo-Kantians with differentiating the social sciences from the natural sciences and accusing the Marburg philosophers of not doing so, Alekseev saw Stammler, who was and is typically associated with Marburg, as more inclined than the Marburgians to make that differentiation.

It must be said, though, that Alekseev immediately qualified his recognition of Stammler's significance. The correct formulation – even were it that – of a problem does not mean that the execution leading to its solution is correct. In attempting to solve the problem, Stammler introduced a number of inadmissible blunders that Alekseev believed could be reduced to two: (1) Stammler, in addition to the problem's somewhat ambiguous formulation, had inadequately grounded the problem in epistemology, and (2) Stammler had provided "an erroneous solution to it within the bounds of social philosophy."[12] Whatever we

---

[11] Alekseev (1909), 14.
[12] Alekseev (1909), 15. This second point is circular. He said, in effect, that Stammler's error was in having provided an error!

may make of this second "blunder," Alekseev held that Stammler's conception of a fundamental unity had a variety of meanings, but more importantly, he assumed there was such a unity without adequate proof. Thus, it remained in his formulation like a mathematical variable that could have different significations depending on the context.

Alekseev found another lacuna in Stammler's social philosophy in his failure to account for how an object's fundamental unity could be referred to from various standpoints of study. The two most important categorical concepts of social philosophy are form and matter, both of which are abstractions from the unity that is human social life. Stammler simply declared that form can, by its essence, be studied scientifically in isolation from its content. He understood this thesis as justifying the study of the form of social law apart from the content of law and that this study formed the sub-discipline of philosophy of law. Alekseev found Stammler's conception of form closer to that of the Marburg philosophers than to those who represent the Baden School. For Stammler, physics today meant mathematical physics, but mathematics can be studied as a discipline independently of its application to the physical world. He held that social philosophy can, likewise, be studied independently of factual social life. Alekseev objected to such a blanket assertion.

> From the fact that something takes place in physics, it is impossible to conclude that it must take place everywhere, particularly in social philosophy. The concern of the fourth heading of Kant's amphiboly is with form and matter in general. However, we firmly declare that there is nothing in Kant's discussion in the *Critique of Pure Reason* to prove the thesis that a general peculiarity of form can be studied scientifically by itself.[13]

Moreover, what is this generality, this feature or peculiarity, that constitutes the concept of human social life and that allegedly can be the focus of sociological investigation? Stammler's answer is the concept of the external regulation of human relations. This concept is that of a social norm, and it serves as the fundamental unity underlying human social life. Alekseev found the basis of Stammler's one-sided social philosophy to be this positing of the concept of social norm as the "constitutive principle of social life in general."[14] This elevation of social norm is reflected in all of Stammler's further investigations. Alekseev remarked that unfortunately the size constraints of an article prevented him

---

13 Alekseev (1909), 19–20. Alekseev certainly had in mind Kant (1997), 369–370 (A266/B222/A268/B324).
14 Alekseev (1909), 25.

from focusing on all of the consequences that follow from accepting this, in his eyes, fundamentally false premise.

Back in Moscow, Alekseev defended his *magister*'s thesis in May 1911. Before turning to it, however, let us quickly note the allegiance he expressed to Kantian idealism in the context of a review-article published in late 1912 of Bulgakov's *Philosophy of Economy*. Alekseev, indeed, found Bulgakov to be indebted to Kantian Critical Philosophy but interpreted through Schelling. The outlines of the transcendental method Bulgakov used are directly borrowed from Schelling and, to a limited extent, the Marburg philosophers. Alekseev stressed that he himself had the deepest respect for transcendental philosophy, that his views were entirely dependent on the ideas of German Idealism, and that, thus, he stood closer to that movement than did Bulgakov in 1912. Nevertheless, the main inadequacy of Kantianism as interpreted by its founder's successors was that they erased the Kantian distinctions between the separate stages of human cognition, developing a peculiar monism that was quite foreign to Kant himself. They adopted some features of his thought, but not the inner sense of Kantianism. They attempted to construct all of human knowledge in terms of a single standard. Alekseev, on the other hand, saw the sense of the Kantian Critical method in its liberation from the dogmatic monism of earlier philosophies and the task of natural science not as an absolute grasp of things in themselves, but as an unending quest in always new problems. Kant concerned himself not only with natural science but also with ethics and aesthetics. He was clearly aware of the difference between the subject-matter in the three *Critiques*, even though he attempted to transfer the architectonic scheme of the first *Critique* to the others.[15]

Alekseev's thesis, published in 1912, on the face of it was to be a study and critique of the role of naturalistic methodology – an explanation of phenomena strictly in terms of natural properties and causes – in socio-political theories. One of the chief questions to be studied, he tells us, is whether and in what sense the methods employed in natural science are applicable to the study of social phenomena. This, he hoped, would lead to a determination of whether sociology is itself a natural science. However, he believed that an examination of the sociological theories that had recently appeared offered a better basis for answering his questions and was more consistent in its approach with the sense of Kantian philosophy than an abstract examination – such as that presented by Rickert, though his name remained unmentioned – of such theories. Alekseev also added in the preface to his thesis that his conclusions agreed with the ideas

---

15 Alekseev (1912b), 714.

of Struve and to those of his teacher Novgorodcev. Although his reference to these names was guarded, they would have been clear to all of his readers.[16] In another veiled reference, this time to Hermann Cohen, Alekseev stated that he rejected all logical and epistemological relativism and that his sympathy was on the side of "pure logic" departing from it when and to the extent that it proceeded into ontology and metaphysics.[17]

Alekseev's thesis, as we might suspect, was a scholarly study and hardly the place to articulate – let alone develop – the author's own convictions. He faulted seventeenth-century naturalistic political theories for attempting to construct social philosophies in a "*more physico et geometrico*" and for not taking into account the non-rational elements that pervade the social fabric and praised the century-later Kant for recognizing that the systematic elaboration of morality could not achieve a high degree of metaphysical certainty. This, to Alekseev, is an indication that Kant opposed the universal applicability of mathematics. Notable also is the fact that Kant wrote of natural law not in terms of what is, but of what should be. The entire issue of natural law belongs to the sphere of ethics, not to that of physics. When dealing with scientific fields encompassing organic life we must allow the admission of the concept of purpose, as a regulative or heuristic principle. The human being is a free, noumenal being in addition to being a natural being.[18]

Alekseev did not proclaim in an unequivocal manner any allegiance to one or another school of neo-Kantianism, though that he was of a neo-Kantian persuasion was clear enough. He combined, in fact, elements from both Baden and Marburg. On the one hand, he appreciated the Baden division of the sciences and, on the other hand, the Marburg aversion to any hint of psychologism. Even such a tepid attitude toward the latter from one of Novgorodcev's students, however, was significant, particularly in light of the Saval'skij case. Regrettably, the promised second part to his thesis topic never appeared. The outbreak of the war surely played some role in this.

---

16 Alekseev (1912a), xiii. Novgorodcev in late 1913 reviewed Alekseev's book. As was customary at the time, Novgorodcev praised the author's efforts, remarking on his student's fresh thoughts and deep understanding of the revolution taking place in the social sciences, which made Alekseev's book of particular interest. However, he added that a book that sets such enormous tasks condemns itself to be the object of a large number of objections, doubts, and misunderstandings. Novgorodcev then immediately sets forth many objections, among which is charging Alekseev with failing to answer clearly his fundamental question concerning the interrelation between the social and the natural sciences. Novgorodcev (1913), 719.
17 This is a clear enough reference to Cohen's 1902 *Logik der reinen Erkenntnis*.
18 Alekseev (1912a), 110.

From 1912 Alekseev taught at Moscow University in the capacity of a *privat-docent* and lectured as well at the Commercial Institute in the city. In 1916, he was promoted to professor in the law department at the University and spent the summer of 1918 on a scholarly trip abroad. Alekseev was appointed as well to a professorship in law at Taurida University, which had just opened in Simferopol, Crimea, which was held at the time by the White forces.

Alekseev's opposition to relativism in both ethics and in law, mentioned above, is again evidenced in lectures delivered during the 1918/1919 year. To tie law to the human individual, that law can have no absolute value transcending the individual, is to ascribe to it only a relative value. "One must object as mightily as possible against this individualism, which feeds ultimately on the anarchist tendencies of our day."[19] Alekseev saw the only recently deceased philosopher Windelband as the one who showed clearly the connection between the world of law and that of norms and obligations. As a study of norms, then, the rigorous discipline of jurisprudence is not concerned with the explanation of actual relations, but with their evaluation.[20] However, Alekseev saw norms and facts not as separate ontological spheres, but as two ways of formulating one expression. Such a simple statement as "Thou shalt not kill" – clearly a norm – can also be formulated, he claimed, as the factual statement "Killing is shameful."[21] However, to be clear, norms and values do not lie within the same sphere, though norms and facts do.

Values occupy a separate sphere from material objects, a sphere of non-natural objects "no less worthy of cognition than nature itself" and which makes possible natural science itself.[22] This world beyond nature was revealed by Kant. "If the negative task of Criticism amounts to a *restriction* on natural science, then its positive task can be said to be more or less the plainly expressed desire for the *intuition of new objects*. Criticism cleared the way for such an intuition, and this is its greatest merit."[23] That absolute values, eternal truths, lie in a non-natural sphere allows and indeed makes necessary a methodological pluralism. Alekseev concluded from this that law and the state can be studied from both the nomothetic and the idiographic viewpoint. In the end, "truth lies in a combination of viewpoints, striving as much as possible to capture the object. Certainly, they do not *completely* capture the object. There always remains some-

---

19 Alekseev (1919a), 133.
20 Alekseev (1919a), 36.
21 Alekseev (1918), 45. Actually, rather than to any of Windelband's writings, Alekseev referred to chapter two of the "Prolegomena" in Husserl's *Logical Investigations*. Husserl (1970), 74–89.
22 Alekseev (1919b), 12.
23 Alekseev (1919b), 12.

thing uncognizable, viz., the uncognizable essence of the phenomenon, the 'thing in itself'."[24] In this way, Alekseev allowed within his neo-Kantianism that concept which many German philosophers valiantly endeavored to banish. Whereas some precursors of neo-Kantianism anticipated the admissibility of a pluralistic methodology in the study of law and political science, those who follow Kant hail each method as forming an independent investigative principle, an autonomous hypothesis that can serve as the basis for a separate branch of rigorous study in the social sciences.

Alekseev already at this time – 1918/1919 – saw an "essential deficiency" in the methodology of the Baden School. How is it, he asked, that a single phenomenon or object with such different underlying principles can still be labeled or grouped under a single designation. How is it that we can look at law or at the political state as a sociological phenomenon as well as a psychological or narrowly juridical phenomenon? Rather than inquiring into the linguistic origin of our terminology, however, Alekseev turned to the notion of essence in phenomenology. We can, as a fact, elaborate a general, fundamental "science" of law or of the state that studies not the forms of each, as in Kantianism, but the object itself with its most general features. The result may not be an epistemology, properly speaking, but, rather, a science of essences, an ontology. Alekseev accorded to our mental faculty an ability to intellectually intuit these essences or universal truths. He remarked, "strictly speaking all of our knowledge follows ultimately from this ability, because only with its help do we cognize what we consider to be most valid and true."[25] Mathematics is, arguably, the most obvious example of such intellectual intuition, but such intuition of fundamental principles lies at the basis of other objects and other spheres of knowledge. This intuition of essences, Alekseev believed, as a methodological device allows jurisprudence to become a science.

After his evacuation along with other White forces from Crimea, Alekseev spent a year in Istanbul, and then in 1922, he became a professor in the law department in the Russian-language school established in Prague for immigrants expecting to return in a reasonable time to their homeland with the collapse of the Bolshevik regime. Several years later he was able to lecture as well in Berlin. With the closing of the law faculty in 1931 and the Nazi accession in Berlin, Alekseev moved on to Paris, where he lectured on law at the Sorbonne. His next stop was Belgrade, where he spent the war years. With the end of hostilities, Alekseev's Soviet citizenship was restored, but the deterioration of relations be-

---

24 Alekseev (1919b), 20.
25 Alekseev (1919b), 25.

tween Yugoslavia and the Soviet Union forced him to move on again. Declining repatriation to the USSR, Alekseev's final earthly destination was Geneva.

As we might expect given his continuing involvement in academic work, Alekseev continued to write on law, but he also became involved at least for a time after 1926 in Eurasianism, a movement among the Russian intellectual émigré community. Although no single set of philosophical ideas characterized Eurasianism as a whole, it saw Russia as a self-contained unit belonging to neither Europe nor Asia but a synthesis, so to speak, of both. Alekseev, in reflecting on his earlier years during this time, wrote that despite learning much in Marburg and Heidelberg he adhered to neither neo-Kantian School. Instead, he borrowed much from both in an effort to create a Russian national philosophy that he could call his own.[26] Alekseev's philosophical jottings in this period are beyond both the thematic and the temporal scope of the present study. Nevertheless, we should note that he continued to keep up with developments in philosophy and in particular with phenomenology without abandoning his essentialism.[27] To what extent he retained in his later years any debt to neo-Kantianism, even if it be "diluted" with phenomenology, remains an open question.

7.3 The Journal *Logos*

A number of young students, in addition to Alekseev, went to Heidelberg and Freiburg to continue their philosophical education with the hope of learning there the latest developments in philosophy directly at their source. Two of them, Sergej I. Hessen (1887–1950) and Fedor A. Stepun (1884–1965) became intimately associated for posterity with the founding of the Russian edition of the international philosophical/cultural journal *Logos*. The idea for the journal arose among members of the "Heidelberg Philosophical Society," who after they had passed their doctoral examinations and with the support of Rickert began to discuss its possibility with publishers in Germany and Russia.[28] *Logos*, as originally conceived, was to have a combination of original material unique to each edition in the respective languages as well as "main articles" that would be published in

---

[26] Dmitrieva (2007b), 200.
[27] Alekseev wrote, for example, "What is this essence, the essence of what is, which Heidegger formulates as '*das Sein des Seienden*'? The most interesting thing is that Being, the essence of being, cannot 'exist' in the sense in which any other being exists. If it existed in that way, it could not be the most universal." Alekseev (1937), 38.
[28] Bezrodnyj (1992), 489. For a first-hand account of the organizational meeting in Freiburg, see Stepun (1994), 101–103.

all of the national editions.²⁹ Although the hope was also to have French and Italian editions – even an American one – appear soon, only the Italian edition came to fruition. And the original hope to synchronize the publication of at least the German and Russian editions was realized only for a short time.

The journal in its Russian incarnation appeared from 1910–1914 in fits and starts, its project of seeking "eternal peace" in philosophy falling victim to the ensuing war fervor.³⁰ Certainly, both Hessen and Stepun were products of the Baden School, but the journal itself attempted to reflect all directions in contemporary European philosophy and was not, nor was it ever intended to be, exclusively an organ of neo-Kantianism.³¹ Rejecting the possible charge of dogmatism, the editorial statement prefacing the first issue of the Russian edition of *Logos* emphasized that the journal was "not an advocate of any particular philosoph-

---

29 "A common goal and mood will be expressed in 'main articles,' which will be published in parallel in all national editions of *Logos*. 'Special' articles will cater to more detailed philosophical problems, more intimate issues of national cultures, and therefore, in general, will not be translated into other languages." Hessen, Metner, and Stepun (1910), 12.
30 Plotnikov sadly but correctly observed, "The tragedy of *Logos* and, at the same time, the splendor of its intention were reflected in the fact that its founders in both Germany and Russia – Stepun, Hessen, Jakovenko, Kroner, and Mehlis – advanced the project of 'eternal peace' in philosophy just before Europe descended into the abyss of international carnage, forcing the editors literally to fight each other. (The artillery officer Stepun fought on the Russian side of the Galician front when his teacher and a force behind the idea of *Logos*, Emil Lask, was killed on the German side." Plotnikov (2006), 7–8.
31 Yet even such an otherwise discerning scholar as Meerson labels *Logos* as a neo-Kantian journal! Meerson (1995), 225. He is succeeded by Poole who sees in *Logos* "the purity of its neo-Kantian direction." Poole (1999), 340. Given that a Russian translation of Husserl's essay "Philosophy as Rigorous Science" appeared in *Logos*, it is difficult to know from *today's* perspective how the journal could be seen, even by simply looking at its table of contents, as advocating a *purely* neo-Kantian direction. Somewhat earlier than both, Abramov, an otherwise astute Russian scholar wrote, "A characteristic feature of this publication was that from the very beginning it was conceived and created to be the theoretical organ for the propagation and development of neo-Kantian philosophical ideas." Abramov (1994), 232. Not surprisingly, Abramov provided no source for his claim. However, there is a source for the claim, one which shows that *Logos*'s desire for "eternal peace" in philosophy was in jeopardy for some time before the start of the Great War. S. L. Frank already in a contribution to a St. Petersburg newspaper *Russkaja molva* in January 1913 wrote, "It is no secret to people versed in the matter – and this can be seen from the contents of the journal itself – that the extensive and neutral task of bringing together Russian and Western thought is combined in the hands of the journal with a much narrower and militant task, with propaganda and apologetics in favor of certain forms of contemporary German Kantianism. … Under a neutral flag, it promotes militant contraband. … One could in general doubt the usefulness for philosophical creativity of such a narrow and fanatical partisanship." Frank (1913), 5. Frank, of course, had by this time moved in a religious and mystical direction.

ical direction. The unifying factor of its activity was a general mood expressed in an awareness of the tasks of contemporary philosophy and of the paths leading to their culmination."[32] We can find in Hessen's letter of 5 November 1909 to Emilij Metner (1872–1936) further confirmation that the editors sincerely meant what they wrote:

> *Logos* cannot accept bad articles to please philosophers. ... I am extremely hostile to Losskij, but it would not occur to us to refuse articles by this author, since they would be intelligent and seriously written. *Logos* must be a non-partisan journal, for otherwise it loses all meaning.[33]

Denying a bias toward any particular philosophy, the editors of *Logos*, nonetheless, held German philosophy exerted in their day an influence comparable to that of ancient Greek thought. "This is shown not so much by contemporary German philosophy itself as by the indisputable fact that every original and significant appearance of philosophical thought of other nations today bears the distinct stamp of an influence from German idealism."[34]

According to the editorial staff of *Logos*, Russian philosophical thought, as exemplified in the works of the Slavophiles, had rushed toward a comprehensive synthesis of all aspects of life under the influence of Romanticism. But being chaotic in nature, Russian philosophy plunged into chaos taking Russian culture, as it was, along with itself. Russian philosophy had not forged its fundamental principles in the fire of careful, deliberate theoretical work. Instead, it simply took them from the "dark bowels" of inner experience.[35] For the *Logos* editors, no one saw more clearly than had Vladimir Solov'ëv the philosophical sterility of the Slavophile immersion into chaos and the need for a genuinely insightful synthesis. But despite consciously realizing the chief contradictions of Slavophilism, Solov'ëv unconsciously remained immersed in the Slavophile framework. No matter how clear the basic concepts of his system may appear, particularly in comparison to his predecessors, he could not grasp the importance of theory. Solov'ëv, in summary then, in the eyes of Hessen and Stepun,

---

32 Hessen, Metner, and Stepun (1910), 10.
33 Hessen (1993), 527. Metner was a music critic and early member of the Russian Symbolist movement. His name is also spelled "Medtner." His later years in emigration and idolization of Adolf Hitler are best recounted by others.
34 Hessen, Metner, and Stepun (1910), 14.
35 Hessen, Metner, and Stepun (1910), 2.

"quite vividly embodied in his writings the fundamental contradiction of the history of Russian thought."³⁶

Despite the opposition *Logos* encountered in the Russian press from neo-Slavophiles, there were some who at one time came to its defense. One such had been Semion Frank, who in an article published in September 1910, and thus before turning against the editorial direction of the journal, wrote:

> Certainly, it is regrettable when young Russian philosophers worship every word of Rickert or Cohen and do not read Solov'ëv and Lopatin or do not recognize their philosophical significance. But perhaps even more regrettable is the nationalistic conceit that in evaluating a national philosophy does not know standards and perspectives and tramples on eternal values of European thought. It must be said in conclusion regarding the journal *Logos* that whether good or bad its editors understand and fulfill its purpose, viz., to assert the development of philosophical culture in Russia in close communication with Western philosophical life, a purpose that deserves unconditional approval and all possible encouragement.³⁷

Others at least implicitly defended the *Logos* position by publishing in it. Boris Jakovenko, whose writings we shall discuss later, published two articles in the journal already with its first issue. He would be listed in 1911 along with Hessen and Stepun as an editor, and he became the most frequent contributor to the

---

36 Hessen, Metner, and Stepun (1910), 3. The spiritual heirs of the Slavophiles, of course, could not countenance such a portrayal of either earlier Russian thought in general or Solov'ëv in particular. The most vehement and direct objections to the direction offered by *Logos* came from Vladimir F. Ern. In his memoirs Stepun wrote, "An irreconcilable enemy of German Idealism, and in particular of neo-Kantianism, Ern immediately after the appearance of the first issue of *Logos* angrily attacked us with a voluminous book entitled *Bor'ba za Logos* [*Struggle for Logos*]. In this work, Ern appeared with constant critical speeches and polemics in print against those of us who were associated with *Logos*. He stubbornly held the idea that as apologists of scientific philosophy, divorced from the ancient Christian tradition, we had no right to concern ourselves with a term sanctified by the Gospel and which had not yet lost its meaning for the Orthodox." Stepun (1994), 200. Ern was not alone in his opposition to *Logos*. Others within the Moscow Religious-Philosophical Society, two of whom we have already discussed – Berdjaev and Bulgakov – already at this time linked the development of Russian philosophy to an Orthodox religious renaissance.

37 Frank (1996b), 112. Frank's article, "O nacionalizme v filosofii" ["On Nationalism in Philosophy"], was in response to Ern's article "Nechto o Logose, russkoj filosofii i nauchnosti" ["Something about Logos, Russian Philosophy and Scientificity"]. Ern replied to Frank's article with one entitled "Kul'turnoe neponimanie. Otvet S. L. Franku" ["Cultural Misunderstanding. An Answer to S. L. Frank"], which evoked another reply from Frank entitled "Eshche o nacionalizme v filosofii" ["Still More on Nationalism in Philosophy"]. All of these articles appeared in 1910.

journal over its relatively brief lifetime. Kistjakovskij contributed an article in the second issue from 1910 and was listed on the journal's title page along with, among others, Vvedenskij, Lappo-Danilevskij, Losskij, Struve, and Frank in 1911 as offering the "closest participation."[38]

Two issues of *Logos* appeared in 1914. Hessen in July departed for Germany on vacation, his wife being then in Freiburg. He related in an autobiographical sketch that at the time no one thought of an impending war. Hessen met up with his wife in Heidelberg, but owing to his son being ill they could not leave immediately. The start of the War delayed their departure, but a month afterward his wife and son were able to board a special train. Hessen, however, was unfortunately detained. Thanks to the intervention of Max Weber, whom Hessen knew through Rickert, Hessen was permitted to travel to Sweden after promising not to serve voluntarily in the opposing military forces. While Hessen was held as a civilian prisoner in Germany, Stepun was drafted into the war. Without its editors, *Logos* simply ceased publication, and Hessen was unable to revive the journal upon his return to St. Petersburg in December. We may confidently presume, though, that given the journal's platform, on the one hand, and the prevalent war fever in 1914, on the other, *Logos* would have ceased in any case even if its editors had been available.[39]

7.4 A Disciple of Rickert's

Given the stated intent by both Hessen and Stepun that *Logos* would not be an advocate of a specific philosophical direction, what were the personal positions of the two editors? Despite their study in Germany under certainly notable representatives of German neo-Kantianism, were Hessen and Stepun able to avoid the philosophical framework taught there? Do we take at face value Hessen's statement expressed in a letter dated 17 February 1913 to Vyacheslav Ivanov (1866–1949): "To consider it [*Logos*] an organ of neo-Kantianism is such a provincial anachronism. As though Jakovenko, Stepun, and I (let alone the Germans) are Kantians. With the same right, we could be called Platonists, Fichteans, and Hegelians!"[40] Even if Hessen was a Baden neo-Kantian in, say, 1909, had he changed his fundamental stand by 1913? On the other hand, in re-

---

38 Apart from the announcement of a future article "On the Other I," there is no indication that Vvedenskij participated in the work of *Logos*. Lappo-Danilevskij, on the other hand, worked on the journal from the start. Hessen (1993), 533.
39 Hessen (1994), 158.
40 Kejdan (1997), 511.

flecting on his attitudes at the time of the February Revolution of 1917 he wrote in his autobiographical sketch from 1947, "Despite my Kantianism, I, like before, felt closer to the legal Marxism of Plekhanov than to revolutionary socialism."[41] Thus, the question of the young Hessen's Kantianism or neo-Kantianism remains open.

Fearing arrest for his political activities, Hessen had departed in 1905 for Germany at his father's request upon finishing his secondary education in St. Petersburg. In Heidelberg, he met Stepun, Jakovenko, and Kistjakovskij as well as, of course, Windelband and Emil Lask, the latter advising Hessen to go for a time to hear Rickert in Freiburg. He remained there for six semesters. Hessen completed the writing of his thesis *Individuelle Kausalität* in the early autumn of 1908 and received his degree in March of the following year, despite by his own admission performing poorly in certain examinations.

Hessen's thesis displays the strong influence of Rickert's teachings and concerns. In contrast to Kant's "transcendental idealism," Hessen, by way of laying some needed theoretical groundwork for Rickert's own neo-Kantianism, termed his own philosophical direction a "transcendental empiricism."[42] The problem of universals, he claimed, was fundamental in pre-Kantian philosophy. Rationalism sought to derive the particular from the universal, whereas empiricism sought to obtain the universal from the particular. For the former, truth lies in the universal; for the latter, it lies in the particular. Nevertheless, the empiricists – especially Hume – still viewed necessity as connected with universality. Therefore, in Hume's eyes, since causality was associated with universality and there is no universality, there is no causality, understood as a universal connection or lawfulness regulating events. Hume could not countenance a conception of individual causality.

Hessen held that Kant discovered a new concept of transcendental universality. He did not realize, however, its full implications for epistemology. Causality, for him, was still associated exclusively with universality. Causality meant fol-

---

41 Hessen (1994), 159.
42 Hessen (1909), 1. Why Hessen termed his position as he did is by no means clear, though Lask had already used it in his 1902 *Fichtes Idealismus und die Geschichte*, with which Hessen certainly was acquainted in 1910 and, thus, in all likelihood already in 1908/1909. If Hessen sought to set his stance against Kant's understanding of transcendental idealism with its ideality of categories and forms of intuition, he surely failed to do so. Moreover, if he had intended to oppose his views against those of Kant, we would hardly expect him to do so by means of advocating a neo-*Kant*ianism. This is not to say that we ourselves cannot find such a contrast, but only that Hessen did not pointedly do so. He makes us connect the dots, as it were, instead of doing it himself.

lowing a universal rule or law along the same lines laid down by the rationalists. But Kant also found a new concept of the universal, viz., *a priori* universality, which he designated by the expression "in general." Universality is, indeed, connected with veracity "not because it includes the particular in itself or because it is derived from the particular, but because it depends on necessary and universally valid presuppositions without which the most certain and most universal propositions of science as well as the most common empirical truths would be impossible."[43] In other words, Hessen sees this "new" conception of universality as consisting in the necessary concepts presupposed by everyone in all judgments – be they universal or particular – about objects. Hessen sought to set his transcendental empiricism against not just Kant's original conception, but also against Cohen's transcendental rationalism. In this respect, then, Hessen saw Cohen, investigating the possibility of mathematical physics, as closer to Kant than was Rickert.

For Cohen, philosophy is a second-order or transcendental reflection on natural science. This method, the "transcendental method," thus cannot conflict with natural science. For Hessen – and presumably then also in his eyes for Rickert – philosophy, properly speaking, uses a different method than the transcendental one. Philosophy is the study or "science" of the *a priori*. It cannot conflict with natural science, since its concerns lie within a different sphere. This sphere, discovered by Kant, is one of presuppositions, a sphere "implicitly recognized by anyone who makes a scientific judgment, performs an ethical act, or feels something aesthetically pleasing."[44] Being presupposed in these varied activities, they are valued. Philosophy discloses them and thereby consciously alerts us to them. The *a priori* universality of these values or presuppositions is distinct from the generic universality of a universal concept. With the former, we now have the opportunity to handle the individual transcendentally by seeking the *a priori* presuppositions of a particular cognitive structure. The individual, whether it be an individual concept concerning history or an individual event or person, can be philosophically grounded, just as well as a universal concept of physics. From the perspective of transcendental empiricism, mathematics and even the most abstract of physical laws are as empirical as are ordinary sensations. Our relation

---

[43] Hessen (1909), 4.

[44] Hessen (1909), 6–7. Hessen fails to inform the reader just where he discovered Kant's discovery. Hessen wrote that already in 1908 he and his friends had organized a philosophical circle, the chief theme being epistemological problems. The circle discussed among other works Husserl's "logic." Hessen (1994), 155. However, such universal normativity posed by Hessen received a direct assault in Husserl's work. As Husserl showed, there is no normativity in even a simple syllogism. In light of this, it is difficult to say whether Hessen understood much of Husserl.

to reality has its *a priori* at the start, not at the end of the inquiry as in the Marburg conception, in which all of science "only" asymptotically approaches reality. All spheres of philosophy utilize a uniform method seeking the *a priori* universal in each. From the historical perspective, the search for the *a priori* universal started with ethics, and as a result, practical reason was accorded a primacy over other philosophical spheres.

Pre-Kantian empiricism held that all there is is the empirical. Thus, there are no ideas or supernatural beings. The "Critical empiricist" agrees that there is no transcendent being, However, all being is grounded on an "Ought," and this "Ought" cannot itself be characterized as a being or as having being. The Kantian dichotomy between the "is" and the "ought" is preserved. Since transcendental empiricism sharply distinguishes the study or science of value from the science of being, a confusion of value with being, which results in psychologism, is averted.

Surely, Hessen's quite youthful thesis – he was only in his early 20s – was replete with many ideas drawn, for the most part, from Rickert and, albeit to a lesser degree, Lask. We cannot possibly look at all of them here. Hessen, naturally, sought to apply what we have seen to historical causality in particular, which he found to be different from the causality physicists seek. Cohen's Marburg conception of philosophy consisted of asking how such causality is possible without direct interference in the scientific work of determining what exists. Hessen's Baden conception too believed that philosophy, as the logic of the study of values, does not interfere with empirical problems, e.g., the determination of the cause of individual historical phenomena. Philosophy cannot talk about a specific cause, but only about causes. But both the Marburg and the Baden conceptions of philosophy agree in their own way that the task of philosophy is to investigate the preconditions of experience.[45]

The aim of Hessen's transcendental empiricism, which he also called "critical empiricism," is to strengthen the empirical sciences and our understanding of morals and beauty by removing all metaphysical prejudices that may deter its investigations. Such at least is its starting point, although there is a temptation to misunderstand its search for the *a priori* as though it were something that exists. Philosophy must ever be on guard that its inquiries do not get misunderstood.

Upon finishing his studies in Germany, Hessen returned to St. Petersburg, where he attended and actively participated in meetings of the Philosophical Society. He also took the first steps toward attaining a *magister*'s degree, which required initially passing examinations in history, mathematics, and physics. He

---

45 Hessen (1909), 48.

also arranged with Vvedenskij the topics that would be covered on his eventual philosophy examination, which would require three years of study. Most important for our purposes here, however, is that at this time Hessen published in the first issue of *Logos* an article "Mysticism and Metaphysics" that further elaborated many of the ideas in his German thesis. Philosophy is to be sharply distinguished from the objects that come into its purview. The branch of philosophy called "logic" – which more closely resembles what today we call epistemology – is neither an empirical science nor psychology. "It is the science that studies what makes the empirical sciences sciences. That is to say, it studies the formal presuppositions of the sciences of being."[46] It does not seek to prove anew any physical law or mathematical formula. Logic seeks the formal presuppositions that make not only materially true propositions possible, but also false ones. In other words, we attach some sense not just to true statements, but also to materially false ones. Otherwise, the latter could not be said to be false. In contrast to natural science, logic is concerned with determining the sense of such statements.

Ethics and aesthetics are interpreted in much the same manner in Hessen's Baden-inspired neo-Kantianism. Morality provides norms of behavior. Ethics, though, is the science of the formal presuppositions underlying morality. Thus, ethics "is the science of the value of the identical sense that laws and individual rules of behavior have, independently of the psychic states of the subjects who act in accordance with or against these rules."[47] And aesthetics, in turn, studies the forms of artistic sense, it being indifferent to the psychic states of the individuals involved. Thus, aesthetics is independent of the material representation of beauty. Its concern is with what makes an artwork a work of art, what makes something regarded as beautiful or ugly to be beautiful or ugly. Although Hessen does not express himself in this way, we could say that aesthetics is concerned with the essence of beauty or ugliness. Philosophy, in short, then involves what Husserl would call an eidetic reduction. It is impossible to derive a particular artistic creation from the forms or *a priori* obtained from aesthetics just as we cannot derive laws of physics or chemistry from logic. Hessen construed philosophy to be a formal science of values that allows us to distinguish one scientific field from another, thereby stemming the rise of metaphysics. This, he claims, is the role Kant himself portrayed philosophy as serving. "Every metaphysics is a confusion of the borders between separate sciences or separate

---

[46] Hessen (1910a), 123.
[47] Hessen (1910a), 124.

fields of culture."⁴⁸ Kant had the proper conception. He discovered the *a priori* as the sphere of preconditions, of values, but instead of following through with it he, under the influence of rationalism, pursued its conception of universality.

Upon his acceptance to sit in due time for the *magister*'s-degree examinations, Hessen left for Marburg. He attended the lectures and seminars of the philosophy luminaries there – Cohen, Natorp, and Hartmann. Although, as we saw, he was critical of the Marburg conception of philosophy, he retained great respect for Natorp, eventually writing an obituary upon the latter's death in 1924. The following year he attended lectures on geometry and physics at Freiburg University as well as those of Rickert. Understandably, his philosophical output was not tremendous, but he did write a number of reviews, as well as some articles of marginal concern to us here and translations.⁴⁹ Of the reviews, particularly notable is the brief one of S. L. Frank's Russian translation of the first volume of Husserl's *Logical Investigations*. Hessen recognized the significance of Husserl's work but faulted him – seeing him as a former psychologist!⁵⁰ – for identifying the object of logic with the object of mathematics. Husserl, thereby, ascribes an ideal being to sense, which as we saw above, Hessen could not countenance.⁵¹ Senses do not have being but are ultimately values.

Hessen passed his initial non-philosophy examinations in 1912 and his philosophy preliminary examinations in 1913. He taught at St. Petersburg University already with the summer semester of 1914 as a *privat-docent*, his first lecture being on Kant's ethics, which, in his own estimation went poorly! Hessen's later years lie outside the scope of the present study. It should be noted, though, that he spent time wandering, one might say, from university to university, from one country to another finally arriving in Poland, where he resided until the end.⁵² He never completed a planned study of Kant's philosophy.⁵³ His major publication from his "wandering years" was his *Foundations of Pedagogy*, originally published in Berlin in 1923 and which he informs us was based on lectures already prepared while at Petrograd University and then at Tomsk University in the tumultuous years of 1917–1921. Its subject matter lies well outside our concern here. Nonetheless, he wrote in it, "We will limit ourselves to a presentation

---

48 Hessen (1910a), 128.
49 Hessen provided a valuable, long preface to his translation of Rickert's *Kulturwissenschaft und Naturwissenschaft*. The preface clearly and directly summarized Rickert's own neo-Kantianism. See Rikkert (1911), 3–28.
50 Hessen (1910b), 185.
51 Hessen (1910b), 187.
52 For further biographical information in English, see Walicki (1987), 407–413.
53 Hessen (1994), 159, 162. He wrote that he had hoped to submit the work as a dissertation.

of our own viewpoint, which in general abuts Rickert's theory. It is deeper than others, in our view, and understands the 'limits of the method of natural science' and the need to supplement the 'sciences of nature' with other branches of scientific knowledge."[54] Whatever path his later thought may have taken, Hessen undoubtedly during his early years in Russia was a disciple of Baden School neo-Kantianism.

7.5 A Neo-Solov'ëvian Critique

If the issue of Hessen's neo-Kantianism is, as some would have it, a debatable issue, that of Stepun's is even more problematic. In his memoirs, *The Fulfilled and Unfulfilled*, written years after leaving Soviet Russia, he tells us in the account's preface that his philosophical views were close to the "Slavophile-Solov'ëvian doctrine of positive unity," that unity being the highest object of cognition.[55] As we shall see, the answer to our question depends largely on our determination of how much or how little of neo-Kantianism one must accept in order to be a neo-Kantian.[56]

On the advice of Boris P. Vysheslavcev, at the time a *privat-docent* in philosophy at Moscow University, Stepun went off to Heidelberg in 1902, where he studied under, most importantly, Windelband, who was seen in Stepun's eyes as, above all, a historian of philosophy rather than as the leader of a school of neo-Kantianism. It should be noted that Stepun, being quite young, went to Heidelberg intending not to study Baden neo-Kantianism, but to study with Kuno Fischer, who, being sick, did not lecture at that time. Unfortunately, no particularly strong bond developed between the pupil and the teacher, Stepun finding Windelband to be for the most part the stereotypical cold and haughty German professor. More telling perhaps for our purpose here, is that, as Stepun related in his memoirs, he went to study in western Europe in search of answers to the mysteries of life and of the world, but found little of that quest for "spirituality" in Windelband the man or in his teaching. For example, whereas he was content in class with expounding the Kantian dichotomous solution to the problem of free

---

54 Hessen (1995), 255.
55 Stepun (1994), 5. There is certainly ample ambiguity here. Was Stepun referring to his views merely at the time of this writing, or did he have in mind also throughout his philosophically engaged career?
56 Lossky placed Stepun among the "prominent representatives of that form of neo-Kantianism which may be called transcendental-logical idealism." Lossky (1972), 318. Let us provisionally accept Lossky's classification and see to what extent Stepun merits the label of a neo-Kantian.

will, Windelband persistently rejected it outside the lecture room, holding that there had to be a monistic "highest truth."[57] Stepun could not abide such a rigid separation of privately held metaphysical beliefs, as Windelband proposed, and scientifically determined conclusions.

Another telling incident concerns the Third World Congress of Philosophy held in Heidelberg in 1908. Stepun described it so:

> The Congress centered around a heated struggle between Anglo-American pragmatism and the idealist tradition of German philosophy. Heirs of Rome but not of Athens, the puritan pragmatists, who were hostile to the Christian contemplative tradition, came to the citadel of idealist philosophy with the goal of imposing on Europe the American conviction that the creed of universally obligatory and timeless truth is the height of logical inconsistency and practical nonsense. ... I, like all the Russian philosophy students, with all my heart, sided with the Germans, to whom American pragmatism appeared to be a dreadful barbarism and amounted to rank ignorance.[58]

What Stepun did not notice was that behind this apparently unified front the German neo-Kantian Schools were vying against each other for influence, which, it was hoped, would yield academic positions for their respective students and which, in turn, would further enhance their own reputations.

To culminate his studies in Heidelberg, Stepun planned initially to write a thesis on the "ascetic ethics" of Vladimir Solov'ëv, the late nineteenth-century Russian philosopher/public intellectual.[59] But although a draft had already been written, Stepun decided, with Windelband's advice and approval, to change the focus of his work to Solov'ëv's philosophy of history, adding an extended discussion of prior Russian thought, viz., the views of the so-called Westernizers and Slavophiles.[60] The final result was actually a treatise on the philosophy of the history of philosophy largely along the lines of Solov'ëv's own such

---

57 Stepun (1994), 80.
58 Stepun (1994), 116. Stepun possibly had in mind Josiah Royce's presentation at the 1908 Congress on truth. Royce's address has been reprinted many times, but for the original see Royce (1909). However, since Stepun did not know English, he, more likely, had in mind F.C.S. Schiller's paper "Der rationalistische Wahrheitsbegriff," which was delivered in German. Stepun's unqualified confidence in American "barbarism" surely must have undergone a test when he observed and learned of the political scene in German and Russia in the 1930s.
59 Stepun (1994), 111. Stepun provided no illuminating details. Since Solov'ëv harshly criticized asceticism in one early chapter of his principal ethical treatise, the *Justification of the Moral Good*, Stepun could hardly have had that treatment in mind. Thus, we must infer that at this time he interpreted Solov'ëv's overall ethical position as an asceticism in spite of Solov'ëv's explicit rejection of it.
60 Stepun generally referred to those whom we now label as Slavophiles by the term Pan-Slavists. Historians today generally distinguish Slavophilism from Pan-Slavism.

treatment in his *magister*'s thesis. Stepun had at one time hoped to expand this work greatly, but nothing came of it. Presumably, the planned work would have included an analysis of the rationalistic aspects of Solov'ëv's system, but Stepun simply lost interest in the topic. He found that there was a deep dichotomy between the rationalistic features of Solov'ëv's thought and his "prophetic spirit." It is this "spirit" that animated Stepun's philosophical reflections throughout his subsequent writings.

Stepun's thesis covered ground over which many would subsequently tread in terms of scholarship on Solov'ëv. Nevertheless, it remains important today for its glimpse into how Stepun viewed both Kant and nineteenth-century Russian thought. We should not be surprised that Stepun saw Kant's focus to be on epistemology. Displaying his education in Baden neo-Kantianism, however, Stepun charged that an epistemological investigation presupposed that its object had *value*. In Kant's case, this object was natural science, and he attached to it primary importance at the expense of other issues that are part of a total worldview, in which he had a general interest already upon arrival in Heidelberg. Thus, "Kant's system stands and falls with this high assessment of science as a cultural factor of the first order."[61] Kant's achievement must, then, be judged in terms of whether what he accomplished in laying the groundwork for natural science is worth all that he had to relinquish in other matters, although these could not be avoided forever.

Stepun conceded the relatively late arrival of Russia on the world intellectual scene. Only at the end of the eighteenth century did Russians start to contribute anything to the development of science. It should not surprise us, then, that the Russian "nation" did not clearly recognize the value of the natural sciences. Concomitantly, Russians could not grasp the magnitude and meaning of Kant's project. The more they recognized Kant's demotion of the ultimate questions in a worldview, the more they pressed for answers. Having just removed their intellectual diapers, so to speak, the Russian people were not about to abandon their own quest for answers in the face of others' frustration with their persistence.

Stepun was not uncritical of what he took to be the Slavophile-Solov'ëvian critique of Kant. That critique viewed Kant's philosophy one-sidedly as a step – though an important one – in rationalism's development. This, in Stepun's eyes, was a misunderstanding of Kant's philosophy, of its originality, and of Kant's eminent stand on the issues of the time. This misunderstanding was, in-

---

[61] Steppuhn (1910), 4–5.

deed, characteristic of the entirety of Russian philosophical development.[62] The Slavophiles correctly attacked the psychologism of Kantian epistemology with its breakdown of the mental faculties into quite separate abilities and states. Kant is to be lauded for granting that one and the same object can be viewed from different angles, i.e., epistemologically, ethically, and aesthetically. What he was unable to realize was that the unity, the integral nature, of one's being must never be lost even from a methodological standpoint. Whereas western European cultural life found consecration in Kantian philosophy's resolute trichotomous separation of viewpoints, Russian thought saw such a rigid separation as running counter not just to the very nature of its way of thinking but, more importantly, to truth.[63] In a sense, this was, for Stepun, a tragedy for Slavophile philosophy, viz., that it did not, perhaps could not, think through Kantianism, and in doing so enhance the Slavophile doctrine of a positive all-unity. Stepun would write in an article "Proshloe i budushchee slavjanofil'stva" ["The Past and the Future of Slavophilism"] from 1913, "Regardless of how one relates to Kant, his enormous moral seriousness and the acute nature of his logical analysis cannot be denied."[64] The first *Critique* had dealt a blow to the concept of all-unity. It drove a wedge into the heart of the concept of all-unity creating a dualism between form and content as well as one between subject and object. Stepun failed to articulate clearly how it did this, but in any case, the concept of all-unity could not regain its ground until it had critically surmounted Kant's Critical Philosophy. Slavophilism never concerned itself with the problems Kant had raised. First, they distorted his ideas and then polemicized with this caricature. None of the Slavophiles – not even Solov'ëv – "ever tried [*pytalsja*] to peep into the vast complex that Kantianism presented."[65] But, as mentioned above, Stepun also held at the same time that the Slavophiles could not engage in Kant's problems, since they ran counter to the Russian mode of thought! Even Stepun, who himself never took up the complexities of Kant's epistemology, would later in his autobiography write that Kant's philosophy was foreign "to all of my inner and mental constitution."[66]

---

62 Steppuhn (1910), 290.
63 Steppuhn (1910), 76.
64 Stepun (2000), 834.
65 Stepun (2000), 834.
66 Stepun (1994), 117. On the same page of his autobiography, he singled out, however, Kant's "phenomenalism" for criticism – thereby showing how well he understood Kant – writing, "I could not believe that the entire world with which I have been familiar since childhood was just the content of my consciousness, that in fact it corresponds to nothing in general or to something of which I am quite unable to have the slightest idea."

Having acquired a degree in philosophy from Heidelberg,[67] Stepun returned immediately to Moscow with the hope of initiating a career teaching philosophy. Stepun inquired concerning this possibility with Lopatin and Shpet, but in each case came away disappointed. Lopatin, undoubtedly, was not interested in Stepun's background at Heidelberg, and Shpet objected to any introduction of religiously tinged metaphysics into philosophy.

> After a half-hour conversation with Shpet about philosophy, art, and Moscow (which Shpet did not know), it became quite clear that he was going neither to assist nor to advise me, since he did not care in the least whether Lopatin's Leibnizianism or Marburg neo-Kantianism flourished at Moscow University.[68]

Stepun also met with Chelpanov and Khvostov but in the end came away thinking that he would not be able to pursue an academic career in Russia.[69]

Stepun, of course, had his editorial work on *Logos*, and he managed to publish several long articles in *Logos* and other Russian journals as well as a number of relatively short book reviews. In one of his earliest outings in Russian, an article entitled "Nemeckij romantizm i russkoe slavjanofil'stvo" ["German Romanticism and Russian Slavophilism"] published in March 1910, Stepun's words on Kantianism – and, for that matter, on Fichte – were basically little more than a translation of the relevant passages from his thesis.[70] He added there, as he did in his thesis, that Fichte's personality should have made him more attractive to a Russian audience, but his popularity suffered as a result of the abstract nature of his writings as well as his alleged atheism and his theoretical solipsism.[71]

Stepun, some two years later, published in the Moscow symbolist journal *Trudy i Dni* [*Works and Days*] an "Open Letter to Andrej Belyj Concerning 'Circular Movement'."[72] Stepun wrote that Kant had in essence transformed what had

---

[67] Stepun, in a letter to Hans Gadamer dated 20 April 1962, wrote that the last thing he did while in Heidelberg was witness the emergence of Husserl's phenomenology. "For a time, I was chairman of a student Husserl circle. This, however, was not my real calling." Quoted in Treiber (1995), 104. Oddly, Stepun never mentioned such a circle in his autobiography to say nothing of serving in some organizational capacity in it.
[68] Stepun (1994), 148.
[69] Stepun in his autobiography also called Chelpanov a "popular, sensible professor, but a thinker with little talent." Stepun (1994), 207.
[70] Compare Steppuhn (1910), 4–5 and Stepun (2000), 39.
[71] Steppuhn (1910), 5–6 and Stepun (2000), 39–41.
[72] Stepun's reference here is to the symbolist literary figure Andrej Belyj's article "Krugovoe dvizhenie" ["Circular Movement"] in the same journal, which Belyj helped found, and in which he called the entire neo-Kantian movement an idiocy, an abyss, "in which almost nothing remains of Kant." Belyj (1912), 56. On the same page Belyj called the neo-Kantians, taken as a

previously been considered transcendent into the transcendental. By doing so, Kant had placed the absolute beyond the "horizon" of philosophical study, and with this transformation he departed from the rationalist movement, to which, as we saw, Solov'ëv had confined him. Stepun also leveled a familiar charge against Kant, albeit far more eloquently than usual.

> Just as evening shadows present only the forms of objects but not their complete being, so Kant's transcendental forms are incapable of generating their own material out of themselves. The unity of form and material, and thereby the integral nature of the world's givenness, is, according to Kant, a philosophically inexplicable fact.[73]

Contrary to Belyj's accusation, though, that as a result of Kantianism's fatal flaws neo-Kantianism could offer nothing new, Stepun asserted that it was progressing along the sacred paths bequeathed to it by Kant. The most significant correction that the neo-Kantian movement introduced into Kant's system was the amelioration of its greatest inconsistency, viz., the thing in itself. All of the neo-Kantian schools presented a united front against this concept, and in doing so moved against this irrational element in Kantianism.

Stepun also saw neo-Kantianism as breaking the tight connection between the Kantian forms and the table of judgments found in the first *Critique*'s "Metaphysical Deduction." Although Kant's "closed circle" of the categories was an illustration of a strictly applied principle of rationality, neo-Kantianism's severance of the connection allowed transcendentalism to embark on "the fertile soil of living cultural work."[74] But how then does neo-Kantianism propose to separate or distinguish the categories, forms, ideal beings, etc., which are reflections of the Absolute, from the *a posteriori*, the empirical, the transient? Stepun sought to highlight in this symbolist journal what he viewed as the non-discursivity of contemporary approaches to the absolute. Windelband, for example, held that the presence of values cannot be proved at the start, but is set as a task to be determined. And in light of Stepun's possible engagement with phenomenology, he wrote, "Husserl is an unconditional and interesting defender of the method of intuition as an aspect of critical philosophy. He is increasingly freeing the phenomenological seeing of the structure of sense from his original

---

collective, to be a monster [*chudovishche*], despite the fact that taken separately the neo-Kantians are sharp and quite rational men. In contrast, Belyj had earlier in 1910 defended neo-Kantianism and, in particular, the approach of *Logos* to promote rigorous philosophy.

73 Stepun (2000), 811.
74 Stepun (2000), 812.

path of descriptive psychology."⁷⁵ Even Hermann Cohen, who conceived the transcendental method as an inquiry into the possibility of mathematical physics, stood, in Stepun's eyes, far from rational discursivity, being under the obvious influence of Plato. Stepun sought in this manner to defend neo-Kantianism against Belyj's "undeserved reproach" that it, being purely rational, was lifeless. Thus, Stepun's view of neo-Kantianism at least in 1912 had little to do with the philosophy of either the natural sciences or the social sciences.

Although Stepun did not write any technical philosophical treatises, he did in 1913 publish in *Logos* what he later called a "programmatic" article "Zhizn' i tvorchestvo" ["Life and Creativity"] that was intended to serve as the first draft of his "system."⁷⁶ For us here, its importance lies in that he saw his article as a defense of the religiosity of the Slavophiles and the German romantics from the standpoint of (neo-)Kantianism. Stepun conceded at the outset that Kant's system is riddled with contradictions, but, nonetheless, it has a fundamental, unifying idea, which, while not yet being universally recognized, is becoming more and more so.⁷⁷ Kant sought neither to limit cognition merely to phenomena nor to reject therefore previous philosophical quests for cognition of the absolute. No, Kant, according to Stepun, realized that philosophy by its very nature seeks just such a goal. However, he rejected the dogmatic attempt to see the absolute directly. Kant thought the absolute lies just out of sight revealing enough of itself so that we know it is there, over the philosophical horizon. All acts of human creativity – artistic works, scientific judgments, and even religious symbols – reveal its presence. This remains the case even in the face of the *Critique of Pure Reason*. Indeed, this is where Kant placed it. Kant's accomplishment was to have seen the absolute not as a metaphysical integral unit, but as the transcendental form that gives all human cultural creations the character of necessity and meaningfulness. Transcendental idealism abandons the quest for an origi-

---

75 Stepun (2000), 813. It is difficult to reconcile this account of Stepun's meeting with Shpet with what we independently know of Shpet's frame of mind at the time. If Stepun had been previously involved in some Husserl circle while in Heidelberg and Shpet was already enamored with phenomenology, as Stepun portrays, one would think Stepun would find talk of phenomenology to be a common topic of conversation. Yet, there is no hint of that in Stepun's recollection of the meeting. Moreover, there is absolutely no evidence that Shpet was particularly enthralled with phenomenology at this early date, viz., 1910.
76 Stepun (1994), 117.
77 Stepun (1913), 71. The text of this article as found in Stepun (2000), 89–126 does not fully correspond to the *Logos* 1913 edition, but rather to the 1923 Berlin edition. In the latter, Stepun altered the text to read that the Kantian system's unifying idea "forms its historical merit and basic systematic essence." Stepun (1923), 121.

nary understanding or acquisition of the absolute and instead turns to concrete human creations in all spheres of our cultural activity.

Those who follow Kant's lead – Stepun failed to mention just who these individuals were or are – seek not the absolute directly but only indirectly through its reflections and hints in terms of the formal aspect of human life and cultural creativity. The task of neo-Kantianism is to examine systematically all spheres of human activity separating an aspect of the absolute, which Stepun held was the transcendental-formal element, from the contingent and transient.[78] Such a goal forced neo-Kantianism to investigate spheres that hardly resemble metaphysics as traditionally understood. However, neo-Kantianism, in both of its Baden and Marburg School incarnations, reduced the role of philosophy to determining *merely* the formal aspects of the absolute as manifested already and increasingly so in human cultural accomplishments, understood in a broad sense. This is done at the expense of the empirically given.

Stepun's objection to pragmatism, regardless of its various national forms, is that it disavows Kant's discovery of absolute or unconditional truths. He accused pragmatism of preaching relativism and psychologism, thus replacing Kant's achievement with new forms of metaphysics that Kant had rejected. Yet, Stepun believed that the positive features of Kantianism still allowed for an extra-metaphysical claim concerning the integral all-unity too hastily advanced by Solov'ëv. "Only by means of an unqualified recognition and affirmation of Kant's great work, only by means of a selfless recognition of his brilliant but *hostile* feat can we attempt to take the principle of integrity and concreteness under our philosophical tutelage."[79] There can be no turning back to pre-Kantian metaphysics by thinking we can directly cognize the absolute either in the form of the object or as the subject of cognition. The result is that there is no place for "Final Truth" – note the capital letters – in philosophy.[80] The absolute lies outside the sphere of philosophy.

Such was Stepun's "first draft" of a system of philosophy. Although he applauded Kant's endeavors, he retained a belief in an extra-philosophical absolute. The troubling aspect of this lies in that whereas he criticized predecessors for not taking into account Kantian arguments he himself showed no evidence of doing so. Nowhere did he offer specific comments on arguments in the first *Cri*-

---

78 Stepun, in the 1913 original, wrote that this was the task of "post-Kantian philosophy," whereas the more specific reference to neo-Kantianism appears in the 1923 edition. In this case, the latter is to be preferred, since this is surely what Stepun had in mind earlier. See Stepun (1913), 72–73.
79 Stepun (1913), 81.
80 Stepun (1913), 82.

tique. In the end, his quite anemic neo-Kantianism, if that is what it is, was less than what we could accord to such a figure as Vladimir Solov'ëv, who sympathized with much of the thrust of Kant's ethics, but neither the epistemology nor aesthetic philosophy.

In 1914, Stepun was mobilized as a lieutenant in the Russian army and served first on the eastern front, which took place in present-day Ukraine and Poland. He was also present and often participated in many events during the tumultuous years to come. He even served in the Red Army, albeit as a stage director for plays such as those of Sophocles and Shakespeare. His collaboration in 1922 with Berdjaev and presumably other idealists may have earned him a one-way expulsion out of Bolshevik Russia. However interesting his later years, they fall outside the scope of the present study. Stepun never managed to write any technical works in philosophy or revise his earlier works to make them either more technical or more indicative of recent trends. He secured a position teaching sociology and then economics in Dresden and undoubtedly passed through the Nazi era under suspicion but came to no harm. After the War, Stepun held an honorary professorship in Russian Spiritual History, specially created for him, in Munich. He hoped for a time to emigrate to the United States, but he knew no English and could not find a "suitable" position – apparently, he wanted his cake and eat it too. He died in February 1965.

7.6 An Amorphous Neo-Kantianism

At the very periphery of our story lies another figure, Aleksej K. Toporkov (1882–1934). He wrote no philosophical works that would distinguish him from others on the basis of either originality or insight. Indeed, he apparently wrote no works of technical philosophy. Still, his fate is instructive for understanding the demise of Russian neo-Kantianism as a whole. Toporkov completed his studies at Moscow University in 1906, with an undergraduate thesis concerning the influence of Leibniz on Kant's theory of cognition, for which he was awarded a gold medal. As such a recipient, he was officially sent for additional study to Heidelberg and/or Marburg for a stay perhaps as short as one semester.[81] Soon after his return to Moscow, he joined to a limited extent with those holding a similar Western orientation around *Logos*, but particularly with those around the new symbolist

---

81 The evidence is slim. Dmitrieva (2007b), 155, referencing an archival curriculum vitae from 1923, puts Toporkov "in one of the semesters – summer 1906 or winter 1907/08 – in Heidelberg." She also writes, referencing a Marburg archive, that Toporkov "studied in the summer [semester] of 1907 in Marburg." Dmitrieva (2007a), 110.

journal *Trudy i Dni*. This alone already shows his drift away from technical philosophizing, a drift we note in a number of other Russian neo-Kantians, particularly those who had been reared philosophically in the Baden School. Toporkov was to have studied further on his return with Lopatin, but for some unclear reason they had a falling out.[82] With that, "the doors of the University were closed to him."[83]

In an essay from 1915, "Ocherki sovremennoj filosofii. Neo-kantiancy," Toporkov gave a succinct account of how he viewed neo-Kantianism. He wrote that under a certain understanding and interpretation Kant's philosophy did not offer a worldview, and as a consequence, one could be an atheist or theist, a revolutionary or a reactionary. What it does do is help one to understand the contradictions between our various experiences, thereby "bringing order to our mind and facilitating creativity and preaching."[84] He held that Russian (neo-)Kantianism, in this respect, is more correct than the German variety, which advocating a method and specific formal tenets, does imperceptibly become a worldview, a fixed doctrine. The former, on the other hand, accepts different opinions provided they do not conflict with Kantianism and science. Thus, "In a certain sense, Western Kantianism is the limit and end, whereas Russian Kantianism is the beginning and represents aspirations."[85] Clearly for Toporkov, the Kantian legacy made for a psychological attitude more akin to the "New Age" movement than to positivism.

We can hardly be surprised, then, that Toporkov with such an amorphous understanding of neo-Kantianism could play a role in the development of Russian symbolism. He had already written in 1912 that idealism was not a provable philosophy. Specifically recalling Fichte, a particular philosophy is an act, a willful commitment by an individual. "Whether one commits to it or not depends on the will of each individual person."[86] Nevertheless, standing, as it were, in stark contradiction to that claim, he also wrote, "Of course, my words should not be interpreted as though they mean idealism is a quite arbitrary and unproven doc-

---

[82] If we believe Andrej Belyj, part of the reason may have been sheer jealousy. Many of Lopatin's students went off to Germany, some if only for a rather short time, and came back "Germanized," even "neo-Kantianized" whereas Lopatin stood alone without disciples even after students had studied under him for years. See Belyj (1990a), 272.
[83] Belyj (1990a), 271.
[84] Quoted in Dmitieva (2007b), 273. Toporkov, actually, wrote that Kant's philosophy helps a Russian to such an understanding. Hopefully, he was not singling out Russians as needing mental ordering.
[85] Quoted in Dmitieva (2007b), 273–274.
[86] Toporkov (1912a), 73.

trine that can exist alongside others just as little substantiated and proven."[87] No, in his mind, a philosophy is not something subjective, for the philosophy one chooses *is* that person. Toporkov rejected psychologism, saying that the psychological is the logical, though without clarifying what he meant. He also wrote that ideas are transcendent – not transcendental – to cognition. "Kant expressed this aspect of an idea most vividly and at the same time paradoxically in his 'Transcendental Dialectic' and also in his *Critique of the Power of Judgment*."[88] Truth is not given, but, as with Marburg neo-Kantianism, is a task for reason. It must be said, though, that Toporkov relished contradictions, saying that an idea is both immanent and transcendent, though he explicitly rejected any disavowal of logical laws. In a subsequent article a few months later in the same symbolist journal, Toporkov again affirmed that how the world is, depends on how we view it. There is in it scarcely any sign of neo-Kantianism. Toporkov did not address how to reconcile conflicting conceptions of the world. Both cannot be true. He gave no indication that he understood what Kant sought to convey with the "Transcendental Dialectic." Toporkov went so far as to assert directly that a genuine theism is a pantheism, since every human being is a manifestation of the Divine.[89]

Toporkov's activities during the revolutionary years are unknown. He subsequently emerged as an author of some popular and quite non-technical articles. Finally, in 1928 he published a work entitled *Elements of Dialectical Logic*, which showed he had managed an accommodation with the official ideology without requiring much effort. His previous idealist and pantheistic understanding of the dialectic was conveniently transformed into a materialist dialectic.[90] Sadly, however, it did him little good. He was arrested at the end of 1932 and sentenced one month later to ten years in a forced labor camp, where he perished.

Undoubtedly, owing to the political activities and writings of Novgorodcev and Kistjakovskij and the emergence of the journal *Logos*, the Russian Baden neo-Kantians have gained more recognition in English-language scholarship than those Russian associated with the Marburg School. We surely can say that the *leitmotifs* of the Baden School, e.g., its concern with values, were closer to the traditional concerns of Russian thought, as represented, for example, by Lavrov, than the *leitmotifs* of the Marburgians. Yet, we saw that Stepun, for one, chafed under even the

---

87 Toporkov (1912a), 73.
88 Toporkov (1912a), 77.
89 Toporkov (1912b), 131.
90 See Toporkov (1928).

meager restrictions of Baden neo-Kantianism. His frame of mind was of a more religious nature than was typical of either the Baden or Marburg philosophers. We shall also see that the Baden-inspired figures we have examined generally moved more easily or seamlessly into other intellectual activities than those inspired by Marburg. Did this, however, mean that within Imperial Russia in its waning years the Baden School was more "popular" than the Marburg School?[91] Can we even say that the former received more attention than did the latter within Imperial Russia? To help answer these questions among others, we must turn now to the young scholars who streamed to Marburg.

---

91 Abramov (1994), 240. Abramov holds that the Baden School was more popular in Russia than the Marburg School but provides neither a definition of "popular" nor on what basis he came to his conclusion.

Chapter 8
The Marburg School's Influence in Imperial Russia

Before turning to a discussion of the Russian adherents of the Marburg School of neo-Kantianism, let us first look at how Cohen's philosophy itself fared in Imperial Russia. Its highly idealistic nature, we might add, would later prohibit a careful and sympathetic reading of Cohen's works in Bolshevik Russia. The hostility, however, evoked by Saval'skij's thesis, as we saw, already made it extremely unlikely that another young scholar would take up the banner of Marburg neo-Kantianism during the last years of Imperial Russia, particularly in the law faculty at Moscow University. Additionally, with Sergej Trubeckoj's premature death in 1905 and the dominant figure in the philosophy faculty being Lopatin, who remained hostile toward all recent Western philosophy, neo-Kantianism had little chance of making inroads in that faculty either. The Marburg School approach to philosophical issues could not receive a fair hearing, and there was, in any case, virtually no interest in the philosophy of the natural sciences, least of all in the way exemplified by Natorp and Cassirer, among philosophy instructors in Moscow. Yet, as we shall see, a number of promising young scholars did travel to Marburg to study with Cohen and Natorp. It is a curious and fundamentally unexplained fact, then, that not one of Cohen's dense philosophical works appeared in print in Russian translation. A number of works by Natorp, on the other hand, were translated before the Bolshevik Revolution. Moreover, Cassirer's *Substanzbegriff und Funktionsbegriff* appeared in translation (1912) soon after its original publication[1] and even his *Zur Einsteinschen Relativitätstheorie* appeared in the Soviet Union in 1922, one year after its publication in Germany. Of the young Russians who flocked to Marburg to hear Cohen, none is on record as thinking of translating any of Cohen's systematic works.[2] Why they all shied

---

[1] The translation was not done by a philosophy professor. Boris G. Stolpner (1871–1937) also translated the bulk of Hegel's works. Raised an orthodox Jew, Stolpner graduated from Moscow University and associated with the Menshevik branch of the Social-Democratic Party. He later worked at the Marx-Engels Institute in Moscow. Pavel S. Jushkevich (1873–1945), also a Jewish Menshevik, studied mathematics at the Sorbonne. He too worked at the Marx-Engels Institute from 1922. Philosophically a positivist, he earned Lenin's sarcasm in *Materialism and Empirio-Criticism*.

[2] One recent commentator wrote, "In order to understand why Cohen's books were not translated before the Revolution … it should be noted that knowledge of foreign languages was higher than it is now. … Translations of philosophy books were done primarily for secondary-school

away from such a task is perhaps an unanswerable question. Even Cohen's three commentaries on the three Kantian *Critique*s received little attention.³

8.1 Cohen in Russia

Whether accurately or not, Cohen was looked upon by Russian philosophers in the Imperial era largely – and understandably – through their own concerns with ethics and by extension the philosophy of law, not epistemology and certainly not philosophy of science. The first significant presentation of Cohen's thought was an analysis of his systematic work *Ethik des reinen Willens*, the first edition of which was published in 1904. The review published in 1905 by Evgenij V. Spektorskij (1875–1951), a graduate of Warsaw University in 1898, is actually quite remarkable for its lucidity and analysis. Spektorskij, as with so many others we have already discussed and shall discuss, was sent abroad for further study. Whether he was ever an actual student of Cohen's and/or Natorp's is unrecorded, but he did work in the libraries of Paris, Berlin, Göttingen, and Heidelberg.⁴ In any case, on his return to Imperial Russia he demonstrated considerable knowledge of Cohen's ideas and critical sympathy for the Marburg "project." At the time of his review, Spektorskij was teaching philosophy of law in Warsaw.

Spektorskij stated that in antiquity, ethics was thought to be a methodic, but descriptive empirical discipline. As such, it could not be erected with mathematical precision. Even Kant thought that ethics could not achieve such exactitude. Cohen, on the other hand, wished to look at ethics strictly "scientifically." This

---

students." Vashestov (1991), 230. If Vashestov's explanation is correct, why was Windelband's *Präludien* translated (by S. L. Frank) in 1904 and Rickert's *Die Grenzen der naturwissenschaftlichen Begriffsbildung* translated in 1903?

3 In one of the earliest mentions of Cohen's commentaries on Kant in the Russian language, Preobrazhenskij wrote, "Cohen believes in Kant like in the Bible. ... There are common inadequacies in all of Cohen's works: a one-sidedness, an intolerance toward the viewpoints of others, particularly toward English philosophy, ... obscure language, and a pretentious style." Preobrazhenski (1889), 96. An early discussion of Cohen, amazingly, is found in an article published in 1900 by a Marxist associate of Plekhanov's, Ljubov Aksel'rod. In it, she calls Cohen "one of the most profound Kantians" and writes that *Kants Theorie der Erfahrung* is "undoubtedly the most serious work in the literature of Critical thinkers." Aksel'rod (1906), 200. She concludes, though, that Cohen's principles lead to dogmatic skepticism as must all theories of cognition that base themselves on Kantianism.

4 Filatov (1999), 756. Spektorskij later taught at Kiev University and then in Prague and Belgrade. He went on much later to teach at St. Vladimir's Orthodox Theological Seminary in New York.

direction of German philosophy was, Spektorskij claimed, mostly unknown in Russia apart from a small number of specialists. Cohen and the Marburg School in general, of which Cohen was the leading figure, rejected the contingency and subjectivity of the results obtained by all branches of philosophy. To remedy the situation, we must, in Cohen's eyes, start with a reformulation of Kant's enterprise, replacing his rejection of metaphysics in each branch of philosophy with an examination of an established field of knowledge in each branch. Cohen, in Spektorskij's understanding, recognized Kant's principal pursuit to be the critical grounding of the possibility of science understood as valid, objective knowledge untainted by subjective characteristics of individual minds. When it came to epistemology Kant relied on the already established science of Newtonian mechanics.

Spektorskij gave every indication that he applauded Cohen's procedure thus far. The heart and soul of Kantian philosophy is the transcendental method, which yields results as firm as the science whose possibility is thereby ascertained. By inquiring into Kant's connection with Newtonian physics, the Marburg School exposed and deepened the connection between Critical Philosophy and mathematical physics. Kant, thus, turns out to be not the theoretician of metaphysical illusions, but of scientific experience.

Cohen, in this picture, then aimed to do for ethics what he had done for theoretical philosophy or "logic," i.e., construct an ethics-based not on any subjective, mental attitude, "not on dogmatic faith in transcendent authorities and sanctions, but on demands that with logical necessity follow from the nature of autonomous reason."[5] To make ethics into a pure science, Cohen believed ethics cannot appeal to any contingency, i.e., it must not invoke or deal with things or objects. It must be an ethics of pure will, as Cohen entitled his work, and of what flows from the pure will, namely, pure action, action exclusively willful, but without particular reference. Such action, ideal action, has nothing in common with psychology. It is this "purity" that Spektorskij contested. Spektorskij recognized that Cohen sought over and over to avoid any suspicion of psychologism. But in the eyes of the former, these efforts were unsuccessful and must be so. However "pure" we conceive the will – and even more so an action – it will always retain some element of the psychological. Therefore, a concern with the will is more appropriate in empirical science than in some "pure" science. Spektorskij proposed that if Cohen wished to pursue a pure ethics he should deal not with the actions stemming from a pure will, but, rather, with the goals that fol-

---

5 Spektorskij (1905), 394.

low from reason. "Only such goals can be completely ideal, and only they can be established with perfect purity and transparent clarity."⁶

Spektorskij saw one goal of the Marburg School to be the determination within the history of philosophy of a single, unitary endeavor. Hence, its concern with monographs dealing with individual figures, such as those written by Cohen himself on Kant and Plato as well as by Natorp (on Plato and Descartes) and Cassirer (on Leibniz). Unlike Hegel, who had a similar interest, the Marburg philosophers sought this unity without introducing metaphysics. They saw the unity as given in the purpose of philosophy. This aspiration for unity was, for Cohen, a theme that binds the three philosophical divisions of logic, ethics, and aesthetics. But whereas for Cohen the theme of scientificity ran throughout philosophy, of course properly understood, for Spektorskij the "thread" connecting the divisions is psychology. In short, these divisions derive their scientific significance from psychology. Since their respective concerns are with the mind, with the will, and with feeling, all ultimately incorporate psychological concepts.[7]

Whereas for Cohen theoretical philosophy finds mathematical physics to be the scientific fact on which to orient itself in its search for ideal principles, the parallel fact in practical or moral philosophy, as we know, is jurisprudence. Spektorskij, a trained lawyer, not surprisingly, differed.[8] He rhetorically asked whether current jurisprudence can claim for itself the mathematical validity we observe in Newtonian mechanics. His answer was straightforward: "It seems to us it absolutely cannot."[9] To base ethics on contemporary jurisprudence will not yield a pure ethics with mathematical validity. "Not being a product of pure reason, jurisprudence represents nothing more than a historically relative, albeit enduring or having endured, phenomenon."[10] Spektorskij, instead,

---

6 Spektorskij (1905), 397.
7 Further on in his article, Spektorskij wrote that action and the will *are* psychological concepts. Spektorskij (1905), 411. We might add that at least from a particular standpoint the will must be psychological. What else could it be?
8 Saval'skij, whom we discussed earlier, knew of Spektorskij's article and attempted to reply on this point. See Saval'skij (1909), 193–195.
9 Spektorskij (1905), 401.
10 Spektorskij (1905), 407. Spektorskij rejected jurisprudence as a "science" of ethics and law owing to the fact that as it then existed it was but a moment in the historical evolution of law. His reasoning, thereby, implied that Newtonian mechanics is itself an absolute and final truth grounded on the tenets of Kant's first *Critique*. It is ironic, then, that just as he was writing his article, the Newtonian paradigm was about to be overturned.

proposed that ethics itself must become a moral mathematics.[11] However, regrettably, he himself did not go on to elaborate a "pure ethics" that could stand as the equivalent of Cohen's *Ethik*.

Spektorskij did express additional criticisms of Cohen's *Ethik*. For example, whereas Cohen believed the state could be justified through his "pure transcendental method," Spektorskij found the state to be nothing more than an empirical phenomenon, a historical fact, which could not be deduced from the principles of pure science. It simply appears as the subject matter in a practical ethics. "In no way is it possible to recognize the state as a necessary postulate of a theoretical, pure ethics. But Cohen does precisely that."[12] Spektorskij concluded his essay, writing that the problem of a pure ethics as a completely valid rigorous science remains a problem.

That Spektorskij's essay was and has been ignored is a great misfortune for the development of neo-Kantianism. Of course, an elaboration of the details consistent with Spektorskij's project of correcting Cohen might itself prove impossible. Nevertheless, the Spektorskij of 1905 certainly appeared to be one of the most promising lights of Russian neo-Kantianism. He subsequently published many works but distanced himself ever further from his 1905 project.

Evgenij Trubeckoj, whom we saw previously as a harsh critic of Saval'skij, presented along with his remarks an overall damning picture of Cohen's *Ethik*. Trubeckoj's central point is that the reader can never determine just what the topic of Cohen's work is. "Instead of glass walls, the reader will find here a complex maze shrouded in a dense fog."[13] Cohen presented his book as a study of ethics and, indeed, of the human being and action, but both fall quickly into the background. Cohen rejected the given in its entirety including the empirical human individual. The concern is purely with the abstract will. Thus, since the object and the subject are unclear, the intention and the goal must also remain at least as unclear. Cohen replaced the living, breathing human individual with an abstract concept, with a method. There is neither a material nor a spiritual

---

[11] Spektorskij in a footnote mentioned a contradiction he found in Cohen's writings. See Spektorskij (1905), 409f. In the *Ethik des reinen Willens*, Cohen claimed that jurisprudence "can be characterized as the mathematics of the *Geisteswissenschaften*." Cohen (1904), 63. However, two years earlier in his *Logik der reinen Erkenntniss* Cohen wrote that "history may be characterized as the mathematics of the *Geisteswissenschaften*." Cohen (1902), 426. Cohen on that page also intriguingly wrote that history is to reality what nature is to mathematical natural science. Gustav Shpet, who had little regard for Marburg philosophy, some two decades later would essentially have the same view of history. Saval'skij attempted to explain Cohen's alleged inconsistency without a direct reference to Spektorskij. See Saval'skij (1909), 247–248f.
[12] Spektorskij (1905), 409.
[13] Trubeckoj (1909), 121.

world; there are only ways or modes of thinking. Such an abstract scheme, in Trubeckoj's eyes, can hardly be called an ethics.

Trubeckoj, as we also saw previously, objected to Saval'skij's erection of jurisprudence as the "science" on which to orient ethics. Needless to say, he criticized Cohen for doing just this but expressed an understanding that Cohen would do precisely that given his emaciation of the subject matter. Cohen searched for an object and found law, but again it is not a specific law or set of laws. It is the fact of law within a juridical system or "science." Trubeckoj faulted Cohen for claiming that there is no morality apart from law.[14] Cohen made no attempt to prove that philosophy must orient itself in a fact and axiomatically takes this – the transcendental method – to be Kant's lasting contribution.

What particularly – and understandably – distressed Trubeckoj is that in Cohen's scheme the only means of liberating the individual from self-absorption, from a moral egoism, is an absolute subordination to law, whatever that may be. The state, as the originator of law, assumes unlimited and complete moral authority over the individual, who is not accorded any special moral self-consciousness or conscience distinct from the state.

For Trubeckoj, morality and the meaning of life itself must have their source in a transcendent realm, which Cohen's philosophy denies. "It is impossible to reject with impunity the very thing by which the human soul lives."[15] All such oppositions as heaven and earth, the spiritual and the corporeal, God and human become in Cohen merely a "method." Trubeckoj's comments in this respect were not unexpected from such a religious philosopher.

More interesting in light of twentieth century and current neo-Kantianism was Trubeckoj's attempt to disassociate Cohen from Kant, as though that alone would be a sufficiently damning indictment. Without providing the textual basis for his assertion – however, none really is needed – Trubeckoj held that the Kantian categories are independent of the empirically given, they being *a priori* conditions of the possibility of cognition and of experience *in general*. Thus, they are logically prior to all cognition. For Trubeckoj, Cohen, however, understood the categories quite differently than did Kant. For Cohen, the categories are dependent on a factually given science. Cohen transformed the universality and necessity that Kant took to be the essential characteristics of the *a priori* categories into hypothetical presuppositions of factually given and mutable human cogni-

---

14 Trubeckoj (1909), 137.
15 Trubeckoj (1909), 164.

tion.[16] In other words, Trubeckoj saw Kant as arguing for twelve immutable categories, whereas Cohen held to the relativity of the *a priori* categories. Trubeckoj, unfortunately, did not venture into the philosophical viability of Cohen's project. Surely, however, the details of Cohen's defense would depart considerably from the argument in Kant's Transcendental Analytic found in the first *Critique*.

In the next decade at the very end of the Imperial Russian era, Trubeckoj published a large study of Kant and neo-Kantian epistemology.[17] He again brought up the transformation of Kant's immutable *a priori* categories into Cohen's relativized *a priori* presuppositions of factual science as the "essential difference between the two philosophies."[18] In this work, *The Metaphysical Presuppositions of Cognition*, Trubeckoj devoted considerable attention to Cohen, concentrating on his 1902 *Logik*. Trubeckoj stressed Cohen's attempt to disassociate the logical presuppositions of cognition from its alleged psychological conditions, a confusion that he believed could be found in Kant, and to disavow the presence of any ontological or metaphysical presuppositions in scientific cognition. Seeking to banish even the slightest hint of psychologism from the philosophical enterprise, Cohen concentrated on what he called "method." Yet, Trubeckoj found no trace of it in its everyday sense in Cohen's work. Instead of understanding "method" as the means of investigating or determining truth, Cohen *identified* truth with method. There is no knowing truth apart from *our* means of determining it. The result is a false anthropologism. Cohen posited, in effect, that there is no higher truth than human thought. Instead of solving the fundamental problem of epistemology, viz., how cognition of the truth, of existence independent of the cognizing human subject, is possible, Cohen simply ignored the problem. Truth, for him, lies solely in the search for it. Whereas science presupposes the existence of an object independent of cognition, Cohen's standpoint dismissed that presupposition as "illusory," as misleading, for there is no such object independent of science.[19]

Despite his efforts, Cohen, in Trubeckoj's eyes, could not prove that pure thought, our pure thought, is the only source of knowledge, that human knowledge is independent of sense data. Even if we should agree that mathematical physics concerns itself with nothing obtained from the senses, are *my* experiences too merely illusory, a product of *my* thought? My sensations are certainly not illusory to me. There is no means of convincing me that the summer sunshine

---

16 Trubeckoj (1909), 159.
17 For a discussion of Trubeckoj's examination of Kant, see Nemeth (2017), 337–342.
18 Trubeckoj (1917), 210.
19 Trubeckoj (1917), 227.

and the sound of Beethoven's symphonies are products of *my* thought alone. Trubeckoj faulted Cohen again as he did in his 1909 essay for one-sidedly basing all of his arguments on a factually given science. "They all start from the quite unfounded presupposition that logic is exclusively or primarily the logic of mathematical physics."[20] To be sure, Cohen was aware of this line of argument and dismissively rejected first-hand experience, saying, "We know only a logical expression of sensations."[21] Trubeckoj in turn – and understandably – was struck with the inconsistency of such a retort. Cohen must know the difference between sensation and thought, for otherwise, his polemic would be pointless. If Cohen were correct, sensations would be merely a function of pure thought. "But then his entire attempt *to exclude sensation as a source of cognition* would be meaningless."[22]

Trubeckoj summed up his critique stating that he agreed with Cohen in undertaking a transcendental investigation of the conditions of cognition. These conditions – and again Cohen, in Trubeckoj's eyes, is correct – are not psychological, but necessary logical presuppositions. The true method to be employed is the transcendental method. However, Cohen errs in not pursuing the method to its ultimate conclusion. For "a completed investigation into the conditions of the possibility of knowledge inevitably reveals the essentially *metaphysical* presupposition on which it rests."[23] This presupposition is the original or first principle (*pervonachalo*) of all that is both real and possible. With this, Trubeckoj introduced as well the Solov'ëvian conception of the all-unity, but at this point, we leave philosophy and pass into the realm of metaphysics.

Trubeckoj's criticism of Cohen's systematic work is representative of the ultimately faith-based approach to philosophical issues taken by a significant segment of the Russian philosophical community in the last decade or so of Imperial Russia. Trubeckoj was by no means the only one. Berdjaev in 1910 writing quite eloquently from his Christian existentialist standpoint called Cohen "the most extreme and, perhaps, the most consistent neo-Kantian epistemologist."[24] However, Berdjaev, as did others, objected to making epistemology parasitic on mathematical physics, or indeed on any factual science. Were one's faith in that science to falter, then, of course, the epistemology based on it would collapse as well. Berdjaev, ever willing to throw the baby out with the bathwater, remained content with holding that epistemology had no relation to a cognition

---

20 Trubeckoj (1917), 232.
21 Cohen (1902), 390.
22 Trubeckoj (1917), 234.
23 Trubeckoj (1917), 240.
24 Berdjaev (1910), 287.

of being. "It neither grounds the certainty of science nor does it triumph over skepticism and relativism."²⁵ Schelling and Solov'ëv recognized reason more than did Mill, Spencer, Cohen and Avenarius.

After retiring from his philosophy professorship at Marburg University, Cohen taught at a rabbinical seminary in Berlin until his death in April 1918. He did visit Russia, however, arriving in April 1914 for approximately four months of lecturing, not as a spokesman for Marburg neo-Kantianism or even as a philosopher, but as a representative of western Judaism, as "an authoritative Jewish thinker in order to comfort and cheer up his brothers in faith and blood."²⁶ While in Russia, he was received by such figures as Bulgakov and Losskij as well as others, but in general, his presence on their soil failed to attract much scholarly attention. Russia's religious philosophers simply ignored him. "Their attitude toward him was one of indifference."²⁷ Nevertheless, in St. Petersburg he spoke in German at the city's Religio-Philosophical Society on the topic "The Essence of the Jewish Religion," after which he traveled on to Moscow. There, at the start of May, he lectured not only on Judaism, but also on "Philosophy and Science" to the Society of Scientific Philosophy.²⁸ These excursions have received little attention both in the past and in today's secondary literature.

Cohen died on 4 April 1918, at a time when the young Bolshevik regime was still consolidating its grip on power but doing so with ever increasing determination. Two obituaries appeared that same month in two weekly newspapers, *Ponedelnik* [*Monday*] and *Nedelja "Narodnogo Slova"* [*Weekly "New Word"*] each surely with a small circulation. In the first of the two, the author, signing his name simply as "Vs. R.," wrote that Cohen was, undoubtedly, the most important of the neo-Kantians but "too independent a thinker to remain a mere interpreter of Kant."²⁹ The obituary is quite brief. In the second of the two, Moisej M. Rubinshtejn [Moses Rubinstein], whom we shall discuss in the next chapter, concentrated on Cohen the system builder rather than Cohen the interpreter of Kant. Rubinshtejn noted that Cohen's practical philosophy had, on the one hand, attracted great sympathy from some involved with the philosophy of law – no doubt a reference to Saval'skij – as well as "opponents among our phi-

---

25 Berdjaev (1910), 288.
26 Belov (2016b), 22.
27 Belov (2004), 333. Nonetheless, Boris Jakovenko, a figure we shall see later, wrote in a 1915 article "O polozhenii i zadachakh filosofii v Rossii" ["On the Status and Tasks of Philosophy in Russia"] that when Cohen "visited Russia, the people poured out, one might say, for his German lectures, and they were enthusiastically received." Jakovenko (2000), 723.
28 Vikhnovich (2014), 204.
29 Kolerov (2018), 60.

losophers of a religious-philosophical persuasion."[30] Most importantly for us here, though, Rubinshtejn rejected being seen as a "friend" of the Marburg School despite the respect that Cohen deserved both as a man and as a philosopher.

Having seen examples of the negative reaction to Cohen's neo-Kantianism, let us now proceed to a discussion of the young scholars, whether ethnic Russians or not, who left Russia to study in Marburg. We have already noticed that some went and never returned to the land of their birth. Others returned for a short period but emigrated amid the political turmoil of their homeland. Still others were evicted. All of their stories together form the story of the rise and the demise of Russian neo-Kantianism.

8.2 Two Ethnic Germans

The first student born in Moscow, indeed in Russia, to study with the Marburg neo-Kantians was Kurt Wildhagen (1871–1949). As we may suspect from his name, Wildhagen was of German origin, his father being a businessman or architect. In any case, the family returned to Germany in 1886. Wildhagen enrolled at Marburg University in 1895 and remained there until 1899 without finishing his degree. Whatever the reason, he took up residence in 1903 in Heidelberg, where he remained for the rest of his life as a perpetual student becoming a fixture in the local cafés and university lecture halls. He made his living giving lessons often to foreign students and from editing, writing and preparing translations. He survived the Nazi era, although he reputedly had no love for the regime. Upon his death, he left behind a veritable mountain of books and papers, which found no interest and were largely discarded. Make no mistake, though, Wildhagen was not a Marburg neo-Kantian, not a neo-Kantian of any particular school, not a career philosopher, not Russian, and not a Slav.

Another figure whom we cannot overlook, even if only in passing, is a prominent philosopher whose name has come to be eclipsed by the rise of Heidegger. Nicolai Hartmann born to ethnic German parents in Riga in 1882 belongs to the history of German, rather than Russian, philosophy. However, he studied almost surely under Lossky at St. Petersburg University "for some time between 1903 and 1905."[31] He then went on to study in Marburg – why he chose Marburg in

---

30 Kolerov (2018), 63.
31 Tremblay (2019), 193f. Tremblay, with good reason, writes, "Early inoculation by Lossky would explain why Hartmann remained unaffected, while in Marburg, by the neo-Kantian formalist legalistic ethics of his professor Hermann Cohen." Tremblay (2019), 201.

particular is not entirely clear – where he impressed Natorp with his interest in ancient Greek thought. In 1907, he won a competition on a theme related to Plato's theory of ideas and received encouragement from Cohen to pursue a career in philosophy. That year, Hartmann also submitted his thesis *Über das Seinsproblem in der Griechischen Philosophie vor Plato* and returned to Russia. During the 1907/08 academic year, he taught Latin and Greek at a St. Petersburg secondary school, while also taking classes at the University and preparing a book *Platos Logik des Seins*. Lured back to Marburg in order to complete his degree (exam and thesis defense) there, "the prospect of becoming a university professor of philosophy proved irresistible."[32] Hartmann rather quickly composed his *Habilitationschrift* entitled *Des Proklos Diadochus philosophische Anfangsgründe der Mathematik* published in 1909. Clearly, Hartmann's interest was not in the promotion and defense of Marburg neo-Kantianism, neither later in his own systematic works in ontology nor even in his earliest writings under Cohen and Natorp.

8.3 A Cohen-Inspired Reading of Kant

Two individuals who undoubtedly studied with the Marburg neo-Kantians and returned to Russia imbued with the spirit of their German philosophy professors were Boris A. Fokht (Vogt) and Gavriil O. Gordon. Both have been the subject of much recent scholarly attention. Fokht (1875–1946) is understandably of great significance in our story not so much for his contributions to technical philosophy as for his sheer perseverance and resolute adherence to Marburg neo-Kantianism. He was born in Moscow, where his father was a professor of medicine at the University. After completing his secondary studies, Fokht enrolled in the natural sciences department at Moscow University and was awarded a first-class diploma in 1899 with an undergraduate thesis entitled "Comparative Anatomical Sketch of the Circulatory System of Vertebrates." However, with an interest in philosophy already in his undergraduate years, he officially requested permission in 1896 to go to Switzerland and Germany owing to ill health from late-April to mid-August. Whether he did go to Switzerland is unclear, but surviving documents show that he attended lectures by Kuno Fischer in Heidelberg on the history of philosophy. Moreover, immediately upon finishing his undergraduate studies in 1899, he went to Freiburg to attend Rickert's lectures on logic and epis-

---

32 Harich (2000), 6. Harich also writes that the appearance in 1903 of Natorp's *Platos Ideenlehre* immediately captivated Hartmann. Harich (2000), 4. Thus, Natorp's book may have been a factor in Hartmann's decision to go to Marburg.

temology. Returning to Russia for the start of the academic year in Moscow, he enrolled that fall in the liberal arts department. A reasonable conjecture is that in these early years he continued to visit Germany when possible, but instead of going to Freiburg or Heidelberg he went to Marburg. Whatever the case, the noted literary figure Andrej Belyj remarked that Fokht already in 1902 was a Kantian.[33]

By the time he had completed a second undergraduate degree, this one in the liberal arts faculty, in 1904, Fokht was a disciple of the Marburg direction of neo-Kantianism and at least to some of his Russian friends the leader of the movement, such as it was, in Russia. The thesis submitted for this second degree was a work concerning Kant's theory of the *a priori*,[34] but he was awarded only a second-class diploma. Although he was officially commissioned to study abroad owing to his previous first-class diploma in science, he would receive no government funding. Fokht, nevertheless, wished to return to Marburg, but for an unrecorded reason, he could not proceed with his wishes in haste. For a time, he taught philosophy at the Higher Women's Courses in Moscow, and in 1905 lectured on pedagogy at a women's secondary school. In the years ahead, he continued to teach at the former, allowing, that is, for trips abroad, particularly to Marburg to hear Cohen and Natorp.

Whatever impression his friends may have had of him, Fokht would publish comparatively little, and it is only on the basis of those few works that we can engage with him today. The brute fact is that whereas one of his close friends wrote to him at the start of 1907 from Marburg that he (and possibly others) "expected much" from Fokht, little appeared in the immediate subsequent years.[35] The "normal" procedure for a young Russian student after spending time at universities in a foreign land was to complete, publish, and defend a *magister*'s thesis. However, Fokht for some still unknown reason failed to produce such a work until the end of the Imperial era. Speculation has abounded that Lopatin, a friend of Fokht's father, objected to Fokht's neo-Kantianism. Was it purely coincidental that his thesis defense took place only a year following Lopatin's death from influenza and was possibly delayed for that period owing to the political tumult?[36] However, if Fokht feared Lopatin in Moscow, he conceivably could

---

[33] Belyj (1990b), 39. Although there is no reason to doubt Belyj's observation, we should bear in mind that Belyj's aptitude in philosophy may not have been outstanding. He remarked, "Kant, Rickert, and Cohen as philosophers are completely alien to me." Belyj (1990b), 451.
[34] This work has apparently been lost.
[35] Letter from 12 January 1907 as in Dmitrieva (2007b), 416.
[36] Belyj wrote concerning Fokht, "Hounded by Lopatin, he would have liked to bite the head off the Lopatins and the religious philosophers. A most magnificent and cleaver teacher and per-

have submitted the thesis in St. Petersburg. Whatever the case, Fokht's interest in philosophy and particularly the Marburg approach never waned. He devoted his time to translations, creating a series entitled "Kantiana" that published works intended to serve as guides to the study of Kant's philosophy. The first publication in the series was a joint translation of Carl Stange's incredibly brief *Der Gedankengang der "Kritik der reinen Vernunft"*. Unfortunately, Fokht provided little information concerning his own position in his "Preface to the Russian Translation" apart from saying that Windelband had presented in his *History of Philosophy* the true sense of Kant's philosophy: "But apart from expressing the authentic sense and significance of transcendental philosophy, Windelband's work does not carry sufficient objectivity and is not an exposition of Kant's philosophy so much as a study of it."[37]

The second volume in the "Kantiana" series appeared in 1910 and was a translation of Johann Schulze's *Erläuterungen über des Herrn Professor Kant Critik der reinen Vernunft*. Fokht again tells us in his "Preface" that the translation was meant "to facilitate the understanding of the genuine content of the *Critique of Pure Reason*."[38] Fokht also prepared translations of Natorp's *Sozialpädagogik* and of the *Philosophische Propädeutik*, but a planned translation of Cohen's *Kommentar zu Kants Kritik der reinen Vernunft* never appeared in print.[39] This translation was intended to appear as the fourth volume in the series. The third volume was to be another translation, viz., that of Alfred Hölder's *Darstel-*

---

son, he was unfairly driven out of the department and so made his home his department, teaching us methodology there." Belyj (1990b), 384. We should add here that the influenza pandemic of 1918–1920 hit Russia particularly hard. Already reeling from the socio-economic effects of the Great War, followed by Revolution, Russia was hardly prepared for what followed with the pandemic. Unfortunately, it has not been the object of much scholarly attention.

37 Stange (1906), ii.

38 Shul'ca (1910), i. Fokht also writes that the translation was undertaken several years previously on the advice of Lopatin! Fokht collaborated with three others in the preparation of the Schulze translation.

39 Fokht specifically mentioned this translation in a 1925 curriculum vitae. See Dmitrieva (2007b), 305. Thus, there are grounds for believing that Fokht did prepare it. We also find in a late-December 1908 letter from another young Russian neo-Kantian, Aleksandr V. Kubickij (1880–1937), to Fokht: "Have you begun your translation of Cohen's *Logik* [*der reinen Erkenntnis*]?" Letter printed in Dmitrieva (2009), 395. On what basis did Kubickij believe that Fokht was going to prepare such a translation? Unfortunately, no additional information seems to be had. In another letter a week later, Kubickij again raised the issue of the translation, saying that Fokht would soon present an excellent translation. Dmitrieva believes this refers to the translation of Cohen's *Kommentar*, whereas in light of the December letter and its early date, viz., 1909, I believe this refers to the idea to translate Cohen's *Logik*.

lung der Kantischen Erkenntnistheorie mit besonderer Berücksichtigung der verschiedenen Fassungen der transscendentalen Deduction der Kategorien.[40]

The journal *Logos* announced in 1911 the future appearance in it of an essay by Fokht to be entitled "On the Principle of the Transcendental Method in Kant's Theoretical Philosophy."[41] Admittedly, the existence of *Logos* was cut short, but Fokht's piece is nowhere to be found in any subsequent issue of the journal. Whether the essay was ever written as such is unknown. Quite possibly, what was written simply became part of his later thesis. The fact is that nothing written by Fokht, the supposed leader of Russian neo-Kantianism, ever was published in the sole allegedly "neo-Kantian" journal *Logos*. We need not engage in pointless conjectures as to the reason, for in the end, the fact remains unaltered.

The Bolshevik Revolution eventually led to a restructuring of the universities and, more immediately, in the manner in which degrees were administered. Instead of a distinct *magister*'s examination, there was the more "proletarian" colloquium. We can infer that Fokht was successful in that he was able to conduct a seminar on Kant at Moscow University from 1919–1921. Finally, he defended his thesis at the end of May 1921 with Chelpanov as one of the two opponents.

Fokht's thesis, *Ob osnovnoj idee, sushchestve i glavnejshikh momentakh transcedental'nogo metoda v teoreticheskoj filosofii Kanta* [*On the Fundamental Idea, Essence, and Chief Moments of the Transcendental Method of Kant's Theoretical Philosophy*], reveals a distinct and massive influence from Cohen's reading of Kant's epistemology. Although given its late date, the thesis "in spirit" belongs to the late Imperial era and demonstrates Fokht's continuing commitment to a philosophical movement that would soon disappear. Fokht tells us at the start that his work was concerned with the conditions and the very possibility of experience, understood not in the quotidian sense, but "experience" as scientific cognition. Such an undertaking, seeking the conditions of experience, is what Kant (and the neo-Kantians) called the transcendental method. Philosophy is distinguished from the other rigorous disciplines or "sciences" in that they presuppose the possibility of experience, i.e., of cognition in the respective scientific discipline. The aim of the exact sciences is to extend the content of such experience. But since philosophy poses a different task, viz., the determination of the possibility and the conditions of cognition, it must employ a different

---

40 The precise fate of this third volume is also unclear. Dmitrieva mentions a typewritten copy of this volume in the library of the "Institute of Red Professors." Fokht stated in the preface to this copy his translation of Cohen's *Kommentar* as having been made "many years ago." Quoted in Dmitrieva (2003), 33.

41 Predpolagaemoe (1911), 239.

method. Inasmuch as its aim and methodology are different from that of the other sciences, it cannot be reduced to any of them. Hence, the specter of positivism is eliminated. And inasmuch as it inquires whether the other sciences are justified in assuming what they do, it is logically more fundamental.

Fokht relied heavily, though not exclusively, on Kant's *Prolegomena* as read through Marburg eyes. His is an attempt at a non-subjective interpretation, in which "experience" is taken as universal natural-scientific experience, i.e., the "experience" or cognition of the impersonal community of scientific investigators. The object of such experience is nature. Fokht remarked that from this Marburg-inspired standpoint, the possibility of natural science and the principles of its grounding as absolutely valid cognition became the fundamental problem of theoretical philosophy after Plato. Fokht then proceeded in his thesis to summarize the history of modern philosophy. He wrote that Galileo ushered in modern science through the idea of the mathematization of nature, that the universe cannot be understood without learning its language, that language being mathematics.[42] And both Descartes and Leibniz were concerned with fundamentally grounding science, but neither took natural science as a fact. It is this that distinguished Newton from Leibniz. The former confronted philosophy with the fact of his system of scientific knowledge, i.e., with the indisputable reality of mathematical physics. We are asked to find an additional hint to the proper approach to philosophical problems in the title of Newton's most famous work – rendered into English – *Mathematical Principles of Natural Philosophy*. The principles that make physics and thereby nature possible are exclusively mathematical. As original factors of our cognition, they play a determinative role in it as *a priori* principles.

Fokht saw this conception that mathematics is the language of nature to be the guiding thread throughout the modern scientific era and for all scientific philosophy. However, the fundamental question remained unanswered. How is mathematical physics possible? How is it that mathematics, as an *a priori* discipline, is uniquely suited to describe the *a posteriori* given of nature and thereby makes theoretical physics possible? The resolution of this question formed the chief task of Kant's "Critical" investigations.

The central concept of Kant's philosophy, the one by which he characterized his entire philosophy is the term "transcendental." He also characterized his method as "transcendental," and his concern was with "transcendental cognition." Kant, that is, was concerned not so much with objects as with our means of cognizing objects in general insofar as such cognition is possible *a pri-*

---

42 Fokht (2003), 60. Cohen quoted Galileo to make this same point. See Cohen (1885), 416.

ori. Such talk by Fokht, surely, will not sound strange to anyone familiar with Kant's own writings or those of Cohen. Fokht wrote, "Clearly, therefore, transcendental cognition has to do with the means of cognition that we call a science of nature, mathematics and physics, in their systematic connection and unity, or, in other words, 'mathematical physics'."[43] Here and throughout his thesis Fokht merely echoed Cohen, claiming that Kant had established a new concept of experience as the science of nature set against the old, pre-Kantian and ordinary sense of the word.

Fokht firmly rejected the assertion that in accepting physics as a finished, complete science, Kant fell into a form of dogmatism, which his philosophy was supposed to abhor. No, Fokht claimed, Kant did not hold contemporary physics to be a closed system. Kant proceeded on the assumption that today's science is the result of its historical development, that its validity is the subject of ongoing investigations, and that on the basis of such investigations it will continue to develop. Whereas dogmatic philosophies take their objects to be fixed "things in themselves," Kant held that the very object of his concern, natural science, is itself evolving. The empiricism of Locke and Hume cannot be reconciled with the fact of physics and pure mathematics. This, in Fokht's eyes, already was an endorsement of Kant's entire transcendental deduction. It also serves as testimony that the content of physics as a science is the laws of nature, and these laws are grounded in synthetic principles revealed by philosophy. "*Nature* is nothing other than *experience*, but this experience, in turn, is nothing other than the *aggregate of a priori synthetic judgments* about its object."[44] In this interpretation, then, the possibility of experience is the possibility of nature itself as cognizable. Newton already understood and used the word "experience" in this sense.

Fokht truly had little to say about the specifics of Kant's presentation in his first *Critique* and precious little concerning the particular arguments found in the Transcendental Aesthetic. He did mention the manifold of pure intuition, the *a priori* forms of which bring order to this manifold and thereby prepare the unity conferred by concepts of the understanding. In answer to the query concerning how do we obtain the pure intuitions of space and time if not from the empirically given, Fokht essentially repeated Kant's position without qualification. Fokht wrote, "If we mentally eliminate from empirical intuition everything that belongs to sensation, which is the material of appearance, we will still have its forms, space and time, from which we can never escape no matter how far

---

43 Fokht (2003), 67.
44 Fokht (2003), 75.

we carry out our abstraction."⁴⁵ However, whereas Kant explicitly concluded from this that space and time are *a priori* intuitions, Fokht did not and instead presumed the reader would draw the intended inference. Moreover, in light of Fokht's debt to Cohen's Kant-interpretation, we, the reader, can ask of Fokht: "Who is this 'we' that has empirical intuition? Eschewing much that we traditionally understand of Kant's discussion of space and time, Fokht not just forsook them as intuitions had by you and me, as individual cognizers, but held sensibility itself to be "not so much a faculty as, above all, a cognitive means or method."⁴⁶ Thus, sensibility is not to be understood as a psychological faculty, but as an abstract condition for the unity of consciousness.

Fokht's reconstruction of the Marburg position vis-à-vis mathematical physics forces us, as it were, to ask whose consciousness is this supposed unity and just who has intuitions, etc. If he did not mean the individual's empirical consciousness, opponents of Fokht's position would surely ask of him whether he meant some collective consciousness. Is it perhaps the consciousness of an absolute, creative entity, say, the consciousness of God? Fokht demonstrated no concern with whether space, time, and the Kantian categories are part or a facet of the human individual's innate cognitive faculty. As conditions of the possibility of experience, i.e., scientific experience, they are formal conditions of scientific consciousness or, to use Kant's own expression, of a "consciousness in general." "The chief aim for Kant always remains the discovery of the fundamental principles of scientific, and not of the individual, consciousness, or in his own terminology, the conditions of experience understood in the sense of a science of nature."⁴⁷ Have we, with this, answered our initial question concerning who or what has this "scientific" consciousness? I, this empirical being born in such-and-such a year, etc., have a consciousness. Thus, do I and my fellow scientists possess and share a unique "scientific" consciousness when engaging in scientific research, but not when going about my other daily routines? For Fokht, "scientific consciousness" is a unity of thought, of scientific reason. As reason, it is a unity of the system, and therefore as an internally consistent construction,

---

45 Fokht (2003), 122. Fokht here refers to the *Prolegomena* §10 rather than to Kant (1997), 158 and 162 (A24/38–39 and A31/B46).
46 Fokht (2003), 122. Fokht also claims there that this understanding of space and time is "not just the interpretation of Kant's doctrine by representatives of the so-called neo-Kantian schools, but … is in complete and strict agreement with the actual texts of Kant's *Critique* and the *Prolegomena*." Fokht certainly may have the right to reconstruct Kantian idealism as he seems fit, but to see it as in "strict agreement" with Kant's texts involves considerable legerdemain at the expense of the actual texts.
47 Fokht (2003), 219.

the "scientific consciousness" belongs to no one. But can a "system" have intuitions, particularly of the kind Kant discussed in his Transcendental Aesthetic?

Fokht ended his thesis stressing that his exposition sought to distinguish the psychological from the transcendental, and thereby answer the questions posed above. A transcendental investigation seeks not just to determine the *a priori* elements in experience, for that would be a metaphysical "deduction." No, the transcendental investigation consists in showing how the formative structure of experience emerges from the basic *a priori* elements.[48] This structure would, then, have the objective significance of scientific cognition. Fokht's thesis presented a generally clear summary of Cohen's Kant-interpretation but marked no advance beyond it.

Regrettably, details concerning Fokht's thesis defense have not come down to us, though it would be interesting to know the questions, assuming there were some, that were posed to him and his answers to them. Whatever transpired, the defense itself took place on 30 May 1921. Whether a sheer coincidence or not, within days the official academic year ended and the philosophy department at Moscow University was closed.[49] In late October of that year Fokht was appointed professor of philosophy at the university in Jaroslavl, where he had been teaching since 1920 in the capacity of a lecturer. He held the position in philosophy until 1925 and became a corresponding member of the State Academy for the Study of the Arts (GAKhN). It is in papers written for GAKhN that we can see his continuing commitment to Marburg neo-Kantianism in the years ahead.

8.4 The "Last of the Mohicans" in Soviet Russia

In March 1924, the predecessor of GAKhN, the Russian Academy for the Study of the Arts (RAKhN), formed a small committee to develop a plan to commemorate the two-hundredth anniversary of Kant's birth. The committee consisted of two individuals, whom we shall see later, Matvej Kagan and Aleksandr Sakketti, and was chaired by Gustav Shpet. Fokht for the occasion read his paper, "O postanovke osnovnoj problem estetiki u Kanta i Kogena v svjazi s kritikoj osnovnykh

---

[48] Fokht failed to show precisely how a transcendental investigation or "deduction" reveals the formative structure of all scientific experience. He also failed to provide the criteria by which the community of investigators can know that they have correctly uncovered the *a priori* elements or have completely revealed the formative structure of experience.

[49] Dmitrieva suggests that this "closing" [*zakryt*] of the philosophy department in the immediate wake of Fokht's thesis defense was not a mere coincidence. Dmitrieva (2003), 29.

ponjatij i principov, primenennykh Kantom k ee resheniju" ["On the Formation of the Basic Problem of Aesthetics in Kant and Cohen in Connection with a Critique of the Basic Concepts and Principles Applied by Kant to Their Solution"]. Given its sheer length, it is unlikely that Fokht read the entire paper at a single session. Perhaps, he expanded his text considerably after delivering it orally or, vice versa, he summarized it at the actual event. In any case, the text, published only posthumously, is itself a testimony to Fokht's unwavering and essentially unaltered commitment to Marburg neo-Kantianism even in light of the ongoing changes in his country's political institutions.[50]

Whereas his thesis concentrated on "theoretical" philosophy with little regard for ethics or aesthetics, on both of which Kant and Cohen had much to say, Fokht's involvement with the Academy gave him the opportunity to address directly at least the latter of the two. Even a casual reader, if there be such, of Kant's first two *Critiques* can easily discern a difference in procedure between them apart from the topics covered. Much like Cohen, Fokht would have none of this. The formulation of the fundamental problem in all three spheres of philosophy can be understood only if one adopts and recognizes that the transcendental method is central to all three. This method proceeds from the (necessary) conditions for the possibility of objective cognition in a specific sphere. These conditions, as we saw in Fokht's thesis, are understood to be the conditions of the objects themselves in the respective cognitive sphere.[51] The transcendental method is "necessary for solving the theoretical problems of nature and no less so for ethical and aesthetic problems, indeed even for all problems of human culture in general."[52] Thus, Fokht "simply" wished to broaden Kant's "Copernican Revolution" to all spheres of human activity.

Fokht himself raised the question whether ethics as a philosophical discipline can be treated purely rationally and its problems solved in the same way

---

50 The effect of the political situation in the country on domestic philosophical thought must have appeared puzzling to those inside. Iosif D. Levin (1901–1984), a graduate of the philosophy department at Moscow University and thus a student during these years of both Shpet and Frank, recalled in his memoirs, "up to 1922 there was no state monopoly on philosophy (nor on literature and especially none on poetry). ... A turning point in the relation to philosophy began to become apparent only at the end of 1922." Levin (1991), 281. Yet, we see that in 1924 open philosophical discussions still could take place within the narrow confines of specialized research institutions.

51 Kant (1997), 234 (A111) – "The *a priori* conditions of a possible experience in general are at the same time conditions of the possibility of the objects of experience." See also Kant (1997), 283 (A158/B197).

52 Fokht (2003), 196. Cassirer wrote, "For Cohen himself this idea – the unavoidable necessity of the 'transcendental method' – became the essence of scientific philosophy." Cassirer (1981), 3.

as in theoretical philosophy, i.e., through use of the transcendental method. "Will the needed conditions and concepts be found for applying the transcendental method, conditions and concepts that, like certain assumptions made by pure thought, could ground and construct the corresponding cognition also from pure practical reason, analogous with the cognitive object from pure reason?"[53] This question was, in Fokht's mind, equivalent to asking whether we can find necessary conditions for grounding and determining the will's conformity to law. Fokht believed that Kant had answered these questions in the affirmative, although the former thought that we must indulge Kant some, since Fokht failed to see an immediate similarity between the starting points on which we must apply a transcendental analysis. In one, the objects of natural science are the starting point, whereas in the other we start with the objects of the will, i.e., human actions and the law-conforming connections between them. Fokht criticized both Kant and Cohen for their respective concerns with laws at the expense of the law-conforming will. "The *Critique of Practical Reason* and even more the *Ethics of Pure Will* should have considered and actually taken as their basic task nothing other than the grounding of rational goal-conforming laws of human activity, or the will. However, that applies even more to the activity of the originally inherent law-conformity of pure practical reason, namely the pure will."[54] To be true to the basic principle of transcendental philosophy's method, we must give an account not of laws and even less of this or

---

[53] Fokht (2003), 205. We should observe that Fokht's writing style now in 1924 bears a closer resemblance to that of his German masters than it did previously in his thesis.

[54] Fokht (2003), 211. Fokht's position raises myriad questions and problems that beg for a detailed treatment. Fokht had little more to say about the will, on which Kant had pivoted his ethical theory: "It is impossible to think of anything at all in the world, or indeed even beyond it, that could be considered good without limitation except a *good will*." Kant (1996), 49 (Ak 4, 393). Thus, one could argue that whereas for Kant the moral good is, to use Cohen's terminology, a *factum*, for Fokht it is a *fieri*. Such an understanding, of course, is consistent with Fokht's Marburgian "theoretical philosophy" where truth is also a *fieri*. We could also criticize Fokht from another standpoint, namely, that of Cohen's mature system, for its failure to take jurisprudence as the analogue in practical philosophy to mathematical physics in theoretical philosophy. Concerning Kant's system, Cohen wrote, it lacks "a precise concept of the social sciences [*Geisteswissenschaften*] as the methodological analogue to the natural sciences. However much we must observe strictly and precisely the methodological difference between the two, we must, on the other hand, also affirm and enforce the analogy. Because Kant did not pose the problem of the social sciences, he did not recognize jurisprudence as the analogous fact to which the transcendental method has to be directed and oriented in order to constitute and ground ethics, just as we do with the fact of science." Cohen (1904), 216. Owing to his failure to recognize jurisprudence as the analogue in ethics for mathematical physics in theoretical philosophy, Fokht was further from the letter of Cohen's teaching than Saval'skij.

that law, but of the concept of law-conformity. The idea of a goal of action, of principles and categories are merely logical rules and determining principles for the will, but by no means are they objectively existing bases immanent to things. If we take the idea of the goal of action to be a logical rule and the action's determining principle to be a condition of the possibility of action, we have sufficiently prepared and fundamentally grounded a transcendental formulation of ethics and the path toward the solution to ethical problems.

Regrettably, Fokht had little more to say on ethics in this essay. He did not formulate, à la Kant, any categorical imperative, and since Fokht did not point to any existing ethical "science" as Cohen did, we are left in the dark where to find the content for our ethical laws. Fokht conjectured that if Kant did not fully and sufficiently reveal the meaning and significance of ethics as a transcendental logic of history and of the *Geisteswissenschaften* in general, this was due to their inadequate foundation at the time.[55] If Fokht, though, believed that he had sketched the path leading toward the solution of ethical problems, he did no more than spot it without stepping a foot on that path. Notably absent in his discussion are any references to freedom or autonomy, so prominent in Kant's own ethical theory.

Just as ethics and logic, in the neo-Kantian sense, are two parts of an overall philosophical system that are grounded by means of the transcendental method, so too can we speak of philosophical aesthetics as grounded by that method. For Fokht, the fundamental question here concerns the conditions for the construction of aesthetics as an integral part of such a system. To do this, we must show that the fundamental sense and significance of aesthetics too is grounded and formed through the concept of law-conformity, as revealed through the transcendental method. As an integral part of a philosophical system, aesthetics is inextricably connected with the law-conformity of cognition in the other two philosophical spheres. Nonetheless, it has a unique law-conformity, the grounding of which is its most fundamental problem. Fokht held that since nature and morality are products of our reason in the form of objectively expressed scientific and moral cognition, so too is art an objectified product of our creative fantasy, "only of a reason that objectifies itself even more fully."[56] Although theoretical and practical philosophy form along with aesthetics a system, material from neither of the former two parts or segments of philosophy are needed for aesthetic judgment or evaluation. Indeed, any such material, if taken as a determinative

---

[55] Is this, then, an implicit disavowal of the specifics of Kant's moral philosophy and his conjectures on history? Fokht was silent, but the implication appears clear enough.
[56] Fokht (2003), 221.

factor in a judgment's pronouncement, threatens to obscure and even reverse the judgment's sense.

However odd given his departure up until this point in his essay from Kant and even from Cohen, though to be sure less so, Fokht returned to Kant, arguably with the recognition that he could not do otherwise, that he had largely exhausted what he could say. He, instead, essentially summarized key points in Kant's third *Critique*, that, for example, we take the aesthetic form of an object as belonging to it, though that form itself is not a thing that has an existence independent of the aesthetic object. The aesthetic consciousness, the realization that something is beautiful, is a consciousness of a feeling. Clinging, as it were, however, to his architectonic, Fokht, unlike Cohen, was not so quick to abandon the view that there should be an exemplary science for aesthetics, to serve much like mathematical physics does for theoretical philosophy and, in Cohen's case, jurisprudence does for practical philosophy.[57] Just as pure practical reason, the pure will, is an expression of the ethical consciousness, so feeling, aesthetic feeling, is an expression of the aesthetic consciousness. However, with the understanding that the moral law is an expression of practical reason, can we know what the comparable aesthetic law, the law of feeling, is? Such a determination is the task of the genius.

> There are, indeed there must be, laws, a law-conformity of beauty. But neither science nor a cognition of the moral, neither logic nor ethics has as its task to discover these laws, this law-conformity. To discover and express this task is the concern of the genius. Only a genius gives a law to one's art, reveals its particular law-conformity in its contact and ultimate unity with the law-conformity of all other arts.[58]

In short, Fokht, by invoking Kant's conception of the genius, absolves himself of any responsibility to provide any aesthetic laws or rules.[59] To do so is the work of

---

[57] How else are we to understand Cohen's aesthetic philosophy except as a glaring resignation that there is no "science" of aesthetics in the least comparable to mathematical physics? That Cohen realized the need for an "aesthetic science" is clear enough. See Cohen (1912), 4–5. That he sought to gloss over this lacuna is also clear enough from Cohen (1912), 6–8.

[58] Fokht (2003), 225.

[59] See Kant (2000), 186–189 (§§46–47). To be sure, Kant does say there, "In the scientific sphere, therefore, the greatest discoverer differs only in degree from the most hard working imitator and apprentice, whereas he differs in kind from someone who is gifted by nature for beautiful art." Fokht, however, forgot Kant's qualification that an aesthetic rule "cannot be couched in a formula to serve as a precept, for then the judgment about the beautiful would be determinable in accordance with concepts." Kant (2000), 188 (§47). Kant explained earlier in his treatise that the determining ground of any aesthetic judgment is not a concept of the object, but the subject's feeling. Thus, there cannot be any objective rules of taste. Since aesthetics involves

the genius in the respective sphere of human culture, be it the physicist, the moral legislator, or the artist, but not the philosopher.

Fokht's paper on the aesthetic problem in Kant and Cohen delivered in 1924 was not his only one that year and undoubtedly not the only one prepared in connection with his work at the Academy for the Study of the Arts. He, additionally, also wrote a separate lengthy piece on Natorp's aesthetic system and yet another dated 20 April 1925 to commemorate Natorp's death the previous year. There is no indication in the surviving papers that Fokht renounced his fundamental allegiance to his early Marburg neo-Kantianism even into the Stalinist era, though we can be sure he kept as low a profile as possible. The end of the State Academy in 1929 also meant an end to Fokht's free philosophical activity. He evidently devoted considerable attention to Aristotle in the 1930s including the preparation of a translation of Trendelenburg's *Elementa Logices Aristotelicae*, which went unpublished, and a German-Russian dictionary of philosophical terms in 1937 also unpublished. Fokht made a living as a teacher of Latin and Greek until 1941. The surely most remarkable fact is that he remained "at large," miraculously evading the Siberian camps. The same cannot be said of his son Kirill.[60]

Boris Fokht died on 3 April 1946, presumably of natural causes.

8.5 *Dii Minores*

A name often associated with that of Fokht in the meager secondary literature on Russian neo-Kantianism is that of his close friend Gavriil O. Gordon (1885–1942), who spent the academic year 1906–1907 in Marburg.[61] Gordon graduated from a Moscow secondary school in 1903 and then enrolled in Moscow University. After the mentioned academic year in Marburg, Gordon returned to Moscow, graduating with a thesis on "Psychological Parallelism in Spinoza and Contemporary Philosophy." He then traveled back to Marburg in 1909 to hear Cohen and Na-

---

no determinate universal rules, originality, not reason, is the primary characteristic of the aesthetic genius.

60 Kirill, born in 1909, was first arrested in January 1933 and sentenced to three years of exile in Tashkent. In 1936, he was freed but re-arrested the following year on a charge of attempting to flee abroad and sentenced to ten years in prison. At the end of August 1945, he was sentenced to an additional ten years in a prison camp. Freed in 1955, he somehow managed to teach in a secondary school but was again arrested in 1962 and sentenced to seven years in prison. He died later that year.

61 Gordon gave his name in Germany as "Gabriel Gordon."

torp. While there he interacted with Hartmann in working out terminology for the translation of classical philosophical texts into Russian. What, if anything, became of this project is unrecorded, but it does show that Hartmann and Gordon were well acquainted with each other.

Gordon left little by which we can now determine his own philosophical ideas or even the strength of his commitment to neo-Kantianism. That he ventured to Marburg and stayed there for a time speaks somewhat in favor of at least a broad allegiance to or deep respect for the teachings promoted there.[62] However, we cannot entirely discount other possibilities as the motivation for initially seeking to study there, such as impressions from others of the quality of the instruction, the promotion there of an innovative interpretation of Plato, or simply to study in a relatively small city as compared to the bustle of Moscow.[63] Gordon, unlike Fokht, never completed a *magister*'s degree. Whereas the delay in the presentation and defense of Fokht's thesis may have been due in some degree – whether small or considerable – to a falling out with Lopatin, Gordon directly attributed his own failure to a falling out with Chelpanov.[64]

Gordon joined the army in 1909 "in accordance with the law at the time concerning military service," but was discharged the next year.[65] He then taught in a secondary school until he was called up at the start of the Great War. Gordon in the immediate years before the events of late 1917 wrote a number of short reviews of philosophical works but nothing more. They provide little insight into

---

[62] Gordon from Marburg wrote in a letter to Fokht from January 1907, "In our long conversations at times with Hartmann about the future of Kantianism and Cohenianism in Russia, for the benefit and flourishing of which we both (he, certainly, to a greater degree) intend to work, we always consider you as the leader and pioneer...." Dmitrieva (2007b), 416. Gordon's invocation of Hartmann here in retrospect is certainly odd, and his use of "we" is indefinite. Did Gordon believe at this time that Hartmann intended to return to Russia to help spread Marburg neo-Kantianism? Gordon's use of "our" here is also somewhat ambiguous. Did he truly mean Hartmann and himself?

[63] Dmitrieva, based on archival documents, writes that already at the time of his first trip to Germany Gordon knew Plato's dialogues "perfectly as well as ancient Greek." Dmitrieva (2006), 471.

[64] To be sure, we have only Gordon's version of his relationship with Chelpanov. Gordon in his memoirs attributed the deterioration of their relationship to his sympathy for the Marburg School. But Gordon also gives us much information to think that Chelpanov's tepid attitude toward Gordon was not purely a matter of philosophical differences. The latter, apparently, was not reticent to express his opinions: "Whether bad or good, I never in my life knew how to subordinate my free thought – even if it was wrong – to outside considerations." Gordon (1995), 87. And Gordon was not above faulting Chelpanov for his "speech defects," i.e., he spoke "with a touch of various South-Russian sayings." Gordon (1995), 86.

[65] Dmitrieva (2007b), 317.

his own positions and are of little significance to us today. Even in their own day, they received scant attention.[66] In 1918, he served in the Red Army and then went on to teach briefly at Moscow University. For a time in the 1920s, he was associated with the philosophy section of the State Academy for the Study of the Arts. He, unfortunately, did not play a significant role therein.

Gordon was arrested in 1929 and sentenced the following year to ten years in a labor camp. The term was changed in 1931 to internal exile in Sverdlovsk for five years and then in 1933 he was freed and allowed to live in Moscow. For the next three years he worked as a book editor until in 1936 he was again arrested, receiving a sentence of five years in a labor camp. Gordon, however, was sent to a camp for common criminals, where the conditions reportedly were better than the ones for political "dissidents," owing to the intervention of Lenin's widow (Nadezhda Krupskaja). Although scheduled to be released in August 1941, the attention of the authorities was elsewhere with the start of World War II. In short, Gordon, the man, was forgotten. He died in the camp from starvation in late January 1942.

Boris P. Vysheslavcev (1877–1954), born in Moscow, studied law at Moscow University and graduated in the same class with Saval'skij. Although he practiced his profession for a few years, his interest in philosophy grew leading him, with an introduction by Saval'skij, into Novgorodcev's circle.[67] After completing his *magister*'s examinations, he went to France and Germany, spending at least some time in Heidelberg, Freiburg, and Marburg. Vysheslavcev clearly impressed those in Marburg to the extent that he contributed an essay to the *Festschrift* commemorating Cohen's seventieth birthday. The essay "Recht und Moral" argues that anarchist theories, such as Tolstoy's, fail to see that the realization of the idea of the moral good involves a two-step process, where law as the first step realizes through legal obligations in organized actions a higher unity that then creates the next stage, i.e., that of morality. The fulfillment of these obligations necessarily involves some coercion. Anarchists mistakenly seek to circumvent the first stage, that of law.[68]

Vysheslavcev believed Kant and the early Fichte had mistakenly separated the two spheres of law and morality. Indeed, they had set the two spheres against each other. Yet, the relationship between the two is one of the most

---

66 For example, Gordon's brief review in *Logos* of Natorp's 1911 book *Philosophie. Ihr Problem und ihre Probleme* is not so much as mentioned in the issue's table of contents.
67 Dmitrieva (2007b), 195.
68 Wischeslavzeff (1912), 201. Natorp critically discussed Vysheslavcev's essay in a piece entitled "Recht und Sittlichkeit" published the following year in *Kant-Studien*. Vysheslavcev transliterated his name in German as "Wischeslavzeff."

basic problems in both jurisprudence and ethical theory. Were we to follow Kant in this matter, that law does not belong in the ethical sphere, the state and its actions would lie outside the sphere of moral evaluation. Still, Kant in his ethical theory could not consistently maintain the separation, and Fichte reversed himself in a later period.[69]

Whatever merits we find in Vysheslavcev's slightly later book-length treatise on Fichte – and there are many – there can be no mistake that he did not view himself and cannot be viewed as a neo-Kantian of any discernable stripe. He applauded the neo-Kantian movement for its systematic reexamination of Kant's ethics in light of contemporary moral issues. But Vysheslavcev lamented that the same had not yet been done with Fichte's ethical philosophy, which is "deeper, more fundamental, and more contemporary than Kant's. Neither Schelling nor Hegel surpassed him in this sphere."[70] We can hardly be surprised that Vysheslavcev, given his training as a lawyer, found Fichte's political and legal philosophy to be of particular interest, finding it to be a true justification of the state and of law. He found that whereas Fichte obtained the basic principle of his ethical system from Kant, a debt the former readily acknowledged, the respective systems each developed upon that principle are markedly different. Kant's morality is cold, pale, and abstract; Fichte's is deeply vital and concrete. Even with regard to the Kantian antinomy between freedom and necessity, Vysheslavcev thought Fichte's resolution to be fuller, more interesting, and more systematic even compared to what we find in the neo-Kantians.[71] Finally – and this is most telling for his future intellectual path in emigration – Vysheslavcev contrasted Kant's alleged subjective idealism against Fichte's overcoming of it, particularly in and for the sake of ethics. "If the neo-Kantians sought to free transcendental philosophy from subjectivism and psychologism, we should not forget that Fichte did this earlier and better."[72] What especially attracted Vysheslavcev to Fichte was the latter's aspiration for ontologism and the Absolute, an aspiration that would later occupy the central place in his writings. Undoubtedly, Vysheslavcev learned much studying under the neo-Kantians in Marburg and in Baden, but he did not come to share either school's fundamental orientation.

Vysheslavcev submitted his work on Fichte as his *magister*'s thesis and defended it in 1914 at Moscow University, after which he taught a course there on the history of political theories. At this time, he also taught at other higher edu-

---

69 Wischeslavzeff (1912), 190.
70 Vysheslavcev (1914), xv.
71 Vysheslavcev (1914), xvi-xvii.
72 Vysheslavcev (1914), xvii.

cational institutions in the city and received an assistant professorship in 1917 at Moscow University. His idealistic views, obviously not finding approval from the Soviet government, led to his involuntary expulsion in 1922 from Russia aboard the infamous "philosopher's steamer."[73] He spent his years afterward in Germany, France, and Switzerland, where he died.

Many other young Russian men with some connection to Moscow went to study for various periods of time in Marburg in the last years of Imperial Russia. Most went to supplement what they had already learned; others surely went for the adventure of studying abroad. Many would contribute little, if anything, to Russian philosophy and even less to Russian neo-Kantianism. A number would in time turn to other occupations or professions. Even in normal circumstances, this distancing of promising students from an early intellectual infatuation is not uncommon, and the Russian cultural scene after 1917 was far from normal. One such young man who ventured to Marburg well before his country was turned upside-down rather quickly realized that professional philosophy was not his *métier*. His is a name that, unlike so many others mentioned in these pages, would be remembered in the years and decades that followed, though not for contributions to philosophy or even academic scholarship, but to literature. The case of Boris Pasternak is particularly instructive for us not so much for what it says about him or his philosophical ideas as for what it tells about the overall impression young Muscovite students had of Marburg philosophy in the years leading up to World War I.[74]

Pasternak enrolled at Moscow University initially as a law student at the start of the academic year in 1908 but did not take his studies very seriously. He viewed the law curriculum merely as an easy option. He switched to philosophy with the 1909/1910 academic year, during which time he studied with Lopatin and Shpet.[75] Precisely how Pasternak became aware of the Marburg philos-

---

[73] This is based on Chamberlain (2007), 161. However, Finkel (2007), 282 writes, "Vysheslavtsev, often listed among the deportees, in fact left at the same time of his own accord."

[74] Belyj provides us with his striking impression. "Cohen and Rickert, without coming to Moscow, reigned within the walls of the University, because their 'students' from Moscow furnished them with youths for all of kinds of processing. A genuine exporting of youths was organized to Marburg and Freiburg, where venerable minotaurs ate them up and had done with them." Belyj (1990a), 272.

[75] Pasternak may have had Shpet for more than one course while he attended Moscow University. He wrote a paper for one of these courses on Hume's "psychological skepticism," which is intriguing in that Shpet published a paper on Hume's "skepticism" in 1911 that grew out of a seminar he led on Hume in the 1909/1910 academic year. Could Shpet's seminar be the course for which Pasternak wrote the mentioned paper? For an English-language translation of Shpet's 1911 essay, see Shpet (2019), 282–293.

ophers is unclear. He may quite possibly have become enamored with the idea of studying for a time in Germany from his friend Dmitrij F. Samarin (1890–1921), who had first traveled there in the summer of 1909 only to find that the university was between semesters. Still, Samarin fell in love with the city and returned the following year in time to study Kant with Cohen and Aristotle with Hartmann. Samarin afterward returned to Moscow, where in a café in February 1912 he talked with Pasternak about Marburg.[76] We cannot say whether this single encounter with Samarin convinced Pasternak to study in Marburg or whether additional testimony from others persuaded him. In any case, he set out in April for Germany. Pasternak, upon arrival, registered for at least two seminars: one with Hartmann on Leibniz and the other with Cohen on Kant.[77] Interestingly, there is no mention of Natorp in Pasternak's autobiographical sketch even though Cohen's name appears prominently in the account of his days in Marburg. This is particularly interesting in that Pasternak esteemed Cohen as an interpreter of Plato, but Natorp's own magnum opus *Platos Ideenlehre* from 1903 gets no mention! Whatever the case, Pasternak surely valued the Marburg School more for its work on Greek thought than for its Kantian interpretation of contemporary mathematical physics.

Whereas there is every reason for us to think that Pasternak took his philosophical studies quite seriously, he quickly and quietly abandoned them without further regret. By mid-July, he announced his plans to bid farewell to philosophy and to Germany. Pasternak returned to Moscow and finished his degree – likely without much enthusiasm – with a thesis, now lost, reportedly on Cohen's philosophy.[78] Undoubtedly while he engaged wholeheartedly in his philosophical studies, he was able to abandon them rather quickly without looking back and without an explicit acknowledgment of their influence on the formation of his own world outlook. One cannot help but notice that however deeply he may have penetrated into neo-Kantianism, he remained essentially uncommitted

---

[76] Dmitrieva (2007b), 187 gives the year of this conversation as 1911. However, Pasternak in his autobiographical sketch *Safe Conduct* certainly gives the impression that it occurred in 1912. Pasternak (1959), 33. Barnes also places this conversation in 1912. See Barnes (1989), 124.

[77] Pasternak alludes to this in Pasternak (1959), 44. For other classes, Pasternak may have had in Marburg, see Barnes (1989), 128. Pasternak was equipped with a recommendation from Gordon for his initial meeting with Hartmann. See Pasternak (1997), 134.

[78] Dmitrieva (2007b), 354–355. Tamara Dlugach reported on this too even earlier than Dmitrieva, but the former remarked that Pasternak "chose Leibniz's philosophy as the theme of his candidate's thesis." Dlugach (1991), 352. Presumably, Dlugach wishes us to understand that although Leibniz was the intended theme of his eventual thesis, he changed it upon his return to Moscow from Marburg.

to it or to any other philosophical standpoint. Like Dr. Zhivago, his most famous literary character, Pasternak remained throughout quite removed from all that happened around him.

8.6 From Neo-Kantianism to the Early Husserl

Lastly, we must turn to a rather peculiar figure in the story of neo-Kantianism. Henry Lanz (1886–1945) was born in Moscow to American parents Caroline and Ernest, his father working as the director of a rail mill. Lanz attended Moscow University, but upon completion of his studies he left, not unlike so many others we have seen, for further study in Germany.[79] If we accept the received chronology of Lanz's early life, he demonstrated a familiarity with Husserl's *Logical Investigations* even before this German interlude, becoming one of the first writings in Russian to set Husserl's early ideas into a historical context. Not surprisingly, Lanz focused on Husserl's critique of psychologism, seeing it as the culmination of the gradually ascending assault on the psychological interpretation of the theory of cognition.[80] But we cannot be surprised that Lanz related Husserl's endeavor vis-à-vis Plato's. Indeed, Husserl is seen as an even more thorough idealist in that whereas Plato held ideas to be "symbols" of the ultimate reality – ideas having the highest degree of being – ideas for Husserl were neither worldly nor merely products of human psychology. They are the object of the consciousness directed at them. The similarities in the respective conceptions are striking. For both Plato and Husserl, ideas are eternal and immutable, but the latter's conception does not carry the "poetic unscientificity" of Plato's doctrine of recollection.[81]

What is of most interest for us in this early work of Lanz's is his observation that whereas Husserl saw science as the discovery and statement of ideal, timeless essences, Cohen saw science as developing in scientific exploration itself. For the latter, science is constructed by "pure thinking" with the help of categories, a position that Husserl unequivocally rejected. But if Lanz found something appealing in Husserl's constructions, he could not see Husserl's early thought as conclusive. Husserl's *Investigations*, despite their status in the war against "re-

---

[79] Popova (2006), 185. Popova's data are based on genuine archival work, rendering previous speculation on Lanz's origins obsolete.
[80] Lanc (1909a), 418. On his German-language thesis, he gave his name as "Heinrich Lanz," and in his English-language writings he gave his name as "Henry Lanz."
[81] Lanc (1909a), 434.

ductionisms," were by no means themselves free of psychologism, precisely in his theory of truth and in his elucidation of the meaning of judgments. "If the sense of a judgment is an idea, i.e., a common concept among separate individual judgments, then how is this different from the psychological concept of a 'judgment in general' or of the 'given judgment in general'?"[82] Lanz admitted that Husserl was aware of such an objection, but in attempting a satisfactory reply he loses sight of the meaning of "idea" as something independent of individual mental acts. Vaguely formulated concepts are the result. Lanz was quick to add, though, that Husserl had merely left a large gap in his construction and not necessarily a mistake. Addressing this shortcoming constituted a task for the future. This problem, viz., the relation between the true, ideal sense of a judgment and our conscious processes, is the most difficult in all of the theory of cognition.

Lanz returned to the problem of the transcendence of "sense" vis-à-vis "real" things two months later in a follow-up article, where he posed the issue in even starker terms. "Just as it would be absurd to seek a musical melody in the *Critique of Pure Reason*, so would it be absurd to seek truth or sense in a real, psychic process."[83] The difficulty in resolving the problem led some to a "universal immanence," i.e., to an utter rejection of the transcendence of sense. It cannot be separated from consciousness, for it is nothing other than pure consciousness. The two are fundamentally identical. Therefore, the problem of transcendence dissolves, since there is no transcendence. This position, which found its original expression in Fichte and was developed, at least along its essential lines, by Wilhelm Schuppe (1836–1913) and Cohen, is the fundamental idea of immanent philosophy. Lanz devoted much of this second article to elucidating and criticizing this alleged identity of being and consciousness, referring frequently to the works of its exponents including Cohen.

Lanz saw the entirety of Marburg neo-Kantianism as an immanentism, which, as a consequence, rejected the very possibility of empirical psychology, a conclusion that he, Lanz, found to be unacceptable. "It is impossible simply to deny the reality of the psychological subject. We must precisely ascertain it and pass it into the world of objects as part of objective reality."[84] The psychological subject, the human psyche, belongs in and to the world of objects, and as such, it can be an object of, and therefore in, consciousness. As such an ob-

---

82 Lanc (1909a), 443. Lanz refers here to Husserl's "Prolegomena" in the *Logical Investigations*. See Husserl (1970), 180 (§46).
83 Lanc (1909b), 760.
84 Lanc (1909b), 788.

ject, consciousness cannot be in it. Thus, being distinct from its objects, which can include the psyche, an object cannot be the bearer of consciousness. Sounding much like the early Husserl, Lanz writes, "Consciousness has no need of any bearer except itself and cannot have any foundation outside itself."[85] It is immanent to the world, but it is impossible to conclude from this, as immanentism does, that the entire world reduces to consciousness. Lanz, to be sure, did write of the "I" without giving a precise determination of it, but he added that it produces itself spontaneously with the help of its own categories. If, however, we objectivize our individual "I," i.e., make it an object in the world of objects, we lose its "noetic" relationship to the world of essences, to truth.[86] Thus, the psychological individual loses the very possibility of cognizing. Lanz ends this portion of his essay – though he continues it by addressing what he sees as objections – by saying that the limitations of a journal article prevent further elaboration.

Completing his work at Moscow University in December 1910, Lanz then went for further study to Heidelberg. He must have been somewhat proficient in German, for he submitted and defended a thesis with Windelband already the following year. To be sure, the dissertation, *Das Problem der Gegenständlichkeit in der modernen Logik*, contained many pages that were translations from the two Russian-language articles we have just discussed. Moreover, the conclusions, such as they were, merely supported those offered in the earlier articles, though without ever hinting that their author hoped eventually to help bridge the seemingly unbridgeable chasm between the logical and the real. For this reason, we can largely forego Lanz's discussion of points previously presented. Overall, though, the thesis stands as an amazing achievement in such a short time.

Lanz's thesis focused on the conscious object, viz., its nature and its essence. This, he claimed, has been since Kant the central issue in philosophy. What is the basis of the object's sense of objectivity? Is it a property of each object that we cognize as "real," or is it a feature of objects that the mind ascribes or projects from itself onto objects?[87] That Lanz viewed this question as tantamount to that of how *a priori* synthetic judgments are possible is clear from

---

85 Lanc (1909b), 793.
86 Lanc wrote that he was using the term "noetic" in Husserl's sense as a condition of cognition and the term "noetic relationship" in the sense of a "cognitive relationship." Lanc (1909b), 794f.
87 Lanz (1912), 1. Framed in this way, we can see Lanz viewed Kant as centrally concerned with the problem, posed in the *Prolegomena* and then in the second edition of the first *Critique*, of how "judgments of experience" are possible. If our judgments were limited to first person perceptual reports, we would be limited to "judgments of perception." But we can and do make judgments of experience. How is that possible?

the fact that he held it to be the basis of Kant's Transcendental Deduction. Surprisingly, though, Lanz thought Fichte had a clearer understanding of the issue at stake than did Kant. Fichte understood that a transcendental deduction of a concept or an intuition is intended to show its objective validity *a priori* in experience and, thus, without that concept experience would be impossible. A transcendental deduction of a concept shows that the concept is a constitutive moment of the object, i.e., that it is a conveyer of an object's sense of objectivity. Fichte recognized the centrality of the "I think" in all of these deliberations and that the I is the basis of every objectivity. This realization, Lanz opined, forms a turning point in the history of philosophy, taking it in a completely new direction, whereas previously the concept of substance, understood as spirit, occupied the pivot of philosophy.

Whether derived ultimately from Husserl's works as they were available in 1911/1912 or not, Lanz interpreted Kant as espousing the view not only that cognition (and therefore consciousness) is a cognition of something, but that objectivity is always vis-à-vis consciousness. "All of reality, the whole of objectivity and reality, is exhausted through consciousness."[88] There can be no objectivity apart from thinking. Objective transcendence is a contradiction. But, the critic will object, "What of the Kantian thing in itself? Is it not an objective transcendence, an objectivity that cannot be thought?" Lanz's reply was that this objection hides a misunderstanding. The crux of the matter concerns not an object without thought, but a thought without an object, a thought coupled with the subjective elements of the understanding but without the conditions of sensibility. It is an object of the understanding without the slightest manifold of sensibility. The thing in itself is an object "in itself," an "object as such." "The pure concept of objectivity, or objectivity itself, apart from the conditions of sensible intuition is possible. It is simply the thing in itself, i.e., an object independent of sensibility given through pure thinking."[89] Although this concept remains problematic, it is not self-contradictory.[90] Lanz saw Kant unfortunately adhering to this transcendental ground.

Whereas many have viewed Kant as proceeding away from an emphasis on faculty psychology in the first edition of the first *Critique* to a more thorough transcendentalism in the *Prolegomena*, Lanz found that in the latter Kant quite frequently affirmed the world to be a product of the psychological organization of our individual minds. Kant was not content merely with a transcendental jus-

---

88 Lanz (1912), 13.
89 Lanz (1912), 28.
90 Lanz's use of the term "problematic" follows that of Kant in the first *Critique*. See Kant (1997), 350 (A254/B310).

tification of objective cognition but tried to explain the cognitive process in terms of the mind's subjective mechanism. In doing so, the cognitive object becomes a function of a metaphysical substance. Kant, thereby, becomes a founder of psychologism.

Make no mistake, Lanz is not opposed to empirical psychology. Lanz, unlike Cohen, had no wish to deny the reality of the psychological subject. But in studying it, the psychological subject becomes an object not fundamentally unlike the other objects of our everyday reality. Thus, it has no logical relationship to the world of truth.[91] Psychologism is this inadmissible comingling of the two worlds. The subject of the determination of truth is not the empirical consciousness, but the transcendental consciousness. The latter has as its object not ephemeral objects, but atemporal cognitive and moral truths. In this scheme, the logical and the psychological lie in quite different spheres. Both Rickert and Husserl attempt to find a way to bridge the two spheres, but in Lanz's estimation all avenues are closed: "Husserl, like Rickert, naturally tries to reconstruct the way back, but we cannot help but see all of their attempts as nothing other than failed compromises with the immanent tendencies of epistemology."[92] One such tendency, of course, in Lanz's eyes, was Cohen's Parmenidian identification of being and thinking, a position that Lanz found to be an unsatisfactory option. And if Rickert and Husserl established two apparently irreconcilable worlds, the quest still remained to bridge these worlds. Lanz offered no solution, but a turn from Kant to Fichte offered the potential of an overlooked approach to the problem. In this way, the rallying cry of the neo-Kantian movement "Back to Kant" became for Lanz the cry "Back to Fichte."

Having completed his studies in Heidelberg, Lanz traveled not to Göttingen to hear Husserl, but to Marburg to hear Cohen, where he stayed for the next three years. There, he briefly encountered Pasternak and others. It was while in Marburg that he wrote an article on the now largely forgotten positivist Richard Avenarius, which is replete with scholarly references to, among others, Cohen and Cassirer, and an article on immortality, which is totally bereft of any such traditional scholarly *accoutrements*. A third article, which appeared in the first of the two issues of *Logos* in 1914, was devoted to Fichte on the one-hundredth anniversary of his death. It demonstrates Lanz's interest in seeking a solution to his cen-

---

[91] Cf. Husserl (2014), 5. Psychology "is a science of *facts*, of 'matters of fact' in David Hume's sense. ... In contrast to this, *pure or transcendental phenomenology* will *be established, not as a science of facts, but instead as a science of essences* (as an '*eidetic*' science), a science that aims exclusively at securing 'knowledge of essences' and *no 'facts'* at all."
[92] Lanz (1912), 149.

tral philosophical concern through a return to Fichte.⁹³ Lanz's recourse to Fichte is most evident in another sequence of articles in *Voprosy filosofii*, particularly the second of the two entitled "Being and Knowledge in Fichte's Philosophy." He unabashedly proclaimed there that both Rickert and Husserl, as representatives of a "transcendent" direction in philosophy, rest on Fichte as does the opposing direction of immanentism. Lanz proffered that Fichte had already previously synthesized realism and idealism to produce a unity. "Therefore, Fichte's philosophy, standing above them both, can look at the truth in each while at the same time recognizing the inadequacy of both."⁹⁴ Both immanentism and transcendentism are constituent parts or two sides of an absolute relation.

Lanz held that although the Marburg School, particularly as represented in the thought of Cohen, may not explicitly depend on Fichte, Marburg served as the philosophical continuation of many of Fichte's ideas. Whether disingenuous or not, Marburg ignored its connection to the speculative tradition that immediately succeeded Kant. Nonetheless, Lanz claimed, many of the questions the Marburg School proposed are indubitably related to those raised almost a century earlier. Cohen formulated his problem as the relation of cognition to natural science, but this was but a manifestation of the broader problem common to Marburg as well as to Fichte concerning the relation of a principle to appearance.⁹⁵

By the time his third article and those in *Voprosy filosofii* had appeared in print, Lanz had returned to Moscow, where he began teaching in a secondary school and a music conservatory. The events that relatively quickly followed must have caught Lanz by surprise and made him reevaluate his options. Following the Bolshevik Revolution, Lanz left Russia, making it eventually to the United States and settled in Palo Alto, California, where he was instrumental in establishing the Slavic Languages and Literature Department at Stanford University, although his principal interest remained philosophy. He did publish several articles during his U. S. residence, but they provided little inkling of his earlier neo-Fichtean leanings. In any case, Lanz's orientation and background was quite out of sympathy with the American philosophical scene at the

---

93 All three articles appeared in *Logos*. Popova concludes from this that "Lanz's collaboration with this journal demonstrates the philosopher's desire to be involved in the neo-Kantian tradition." Popova (2009), 211. This conclusion assumes: (1) that *Logos* was itself a manifestation of the neo-Kantian tradition; (2) that Lanz saw it that way; and (3) that he saw his involvement was motivated by a desire to contribute to the neo-Kantian movement. By no means are all three assumptions beyond questioning.
94 Lanc (1914), 228.
95 Lanc (1914), 274.

time.⁹⁶ Sadly, he died unexpectedly in November 1945. His philosophical potential and original project remained unrealized, though we should add that he did eventually acquire an appointment in philosophy at Stanford.

---

96 Lanz initiated an invitation to N. O. Losskij to come to Stanford as a visiting philosophy professor for a single term of ten weeks in 1933. We can, thus, take for granted that Lanz and Losskij had much to discuss during Losskij's relatively short stay in northern California. Vanchugov (2016), 192–193.

Chapter 9
One-Time Neo-Kantians Who Stayed: Sakketti, Two Rubinshtejns, Kagan

However sad may be the fate of so many of those associated with Russian neo-Kantianism, not all of them ended their years virtually pennyless abroad. Some opted not to flee but chose to tie their personal fate to that of their country and ended up pennyless in their homeland. In this chapter, we look, in particular, at several individuals, two of whom, though having studied at neo-Kantian centers in Germany, survived the horrific Stalinist years and emerged, though battered, with some sense of dignity in the eyes of history. The caveat for our purposes here is that they largely abandoned their early focus on philosophy and are remembered today – insofar as they are remembered – for their contributions in other fields. Before turning, however, to the early works of these two central figures, let us briefly look at the career of one individual who bore a distinctively non-Slavic family name.

9.1 The Marburgian with Italian Lineage

Antonio Sacchetti, the grandfather of Aleksandr L. Sakketti (1881–1966), was arrested in Rome in the late 1840s in connection with his alleged support for Garibaldi. He was released at the request of the influential Princess Zinaida A. Volkonskaja, then resident in Italy, and took up a position, undoubtedly through Volkonskaja's connections, with the opera house in Odessa. His son Liberio (Liveri) (1852–1916) studied at the St. Petersburg Conservatory with Rimsky-Karsakov among others. After graduation, Liberio held a position at the Conservatory teaching aesthetics and music history, writing several well-regarded texts. His wife and Aleksandr's mother, also a graduate of the St. Petersburg Conservatory, was a singer with the Mariinskij Opera House in St. Petersburg.[1]

Aleksandr attended secondary school in the capital and then studied law at the University, graduating in 1908. His first publication, already in May of that year, was a positive review of Saval'skij's book on Cohen's philosophy of law. While at St. Petersburg University, he participated in Lappo-Danilevskij's seminars. In the fall of 1908, he went to Germany for further study at both Heidelberg and Marburg. On his return to Russia, he worked along with Lappo-Danilevskij

---

[1] The basis of this biographical information is Ladyzhenskij (1966), 154.

on Kant. Sakketti penned no theoretical works of the sort that would allow us unequivocally to establish his allegiance to or departure from neo-Kantianism. His affinity to it rests largely on his associations with it. He did reportedly compose a tract on Cohen's philosophy upon his German professor's death, but this, if it survives, has remained unpublished. He received an appointment as a professor at the University in 1918, but the Bolsheviks soon closed the university law departments throughout Russia. In quick order, he was offered and accepted a position at the new State Workers' & Peasants' University in Kostroma. While there in 1920/21 he conducted "seminars" on Kant, and he proposed under University auspices the publication of a work provisionally entitled "Hermann Cohen's Philosophy" for student use. For unknown reasons, this plan was never realized.[2] In any case, the University underwent various reincarnations but eventually suffered severe financial difficulties, and Sakketti left for Moscow. From 1924, he was an active member of the philosophical section of the State Academy for the Study of the Arts (GAKhN) and from 1926 a corresponding member. It is largely on the basis of his comments, such as they are, and his reports during this time at GAKhN that we can affirm an abiding interest in, if not a continuing commitment to, a Marburg-inspired cultural philosophy.

We cannot with complete confidence ascribe to Sakketti the views of Cohen and Natorp expounded in his lengthy report entitled "Cohen's Theory of the Musical Form" on 4 December 1924. But clearly, in the absence of any disavowal of them, of which there is none in this report, Sakketti valued Marburg neo-Kantianism as at least a valuable approach to philosophical, particularly aesthetic, problems. His report also serves as evidence of his willingness to disseminate the Marburg position at a comparatively late date.

Sakketti held Cohen's neo-Kantianism to be both a deepening and a surmounting of Kantianism in the direction of a pure transcendentalism. Philosophy itself is not the cognition of things or substances as they really are, but of the facts of cultural consciousness. "The problem of transcendental philosophy is the unity of experience, where experience is taken in the sense of the manifold content of cultural consciousness. Its method is related to that of exact mathematical natural science and consists of conducting the mentioned manifold to an aggregate of harmonious assumptions, *hypotheses, basic presuppositions, theses.*"[3] In other words, cognition is itself an activity, but an activity that never attains completion.

---

2 Savenko (2002), 238.
3 Sakketti (2017), 413.

Largely summarizing Cohen's view of science and its cognition as found in the 1902 *Logik*, Sakketti then turned in his report to Cohen's 1904 *Ethik*, finding in it an inquiry into the basic presuppositions of human activities and interactions, such as freedom and autonomy. However, Sakketti's central concern is to demonstrate the unity of Cohen's philosophical system with an emphasis on situating aesthetics with its varied forms within it. As Sakketti wrote, "The variety of cultural directions poses the problem of *the unity of cultural consciousness* as a systematic consciousness, i.e., the unity of a system."[4] Both theoretical and practical philosophy affirm the distinction between what is and what should be. However, Sakketti saw Cohen as posing the question in the opposite direction, so to speak. Cohen asked, in effect, how the unity of these varied spheres is preserved. The sought unity is created by art, which is nothing other than the embodiment of an abstract or spiritual idea in a sensible form. Before engaging in a systematic exposition of aesthetics, we must first turn to the philosophical problem of revealing the unity of natural science and ethics as preconditions for aesthetic consciousness. This line of thought reveals the logical and temporal order of Kant's and Cohen's three systematic works.

Sakketti developed this outline of Cohen's presentation in considerable detail, referring also to a variety of other authors. In the discussion following the report, Sakketti remarked that Cohen's central problem was that of the unity of cultural consciousness, that his concept of pure feeling was close to that of Hegel's "absolute spirit," which unites the objective and the subjective, nature and morality.[5]

Sakketti's activities in subsequent years have received little attention. He, presumably, continued during the years following the closure of GAKhN to serve in a consulting capacity with the People's Commissariat of Justice, a position he first acquired in 1923. From 1942, he taught Latin in the law department of Moscow University and wrote on Roman law. He also wrote on, and translated a number of works by, Hugo Grotius. Of course, he had completely distanced him-

---

4 Sakketti (2017), 417.

5 Sakketti (2017), 445. Sakketti did present additional reports on topics in aesthetics during his association with GAKhN. Unfortunately, the published record fails to reveal much. For example, in the discussion following Aleksej Losev's report in late-1926 on Cassirer's *Philosophy of Symbolic Forms*, Sakketti merely remarked that he was surprised by Cassirer's position, that he saw Cassirer approaching Schelling's transcendental idealism. Sakketti (2017), 735. Of course, at this time only the first two volumes of Cassirer's three-volume work had been published. Still, it is surprising that Sakketti likened Cassirer to Schelling rather than Hegel. For Cassirer himself invoked Hegel, not Schelling, in the Preface to the second volume of his work, that dealing with *Mythical Thought*. Sakketti may have been diverted by associating the subject matter of that second volume, myth, with Schelling's own 1842 *Philosophie der Mythologie*.

self from philosophy and neo-Kantianism in particular. He died virtually unnoticed at the age of 84 in Moscow.

9.2 From Marburg to Soviet Psychology

Like Sakketti, Sergej Rubinshtejn (Rubinstein) (1899–1960) survived the Stalinist years in the Soviet Union and without internment in a Siberian labor camp. Unlike Sakketti, Rubinshtejn was acclaimed after his death for intellectual contributions to his chosen field, though these contributions were not to philosophy. Sergej Rubinshtejn was born into a Jewish family in Odessa, where his father, an important lawyer, had a considerable practice. After his secondary school education, which he completed in 1908, he spent two semesters in Freiburg and then went to Marburg to hear, among others, Cohen and Natorp.[6] He passed the formal state final examination in June 1913 and defended his thesis submitted to Marburg in July 1914, spending much of the interval in Berlin. It was in the German capital that he apparently became friendly with Ernst Cassirer, although they may have known each other previously from Marburg, which Cassirer, fifteen years older than Rubinshtejn, would visit on various occasions. Rubinshtejn left Germany in late July 1914, taught at a secondary school and then was appointed *privat-docent* in philosophy at the university in Odessa. With the death of Nikolaj N. Lange (1858–1921), Rubinshtejn was appointed to Lange's professorship at Odessa University.

Sergej Rubinshtejn[7] had a rather illustrious career, being elected to the Academy of Science, the first Soviet psychologist to be so honored, and received the Stalin Prize. However, in 1947, he was accused of "cosmopolitanism," i.e., antipatriotism, for underestimating the contributions of Russian science and culture through his attention to foreign scientists. As a result, he was stripped of his po-

---

[6] These bare biographical facts are largely derived from Rubinshtejn's Marburg thesis. His later reflections on his attitudes at the time cannot be taken verbatim. He wrote, "The sources of my philosophical development were the struggle of early-twentieth-century decadent, fragmenting idealism with Marxism. ... Departure for Freiburg. Dissatisfaction. Transfer to Marburg ... Here the same dissatisfaction and my own philosophical conception." S. Rubinshtejn (1989), 414. Based on his surviving writings from this time, there is no sense of "dissatisfaction" with either Marburg in general or with the neo-Kantianism taught there. As we shall see, his attitude appears to have been just the opposite.

[7] Since this chapter deals with two individuals, both of whom have the same family name possibly leading to confusion, they will be distinguished when necessary by the addition of their first name.

sition, but he was gradually restored to good graces after Stalin's death and hailed afterward for his psychological work.[8]

Rubinshtejn's slim German-language thesis, *Eine Studie zum Problem der Methode*, is an odd tract. Employing throughout the terminology of Hegel's logic, it never directly cites any of Kant's writings and mentions Kant by name only sparingly and in passing. Rubinshtejn proceeded from the claim that all "true" philosophy seeks cognition of what exists. He understood this, without substantiation, to mean that true philosophical cognition is the unity of being and thought, that the historic quest for such unity has always been presupposed but unacknowledged until the advent of transcendentalism. The logic of this "ism," i.e., transcendental logic, seeks to be the logic of what is, the logic of being. Thus, transcendentalism is sharply opposed to those subjectivisms, such as psychologism, and all of subjective idealism, which set thinking as the subject against objective being. This repudiation of subjectivism of any sort must not be taken as a return to the ontology of earlier times. The formal objectivity of the "old" ontology was only a pretense; its apparent objectivity was just that, a deception. It took being as something absolute and above logic.

Sergej Rubinshtejn, however, sought in his thesis to show not only the defects in the ontology of pre-transcendental philosophy but also and particularly in the conceptual scheme of Hegel. The dualism of thinking and being must be overcome, but it cannot be done through some absolute identification of the two, through speculation in a system of identity. Hegel's absolute rationalism too falls into a dualism. The form of "other" being with which it deals restores a dualism. A rationalism that prioritizes the "concept" to be true being, that takes the essence of being to be the ground of being, becomes a dualism.

As mentioned, Kant's thought does not appear prominently in Sergej Rubinshtejn's dissertation, but this is not to say that it is entirely absent. In addition to his general critique of any dualistic philosophy, of which Kant's is one example, Rubinshtejn charged Kant with resorting repeatedly to techniques that were incapable of realizing his general task and instead proceeded down roads that led him astray. "However, Kantianism depicts correctly if it establishes the relationship between logic and science to be that of the ground and the grounded, conferring to logic a fundamental character vis-à-vis the scientific disciplines." [9] Post-Kantian philosophy merely reversed the previously conceived relation between thinking and being but without intrinsically transforming it. Thus, the same systematic contradictions of old were restored, viz., the rationalistic system

---

8 Dmitrieva (2016c), 11.
9 S. Rubinstein (1914), 6 f.

of identity with its divisive dualism was now placed within a supposedly transcendental philosophy. Post-Kantian idealism, in short, by no means overcame the problems latent within Critical Philosophy. The absolute rationalism of post-Kantian philosophy falls back into dualism.[10]

Sergej Rubinshtejn had intended his thesis to be merely the first of a two-part work, the second part of which would focus on "dualistic rationalism." Although it was never written, Rubinshtejn included a "sketch" of this proposed continuation as a supplement to his thesis. The absence of a substantial discussion of Kant's thought in the body of the text was mitigated in part owing to Rubinshtejn's contention that Kant – and Plato – were representatives of dualistic rationalism. In this conception, the fundamental concepts of thinking appear as mere principles that are taken to be presuppositions or conditions of being, but not the foundations or grounds of being. Although unstated either in the body of his work or in his "sketch" of the proposed continuation of the study, Rubinshtejn had in mind through his criticism of post-Kantian philosophy to justify Cohen's systematic philosophy. Admittedly, neither Cohen nor Natorp is ever mentioned so much as once, but Cohen's efforts were clear enough to Rubinshtejn. Hegelian idealism erred by taking the wrong path in its attempt to develop Kantianism. It unsuccessfully hoped to correct the latter's dualism, but it only succeeded in ultimately producing a new version of dualism disguised as an absolute rationalism. For Rubinshtejn, Kant's attempt to ground the natural sciences was worthy of continuance, and this was what the Marburg approach provided.[11]

If Sergej Rubinshtejn had written nothing else apart from his German thesis, he would merit no more than a brief mention in a footnote. However, in the immediate years following his return to Odessa, he wrote several items that demonstrate a continued interest in, if not a wholesale commitment to, Cohen's neo-Kantianism. In particular, in the years 1917–1918, Rubinshtejn wrote, for unknown reasons, a long summary of Cohen's mature system, "O filosofskoj sis-

---

10 S. Rubinstein (1914), 65.
11 Given, on the one hand, the thesis's terminological abstruseness and, on the other, its provenance, it is quite impossible to give any credence to the claim in Abul'khanova-Slavskaja (1989), 17–18 that the thesis was written by a "socially-minded scholar" who opposed the basic doctrine of neo-Kantianism. A far more judicious view is that Rubinshtejn in this early work presented "the basic direction of his further scholarly investigations – the problem of philosophical method as the methodology of the concrete sciences." Levchenko (2012), 111. Moreover, Sieg has observed, "Rubinstein in his thesis 'Eine Studie zum Problem der Methode,' set Marburg neo-Kantianism against Hegel's philosophy. ... In a letter to the faculty dated May 16, 1913, Cohen ... praised, above all, the 'clarity and terseness' of Rubinstein's writing." Sieg (1994), 383.

teme G. Kogena" ["The Philosophical System of Cohen"]. Possibly it was in conjunction with a proposal at this time from a Russian publisher to Rubinshtejn for a translated collection of short religio-philosophical works by Cohen.[12] Rubinshtejn may have intended this piece to serve as the translator's introduction to the collection. Whatever the case, nothing came of the project. However, Rubinshtejn's substantial summary of Cohen's thought is itself a testimony to his high regard for the latter. We need not go into it in any detail, for the Marburg professor's works are presented straightforwardly and without criticism, much as we would expect of an introductory essay to a translated collection. Suffice to say that S. Rubinshtejn also sought to situate neo-Kantianism in its historical context. Whether a gross exaggeration or not, Kant, in this interpretation, recognized modern science, headed by Galileo, as introducing the idea that reason can grasp only what it, by its own design, produces.[13] Only what thought itself constructs can be rational and understandable. This, in Rubinshtejn's interpretation of Cohen, entails the rejection of any givenness in the cognitive process. Kantianism true to its own principles must rid itself of givenness, regardless of whether it be taken as the givenness of sensibility or of the thing in itself. However, it must also and above all rid itself of the static givenness of the categories. Givenness is an infinite cognitive process, and, thus, the Kantian conception of the thing in itself is the ideal of objectivity. It is an infinite task set at the end of the infinite cognitive process.

Sergej Rubinshtejn accepted Cohen's own self-interpretation of his independence from post-Kantian idealism. Indeed, the former too at this time saw the error of that idealism in its abandonment of the intimate link between philosophy, properly understood, and natural science and its failure to employ the transcendental method. "According to the general scheme of the transcendental method, Cohen starts from a certain givenness, and transforming it into a problem, he seeks the necessary and sufficient presuppositions for its grounding. We ask, however, what this givenness is from which logic must start. Cohen's answer is science."[14] Employing the terminology of the day, the "logic" of cognition is the logic of science, and the most precise formulation of science is mathematical physics. "Therefore if we sunder the connection between logic and science and thereby break the unity of cognition, we will have cognition neither in logic nor

---

12 This surmisal is based on a letter from Cassirer to Natorp dated late June 1918. S. Rubinshtejn in Odessa had written to Cassirer concerning this, but the former's letter has not survived. Dmitrieva (2016b), 128.
13 S. Rubinshtejn (2003), 430.
14 S. Rubinshtejn (2003), 440.

in science."[15] Cohen holds this unity to be the fundamental condition of cognition, and it is *"das Ewige"* in Kant.

Sergej Rubinshtejn did not limit himself merely to a summary of Cohen's "logic," but also briefly discussed Cohen's ethics as an integral part of a total philosophical system. Without an ethical dimension, the human person does not exist. But, contrary to how Kant has traditionally been understood, the moral law is not something that the ethical person enunciates. It is not a product of or legislated by the "self"; it is the content of the ethical person. Contrary to Kant, the subject is not given before and in addition to the moral law. For in this picture, the "self" of the ethical subject would be logically independent of the law, and the autonomy of the ethical subject would thereby destroy itself. In Rubinshtejn's depiction of Cohen's ethics, "moral legislation is not the product or the manifestation of the self. It, for the first time, determines the content of the latter."[16] Thus, the ethical subject does not exist at all until it manifests itself. However, even such an expression is imperfect. Rather, the ethical subject generates itself in moral actions. This contentious claim will be radically modified and adapted in Rubinshtejn's Soviet-era psychology.

In the early 1920s, Sergej Rubinshtejn wrote several pieces on scientific methodology that most likely were written in connection with a planned work provisionally entitled "O zakonakh logiki i osnovakh teorii znanija" ["On the Laws of Logic and the Foundations of the Theory of Knowledge"]. The first of these, "Nauka i dejstvitel'nost'" ["Science and Reality"] states that the task before it was "to elucidate the structure of exact knowledge (mathematics and natural science) and its relation to reality."[17] The goal of the first chapter of this planned treatise was to establish what science is and to clarify the objectivism of scientific knowledge in relation to the subjectivity of consciousness. Such a bald statement as this by Rubinshtejn could not allow us to infer his allegiance to any particular philosophical school. However, he continued by affirming that the world of physics, as a science, is not given to us, as the positivists hold, but is constructed. This construction of the content of science is not the work of a demiurge; the content is constructed by knowledge itself. Rubinshtejn understood this to be a rigorously objectivistic conception of knowledge, which avoids the original subjectivism of Kantian idealism. He defined science as a system in which each element is determined by its relation to the other elements of the system. We find in this a distinct echo of Cohen, particularly the Cohen that Rubinshtejn

---

15 S. Rubinshtejn (2003), 441.
16 S. Rubinshtejn (2003), 444.
17 S. Rubinshtejn (1989), 337. The planned title and the title of this first piece were apparently quite provisional.

had depicted in his summary of the Marburg philosopher's thought. Rubinshtejn – along with Cohen – had proposed his own "Copernican revolution," whereby it is not our point of view on the content that must change, but the very content itself, as Einstein, in Rubinshtejn's reading, had proposed.

We could continue our discussion of Sergej Rubinshtejn's reflections in "Science and Reality." However, since he himself failed to publish them, there is always the possibility that they do not genuinely and completely reflect his considered thought at the time. Although they appear to be written from a strictly Cohen-Marburg viewpoint, there is the possibility that he merely was "toying" with those ideas, rather than that they reflect his personal adherence to them. Nonetheless, he did publish in 1922 a short article "Princip tvorcheskoj samodejatel'nosti. K filosofskim osnovam sovremennoj pedagogiki." ["The Principle of Creative Self-Activity. The Philosophical Foundations of Contemporary Pedagogy"] in Odessa. It, evidently, was intended to be a fragment of a second chapter entitled "The Idea of Knowledge" for the planned book.[18] The article contains no references to Marx, Cohen, or Natorp. However, there is nothing in it that prevents us from seeing its author as deeply indebted to Marburg neo-Kantianism.

Sergej Rubinshtejn began his article by distinguishing his own position from Kant's through invoking the latter's definition of transcendental idealism in the "Transcendental Dialectic," where he wrote that all of existence has no ground in itself "outside our thoughts."[19] Thus, the object of knowledge is grounded in thought, but Rubinshtejn also added that it is limited by the bounds of sensory experience. Kant's caveat was important to Rubinshtejn, for it meant that being is the given of such experience, which is given logically prior to and independently of the synthesis of the understanding. Kant held a negative conception of objectivity, whereby being is defined in terms of its independence from cognition. Such is a dogmatic objectivism. However, Rubinshtejn also inferred from this that any unification of the given manifold reflects an element of construction on the part of the cognition and cannot be attributed to reality. From this, we can come to a positive conception of objectivity, not one that merely states an independence from something else. Objectivity, instead, must be sought in the completion of its content. That is, the objectivity of contents must be determined by the interrelations of those contents. "Not what is given is objective, but what is complete."[20] Rubinshtejn's conception of objectivity, then, depends on

---

18 S. Rubinshtejn (1989), 438.
19 S. Rubinshtejn (1986), 103. Cf. Kant (1997), 511 (A490–491/B518–519).
20 S. Rubinshtejn (1986), 104.

an analysis of completion. We, in turn, can ask of him just how we are to understand, though, the term "complete."[21]

Sergej Rubinshtejn stressed that every objectivism based on pure receptivity leads both to dogmatism and concomitantly to subjective idealism.[22] In effect, such receptivity as found in the original form of Kantian Criticism has before consciousness only appearances, which have no grounded existence in themselves. Rubinshtejn maintained that the recognition of constructivity in the acquisition of knowledge does not lead to subjectivism. Yet, such constructivity grants full recognition to the creative genius of the scientific spirit without lapsing into any metaphysical dogmatism. Rubinshtejn repeated many of the themes we have previously seen concerning the constructivism of science, whereby the scientific endeavor itself creates both the object and the subject, themes associated with Cohen. By not accepting that the subject of cognition is itself created through its activity in the cognizing process, that the subject is given beforehand and independently of its actions, Kant's conception, albeit transcendental, returns us to Hume's empirical conception.[23] We see in this that although Rubinshtejn provided no specifics, he repeated the central theme of Cohen's neo-Kantianism at the expense of a static interpretation of Kant's transcendental idealism.

Nothing more was heard from Sergej Rubinshtejn – whether this was a conscious decision or due to a the lack of opportunity – until his emergence in 1934 as a pioneering Marxist psychologist. His "Problemy psikhologii v trudakh K. Marksa" ["Problems of Psychology in the Works of Karl Marx"] from that year was pioneering not only in the sense that he dared to write on a topic that could prove perilous for him personally owing to its promotion of a complex dialectical relationship between consciousness, human activity, and the objective world. As a Marxist, though, Rubinshtejn leaves our story of Russian neo-Kantianism. Still, his clever adaptation in his Marxist maturity of themes embraced during his "idealistic" youth remains striking. Whether a result of pressure from external circumstances or of an inner intellectual development – probably

---

21 Unfortunately, Sergej Rubinshtejn provided no reply to this rather obvious request. To speak of "completion" is to have an idea in advance of the finished product. But then the question turns to the source of that idea. The danger, so to speak, is that we move closer to the Baden conception of value.

22 Implicit here is a rejection of the Leninist "copy theory of knowledge" that would become philosophical dogma during the Soviet era. Rubinshtejn gave not so much as a hint that he was familiar in 1922 with the naïve realism Lenin presented in his 1909 polemical *Materialism and Empirio-Criticism*.

23 S. Rubinshtejn (1986), 106.

both – Rubinshtejn's Cohen-inspired philosophy melted into a subtle Marxist psychology deftly utilizing ideas drawn from Marx's early writings that had first become available only a short time earlier.

9.3 In the Shadow of Baden

Moisej (Moses) Rubinshtejn (1878–1953), born into a merchant family in a village in the Trans-Baikal region of Russia, graduated from local schools and entered Kazan University in 1899.[24] Attending that university for merely one year, he, accompanied by his wife, went on an internship to Berlin and then to Freiburg in order to study medicine. He quickly became fascinated with the philosophy taught there and abandoned his medical studies in 1901 for philosophy, ultimately defending a thesis in 1905 entitled *Die logischen Grundlagen des Hegelschen Systems und das Ende der Geschichte*, which was published the following year.

The thesis itself at the start presents a brief, but splendid, overview of Hegel's thought with an emphasis on Hegel's attempt to distinguish his system from Schelling's by showing the world's rationality. Moisej Rubinshtejn's concern, however, was with Hegel's philosophy of history, particularly its claim that world history has come to an end with the Germanic world, in which the consciousness of freedom has been realized. This was quite an audacious contention, as he fully realized, and where he located the fundamental inadequacy of Hegel's philosophy most keenly.[25]

Hegel adamantly proclaimed that no part of his system, no matter how small it may be, can be removed without disturbing the whole. For Moisej Rubinshtejn, this, in particular, must apply to Hegel's philosophy of history. "If we conclude that there is no room for history in his system, we hit the nerve of his thought and throw overboard the entire system."[26] Rubinshtejn sought to show that Hegel's system cannot accommodate an end to history. He countered that Hegel's scheme necessitated an infinite development or enrichment of value (*Wertbereicherung*), a conception that stood in direct contradiction with Hegel's idea of an end to history. There is no justification in Hegel's conception to single out a particular temporal moment, in particular Hegel's own time, as absolute. The goal

---

24 Moisej Rubinshtejn gives Irkutsk as his town on the title page of his thesis.
25 M. Rubinstein (1906), 16; M. Rubinshtejn (1905), 706, 764. The German thesis and the Russian-language article are essentially the same with differences in the respective introductory materials. Although one appears, for the most part, to be a translation of the other, it is impossible to be certain which was the original.
26 M. Rubinstein (1906), 21; M. Rubinshtejn (1905), 710–711.

of history is the realization of freedom, but Hegel's scheme provided no specific means to determine when this end had been attained.[27] Hegel wished to have it both ways. His system was built on the possibility of an infinite development of the absolute spirit, but it also allowed for the termination of that development as a necessary consequence. This, Rubinshtejn held, must be seen as an inconsistency in Hegel's philosophy.

We can readily see Moisej Rubinshtejn's concern with the idea of infinite progress as stemming from his immersion in German neo-Kantianism. Additionally, we can also see the influence of Baden neo-Kantianism in Rubinshtejn's conclusion, wherein he stated that Hegel proceeded down a dead end with his understanding of the end of history. The construction of Hegel's system with its conception of historical development had to be from the viewpoint of the absolute. However, that construction made no use of the concept of moral obligation. There was no accommodation in it for the free, autonomous human individual. Only a human point of view, not the absolute, can speak of duty and the good, of endless striving.[28] There is no allowance in Hegel's system for morality.

After the defense of his thesis, Moisej Rubinshtejn spent some time at the universities in Berlin, Dresden, and Heidelberg. He returned to Russia in 1907 and lectured at the Higher Women's Courses with hardly any cessation in his publishing activities. Already in the first issue of *Voprosy filosofii* in 1907, Rubinshtejn published a long article entitled "Heinrich Rickert. A Sketch of His Epistemological Idealism" and thereby showed his deep esteem for his former teacher and *Doktorvater*. In it, Rubinshtejn took note of Rickert's "anthropomorphic" viewpoint on cognition, which we saw he contrasted to Hegel's "absolute" viewpoint. Rickert, in Rubinshtejn's understanding, saw knowledge as a human accomplishment and thus as limited by our human capabilities and resources. Such is, Rubinshtejn remarked, the only possible and fruitful point of view. That Rickert's philosophy stemmed from an anthropomorphic viewpoint did not mean or imply any psychologism. The expression "consciousness in general" in epistemological idealism is not to be understood as a generic consciousness,

---

27 M. Rubinstein (1906), 65; M. Rubinshtejn (1905), 756.
28 M. Rubinshtejn (1905), 764. Regrettably, the copy of Moisej Rubinshtejn's German-language thesis accessible to me lacked the "*Schluss*" that was stated in the "*Inhaltsverzeichnis*" but which I found in the Russian-language version. In the presumably official statement of his opinion of Rubinshtejn's work to the University, Rickert determined that his student was very familiar with Hegel's system and with German Idealism in general. However, he also stated that Rubinshtejn's arguments here and there were contestable, and in particular "some elements are emphasized too one-sidedly. As a result, it is possible that Hegel could escape the indicated absurd consequence." Rickert (1905), 355.

i.e., the conception of consciousness that is common to all individual cases. For such a consciousness still refers to a real consciousness minus its specific features.

In his discussion of consciousness, Moisej Rubinshtejn saw Rickert – and presumably himself – as working with a different conception of it than that offered by Struve and Berdjaev in their early neo-Kantian phase. Rubinshtejn also objected to Struve's rejection of the normative character of logic. Were Struve correct, epistemological idealism built on that understanding of logic would be a "complete fiasco."[29] As we saw in Chapter 4, Struve rejected the Baden view of logic, arguing that if the laws of logic were norms we could think contrary to them just as we can act contrary to moral laws, making them normative laws. However, we cannot think A and not-A. Errors can occur in the content of our judgments, but not in their form. Struve, then, held that logical laws are not unlike laws of nature; they both have the character of natural necessity.

Moisej Rubinshtejn disagreed with Struve's conclusion and disagreed that logical laws operate with natural necessity much as do natural laws. Certainly, Rubinshtejn continued, it is impossible to dispute with the latter. For example, one cannot truly wish to violate conservation laws. However, "the situation is different with the laws of logic. They tell us only, if you want the truth, you must be guided by such and such rules, but whether you follow them is a matter of your logical conscience."[30] Rubinshtejn clearly followed Rickert in this matter.[31] Unlike with representations, judgments can only be made with the conditional "should." The laws of logic that are employed in making judgments are imperatives. As a normative anti-psychologistic neo-Kantian, Rubinshtejn charged the realist anti-psychologistic neo-Kantian Struve with overlooking that people can stubbornly resist recognizing truths even when presented with overwhelming evidence. A person must be open to receiving a truth and have a will or wish to do so in order to acknowledge it as a truth.

Moisej Rubinshtejn also took exception with Struve's view that since truth is fundamentally "transsubjective," the concept of value, being a relation of a subject to an object, has no place within epistemological idealism. Rubinshtejn had no issue with the "transsubjectivity" of truth, but he affirmed the Baden position that truth is a result of a cognitive process. An affirmation of truth is the role of a

---

29 M. Rubinshtejn (1907), 23; M. Rubinshtejn (2008), 30.
30 M. Rubinshtejn (1907), 23; M. Rubinshtejn (2008), 31.
31 Moisej Rubinshtejn here quoted Rickert, who wrote, "Therefore, it is not the value of truth that grounds the value we place on the consciousness of duty, but on the contrary the value of truth is based on our concept of duty." Rickert (1902), 700.

judgment, which, as we saw, requires a desire or will to truth. In this sense, then, truth, contrary to Struve, does stand in a direct relationship to the cognizing subject.

Two years later in 1909, Moisej Rubinshtejn published in *Voprosy filosofii* an article in which, while condemning certain elements of Descartes' philosophy, he saw in it an anticipation of Baden neo-Kantianism. On the one hand, he saw Descartes' dogmatism as resting largely on his conviction that there is an objective reality independent of cognition and that truth lies in cognition agreeing with it. As a rationalist, however, he could not follow the path of experience to find this agreement. He, instead, found the way to truth through clarity and distinctness. The system that one obtains through those criteria will be the most faithful depiction of reality.[32] Seeing scientificity only in unconditional and necessary knowledge, Descartes thought that there is only one *a priori* path forward, namely through reason, with experience in the everyday sense holding only a modest place.

On the other hand, Descartes found that representations as such cannot be true or false. Those predicates are applicable only to judgments. Descartes reduced all human thought to two categories: perception, broadly understood, and volition, or operation of the will. He referred sensing, imagining, and pure understanding to the first. To the second, he referred desiring, affirming, denying, and doubting. We obtain a true or false judgment through a correct or incorrect combination of them. Moisej Rubinshtejn found in this the point that the Baden philosophers would make some 250 years later. With it, Descartes clearly and distinctly outlined that volition participates in judgments and sees it in exactly the manner that Windelband and Rickert emphasized. They, like Descartes, referred affirmation and denial to volitional acts. Volition, thus, is a decisive factor in assigning veracity as a predicate to a judgment. Descartes, in brief, appears as a champion of the primacy of practical reason. "Because only with an affirmation or a denial do we obtain the cognitive product that we call a judgment."[33] Descartes, as every beginning philosophy student knows, began his deliberations with a doubt, which is a volitional act. Rubinshtejn took this as an

---

[32] M. Rubinshtejn (1909), 149; M. Rubinshtejn (2008), 155.
[33] M. Rubinshtejn (1909), 150; M. Rubinshtejn (2008), 156. Cf. Descartes (1982), 16 (§34) – "And in order to judge, the understanding is required (because we can make no judgment about a thing which we in no way perceive); but the will is also required in order that assent may be given to the thing which has been perceived in some way."

additional affirmation of Descartes' adherence to the primacy of practical reason and the central role of volition in all judgments.[34]

9.4 From Baden to Lebensphilosophie

Moisej Rubinshtejn in 1910 became dean of the education department of the Moscow Higher Women's Pedagogical Courses, a position he held while teaching numerous other classes. His appointment also signaled a turn in his interests toward educational theory, one that would persist throughout the rest of his life – and one that would not be so fraught with ideological perils as philosophy. However, he had by no means totally divorced himself from technical philosophy. For example, in an address delivered at a meeting of the Moscow Psychological Society in April 1910, but published only the following year, he spoke on "The Question of Transcendent Reality." He alleged in it that the question formed the basis of the disputed issue between supporters of metaphysics and its opponents. Kantianism had up to then, in Rubinshtejn's opinion, directed itself to the investigation of the possibility of metaphysics, but it was not now a matter of going *"Zurück zu Kant."* It was a matter of how to go forward with Kant from the standpoint of epistemological idealism.

Rickert, in his *Der Gegenstand der Erkenntnis*, asked whether there is a reality independent of the cognizing consciousness, understood not as the consciousness of this flesh-and-blood empirical individual, but as the ideal epistemological subject. The issue, then, amounted, in effect, to the tenability of the Kantian thing in itself. Rickert and other neo-Kantians had rejected it, whereas metaphysicians held that the theory of cognition is inconceivable without ontological presuppositions. This requirement, on the part of metaphysicians, indicated the need for assuming a transcendent reality. The proponents of the primacy of ontology over epistemology, therefore, have historically accorded primacy in philosophy to metaphysics. Their reasoning has been that cognition can be had only if there is something to cognize, something fully independent of the cognizing subject. The minimal condition for cognition, therefore, is the presupposition of an "I" and a "non-I," i.e., a subject and an object. Epistemology, in

---

34 It is interesting to compare Moisej Rubinshtejn's portrayal of Descartes with that of Windelband. Whereas the latter, no doubt, recognized that one "cannot think without at the same time willing," this was not Descartes' central message. Windelband made much of the Cartesian criteria of "clearness and distinctness," which lend themselves to a mathematical treatment of the finite objects of perception. "On this account metaphysics and the theory of knowledge terminate for him [Descartes], too, in mathematical physics." Windelband (1905), 394, 393.

starting from this presupposition, cannot account for this dichotomy. Metaphysics too assumes the subject and the object are given. Moisej Rubinshtejn held, therefore, that the source of this presupposition must lie in another sphere, viz., in practice, in life.

A person is first and foremost an active being and then in addition a cognizer. Moisej Rubinshtejn tells us that he takes these concepts in a naïve-realistic sense with the "I" and the "non-I" understood as practical life presents them to us. We leave aside for now any and all metaphysics, such as whether the non-I is a mere appearance or a thing in itself, and all contrasts, such as that between pure thinking and being. There is just one immediate, empirical reality in which there is an active relationship between the subject and the object. Implicit in this is a distinction between an active and a theoretical attitude toward the world. But in both attitudes values are present. Even in a theoretical attitude toward the world, we rely implicitly on a value, the value we ascribe to knowledge, to learning the truth.

Moisej Rubinshtejn contended that how we understand truth itself depends on how we resolve the issue of a transcendent reality. Metaphysicians, for example, ascribe, as we would expect, a metaphysical sense to truth. They see it as something real, as having the character of being. Thus, the truth for them is a transcendent essence. Truth for Rubinshtejn, however, was not associated with being, but with validity. Truth is a characteristic not of being, but of a judgment. If, contrary to such metaphysicians as Lopatin, we refuse to accept the possibility of immediately grasping essences, we must begin an epistemological inquiry in order to understand cognition. And when we do begin to seek the essence of things, we divide the world into appearance and essence. "In this way, we face the presupposition demanded by the very concept of cognition and the need to begin with a theory of cognition."[35] It is quite natural, then, in speaking of appearances to ask to whom things appear. Such was Rickert's starting point, and in this way, we see that philosophy, above all, presupposes that the theory of cognition plays a fundamental role in it.

Moisej Rubinshtejn wrote that he could not fully accept all the conclusions drawn from Rickert's conception of the epistemological subject, some of which remained obscure to him owing to Rickert's evolving position over the years. However – and this must be stressed here – as the "orthodox disciple of Rickert" that he proclaimed himself to be, Rubinshtejn found Rickert's treatment of transcendent reality from the standpoint of the epistemological subject to be of the

---

35 M. Rubinshtejn (1911), 32; M. Rubinshtejn (2008), 83.

greatest value.³⁶ Against the possible reproach that it is impossible to pass from the physical to the psychological and then from the psychological to the epistemological subject, Rubinshtejn denied that Rickert's neo-Kantianism has any such intention. As we saw in his earlier discussion of Rickert, Rubinshtejn denied that the epistemological subject was a generic consciousness. The former is not a reality, but an ideal point of view, a task, whereas the generic consciousness is taken as a being, the essential characteristics of consciousness found in each of us. Since it is ideal, the epistemological subject does not temporally arise or die. There is no *real* transition from the psychological to the epistemological subject, but there is a *logical* transition. The epistemological subject "is nothing more than a *Standpunkt*, a point of view, the ideal for a cognizing person. Therefore, it is not the epistemological subject that makes judgments. We, being living, concrete people, make them. We should strive for this point of view. Our judgments should come from the point of view of the epistemological subject."³⁷

The immediate unity of pre-reflective experience is shattered with cognition. With it, we confront a dualism of subject and object, and in the latest philosophical reflections, we confront a separation and antagonism between the cognitive object as pure content and the cognitive subject as pure form. Moisej Rubinshtejn stated that he saw philosophy as the "science" of absolute values, norms, and the forms of their recognition. Philosophy can exhaustively reveal the forms by means of epistemological inquiries, but it is quite powerless given its very nature to determine the content. If humanity's ultimate aspiration is to know the world in its entirety not just formally, but also in terms of specifics, we postulate the possibility of achieving this goal, relying on faith. Rubinshtejn quoted Rickert: "Faithful adherence to logical duty and the recognition of the logical 'should' bring us ever closer to the realization of the goal to which science strives."³⁸ To his credit, Rubinshtejn recognized that this faith of which he wrote sets him outside the bounds of scientific knowledge, but he also held that a faith in our values can provide us with the power or strength to persist in the realization of the ultimate goal, it being asymptotically acquired knowledge of "transcendent reality."³⁹

---

36 M. Rubinshtejn (1911), 35; M. Rubinshtejn (2008), 86.
37 M. Rubinshtejn (1911), 39; M. Rubinshtejn (2008), 90.
38 Rickert (1904), 243; M. Rubinshtejn (2008), 94.
39 To be precise, Rubinshtejn wrote that we find the possibility of moral faith in a "transcendent reality." M. Rubinshtejn (2008), 94. However, contra Rubinshtejn, Rickert objected to those who construed his talk of values as meaning "moral" values alone. The scientist, qua scientist, values truth, but this is not a *moral* value. It is a theoretical value.

We see, then, that Moisej Rubinshtejn in 1911 saw himself as adhering to the basic thrust of Rickert's Baden neo-Kantianism at least in the form in which it was enunciated at the beginning of the previous decade. His paper "The Question of Transcendent Reality" was to be the fullest elaboration of his own philosophical stance. Rubinshtejn did write several additional pieces on philosophy that decade before the Bolshevik Revolution, but they were largely historical expositions. One article on Fichte from 1914 contained nothing that further illuminated his own thoughts, even though it held the promise of a direct confrontation of Baden neo-Kantianism with post-Kantian Classical German Idealism. The promise went unfulfilled, however. And then there was a rather lengthy article in *Logos* – Rubinshtejn's only contribution to that short-lived journal – on his Moscow University colleague Lev Lopatin. It presented a thorough summary and criticism of Lopatin's philosophy, but Rubinshtejn's own thoughts scarcely emerged except for his brief self-identification as a "Critical" philosopher. Distinguishing Lopatin's speculative and quite thinly-veiled religious philosophy from that of "Critical" philosophers vis-à-vis their conceptions of the *a priori*, Rubinshtejn wrote, "for us the *a priori* in its necessity and universality is understandable only as the organization of reason as a form independent of us."[40] He, then, went on to write that all parties agree that the *a priori* is independent of the subject. However, Lopatin saw it as a bridge to the traditionally understood metaphysical world of things in themselves, whereas he and his fellow "Critical" philosophers saw the *a priori* as explicable epistemologically.

Apart from an essay on Kant's view of the meaning of life, Moisej Rubinshtejn was largely silent on narrow philosophical issues during the tumult of the immediately successive years.[41] His interests already at that time were largely dictated by his career prospects in pedagogical theory for which he would become an established figure in the Soviet Union. He did, however, emerge from his silence first in 1923 with a separately published essay entitled *The Problem of the "I" as the Starting Point of Philosophy*, which was included as a chapter in his later two-volume work *On the Meaning of Life* from 1927.

By the time of the 1923 essay, Moisej Rubinshtejn had abandoned his earlier stand on the primacy of epistemology in philosophy. He wrote that however interesting and important epistemology may be, "its recent elevation to the rank of ruling the destiny of philosophy on the whole seems to us to be mistaken."[42] The

---

40 M. Rubinshtejn (2008), 113. He was by no means clear on how he understood this independence of the *a priori*. Presumably, he wanted to convey that *a priori* forms are not psychological and thus not an attribute of the empirical consciousness.
41 For a brief discussion of this, see Nemeth (2017), 350–351.
42 M. Rubinshtejn (1923), 4; M. Rubinshtejn (2008), 288.

issue with any starting point in philosophy is that regardless of what is chosen it is a judgment that presupposes a definite content that then dictates what will follow. Philosophers, realizing this, have sought a presupposition-less starting point only to find failure after failure, one of which is psychologism. Rubinshtejn at this time endorsed the thesis of Hegel and Jurkevich that in order to know it is unnecessary to have knowledge of knowledge itself. That we cannot proceed in philosophy without first examining how we know was, for Rubinshtejn, the fundamental creed of Kant and the neo-Kantians. So, how are we to begin? We already saw Rubinshtejn's answer in 1910, an answer that now comes to the fore, viz., in the very activity of living, from which all questions originally arise. As soon as we tear asunder the "I" from the world we commit a fundamental mistake, which can fundamentally distort all that follows. Although we can emphasize the "I" in its intimate and inextricable relationship with the world, psychologism is merely one example of what follows from dismembering the living person from that relationship. The specter of psychologism will disappear only with the restoration of that connection.

True, Moisej Rubinshtejn found that the theoretical dismemberment of the "I" and the real world has led to innumerable scientific discoveries, and in this sense, it has been both fruitful and justifiable. But it also led to distortions in the philosophical sphere. Both psychologism, on the one hand, and materialism, on the other, resulted. However, Rubinshtejn's focus, which is of greatest interest to us here, is not on reductionism per se, but on his distancing from neo-Kantianism. For Rubinshtejn, philosophy recognized the danger inherent in psychologism. But instead of reverting to philosophy's starting point, the unnamed neo-Kantians attempted to go in the other direction, further refining the cognizer in the direction of the "pure subject," from which they hoped to derive not only all living things but also the entire psychic sphere. Rubinshtejn characterized this historical path as a "psychological dissolution of the living, actual thinker."[43] Hanging everything, even being itself, on the epistemological fiction of the "pure I," the neo-Kantians were left with abstractions alone with no vibrant content. Rickert, in characteristically elucidating the object of cognition, placed into that sphere not only the material world and the cognizer's physical body but also the psychological subject and the spirit. He, thereby, excluded everything from the "I" until there was only the bare, abstract idea of the subject of cognition. As for Cohen, his "pure thought" was supposed to be omnipotent both in form and content.[44] With these expressions, Rubinshtejn essentially leaves our

---

43 M. Rubinshtejn (1923), 14; M. Rubinshtejn (2008), 297.
44 M. Rubinshtejn (1923), 17; M. Rubinshtejn (2008), 299.

story. He abandoned neo-Kantianism by 1923 in favor of a turn, as he surely would have expressed it, to life. He concluded the second section of his essay, stating "for philosophy there is no, and cannot be, an abstract subject or an abstract object; there can be only a living cognizer, a concrete 'I' and a living object of knowledge."[45] Philosophy's starting point can and must be the entire living human individual.

Moisej Rubinshtejn's 1927 work built on an unequivocal severance from his earlier neo-Kantianism. The turn away from what he considered the abstractions of idealism to life meant the adoption of a basic metaphysical naïve realism. He found that the problem of other minds – which Vvedenskij and Lapshin found so intractable – had already been dismissed handily by such idealists as Kant with his categorical imperative and Fichte through an act of faith. The problem, not unlike so many others, appears insurmountable when we forget how we arrive at it. We view it as if we were on one mountain top looking toward another, ignoring the paths we took through the valleys to get to that mountain top. Solipsism is a "philosophical disaster caused by an abstract starting point in philosophy."[46] Sadly, Rubinshtejn did not go on to map the valleys.

He also offered in 1927 a brisk dismissal of the problem of personal identity in much the same manner. The problem of interruptions (*pereryvu*) of consciousness, e. g. between periods of sleep, that plagued the empiricists can receive a "simple answer."[47] The body and corporeal processes remain in continuous connection during periods of sleep. A diminished (but not totally absent) consciousness is present throughout and reveals itself through our recognition of our personal identity despite these apparent breaks. Of course, the empiricist might well reply that Rubinshtejn's "simple" solution hides the formidable problem of determining just what this continuous connection is.

At the end of the 1920s, Moisej Rubinshtejn was labeled an idealist and a bourgeois reactionary. As a result, he was arrested, spent a half-year in solitary confinement, and exiled to Alma-Ata in Kazakhstan. He was allowed to return to Moscow in mid-1933 but was unable to secure employment. Although he himself was untouched by the purges of the late-1930s, one of his sons was arrested and executed in 1938. After the war, Rubinshtejn was charged with being a "cosmopolitan," a charge we have observed before. The attacks on him had their effect; he was forced to cease what teaching activities he was able to secure. He died on 3 April 1953.

---

45 M. Rubinshtejn (1923), 21; M. Rubinshtejn (2008), 302.
46 M. Rubinshtejn (2008), 323.
47 M. Rubinshtejn (2008), 323.

9.5 Excursus on the Nevel School

The Nevel School, better known for literary criticism than for its tracts in technical neo-Kantian philosophy, has received considerable attention in English-language scholarship, allowing us here to be comparatively brief. Its most well-known member was Mikhail Bakhtin (1895–1975), who has in recent years become virtually a cult figure among some in the field of literary criticism. Bakhtin was neither trained as a professional philosopher nor did he study in Germany nor did he apparently even complete an undergraduate degree. His surviving writings bear little resemblance to modern philosophical texts, but he did know how to insert names, terms, and brief quotations, though what he made of them was quite idiosyncratic. He took what he wished from various sources when they proved useful for his own purposes without particular concern for the consistency between the use he made of a source with his vision.

Bakhtin was born in the provincial town of Orel, south of Moscow. He enrolled first at the university in Odessa in 1913 but transferred to St. Petersburg the following year to study in the classics department, not philosophy. Although he attended the University until 1918, the destitution that followed the Bolshevik Revolution in the cities led Bakhtin to head out to the countryside where calm and food were more plentiful. He settled for two years in Nevel, a town in the western regions of Russia, located in what had been the Jewish Pale of Settlement, where he taught at the secondary school level. While there he and several others who had also recently arrived formed a discussion circle. Contrary to what one might first think, Nevel at that time was a seat of considerable intellectual and cultural activity. Others were drawn there for much the same reason that Bakhtin was. One circle in which Bakhtin participated was a so-called Kantian seminar. Exactly which of Kant's writings and at what level they were discussed is unclear, but if the history of such discussion groups in Russia is any guide, the talk was animated more by passion than by analytic rigor. Slowly members of the circle moved to the larger town of Vitebsk, approximately 100 km from Nevel, including Bakhtin in 1920, and there the discussions continued. While in Vitebsk, Bakhtin married and unfortunately developed a bone disease that would eventually cost him a leg. He moved to Leningrad in 1924 and took a position there with the State Publishing House and with the Historical Institute.

In one of his earliest surviving works, an extended essay entitled "Author and Hero in Aesthetic Activity," dating from the early 1920s, Bakhtin mentioned Kant and Cohen only in passing. Little of the essay is of interest to neo-Kantian themes, although it does contain extensive observations that belong to what undoubtedly could be held to be of interest concerning the problem of other minds. Bakhtin wrote, for example, "When I contemplate a whole human being who is situated outside and over against me, our concrete, actually experienced horizons do not coincide. For at each given moment, regardless of the position and the proximity to me of this other human being whom I am contemplating, I shall always see and know something that he, from his place outside and over against me, cannot see himself."[48] Such observations, which we may broadly characterize as "phenomenological," are plentiful in this work and as such are akin to those we find in both Husserl and Scheler on the topic.[49] What is missing in terms of our concern here is any discernable connection of these descriptions, however correct we may view them, to Kantianism or neo-Kantianism of either the Marburg or Baden variety.

Another early manuscript of Bakhtin's but first published only posthumously in 1986 was left unfinished. It may reasonably be inferred from the text itself as well as other accounts from the presumed time of its writing that Bakhtin had intended the work to be part of a book on moral philosophy.[50] The treatise itself bears no title, but the editors based on Bakhtin's description of its thesis have named it *Toward a Philosophy of the Act*. There is much in this work that speaks against seeing Bakhtin as either a Kantian or a neo-Kantian of any discernable stripe despite his own much later claims to the contrary.[51] Bakhtin's interest in Kantianism and neo-Kantianism had little to do with epistemology or philosophy of science but revolved around ethics and the centrality of the individual human being as an agent of actions. He objected to viewing the moral ought as based on rationality and universality. He held that any ethics built on such a foundation is defective. The form of universality is "illegitimately appropriated" from scientific laws and assumes that the moral ought can be applied to all without exception.[52] The scientific or theoretical attitude itself can provide no practical orientation for how to live our lives responsibly. We obtain that attitude through an abstraction and elimination of the individual's unique being. Instead of a description of how the world is to me, the theoretical attitude toward the world is an attempt to view

---

[48] Bakhtin (1990), 22–23.
[49] From a Husserlian perspective, all of Bakhtin's descriptions are uniformly taken from within the natural attitude and, as a consequence, no connection is given to either the static or the genetic constitution of the sense of the "Other."
[50] Sergej Bocharov quotes a Vitebsk periodical from 1921 to this effect in his editor's introduction to Bakhtin (1993), xxiii.
[51] Bakhtin related in several long discussions from 1973 that he had a "partiality [*pristrastie*] for the Marburg School." Bakhtin (1996), 39. He also claimed that Cohen "was a wonderful philosopher who had a great influence on me." Bakhtin (1996), 277.
[52] Bakhtin (1993), 25. Certainly, one could argue that Kant *assumed* – and thus did not logically prove – that the fundamental principle of morality must be universal. This is not the point, which is that Bakhtin sharply opposed Kant's conception.

the world "as if I did not exist."⁵³ Here, according to Bakhtin, we see how "formal ethics," the ethics of pure form, which was developed by Kant, is linked to the theoretical attitude to become a category of it. The theoretical attitude theorizes the ought, thereby losing sight of the individual deed. The moral ought in (Kantian) formal ethics is to be applied to individual acts much like a natural law can be applied to individual instances in the real world.⁵⁴ The physicist sees particular objects not as such, but only as instantiations of universal mathematically-formulated laws. Similarly, the (Kantian) formal ethicist evaluates particular human actions only in terms of universal rationally-formulated imperatives.

The young Bakhtin saw the prescription of the moral law by the human will as a second shortcoming of a formal ethics. Kantian ethics would have the will produce the law, that it holds to be universally valid, and thus that it itself must submit to that law. The formal ethicists [Kantians], then, would have the will die "as an individual will in its own product."⁵⁵ Bakhtin saw the primacy of practical reality preached by formal ethics to be a sham. "The primacy of practical reason is in reality the primacy of one theoretical domain over all the others. ... The law of conformity-to-the-law is an empty formula of pure theoreticism."⁵⁶ An unfortunate trait of rationalism that Bakhtin saw persisting in Kant's moral philosophy – though we must honestly say that Bakhtin does not mention Kant by name – was to have the criteria of moral truth be the same as the criteria of cognitive truth. Bakhtin, in all this, again feared the loss of individuality in formal ethics, but he offered no guidance at all concerning how individual deeds, in his opinion, are to be evaluated. He extended criticism but without critique. The young Bakhtin, despite his much later pronouncements of his earlier allegiance, belonged far more to the Christian existential camp than to either the Kantian or the neo-Kantian camp.⁵⁷

Of far more interest in a study of Russian neo-Kantianism is Matvej I. Kagan (1889–1938), who was particularly close to Bakhtin during their time in Nevel

---

53 Bakhtin (1993), 9.
54 Bakhtin (1993), 25.
55 Bakhtin (1993), 26.
56 Bakhtin (1993), 27.
57 In an effort to reconcile Bakhtin's expressions of solidarity with Kant and neo-Kantianism, Sandler sees Bakhtin as professing "a very strange" kind of neo-Kantianism. Sandler would have it that Bakhtin attempted to radicalize Kant's "Copernican revolution" by extending it to the extreme. The individual human being constitutes one's world by one's individual deeds, "not as a Kantian generalized subject, but as a *unique* individual, as the *only* I in the world." Sandler (2015), 176. Sandler can be applauded for trying to square the circle, but in the end the circle remains a circle. Such a "radicalized" Copernican revolution results not in neo-Kantianism, but existentialism.

and Vitebsk. In fact, Kagan's presence in Nevel was not terribly unusual, since it was where he had spent his childhood. After obtaining his secondary school diploma, Kagan went to Germany for further study. He spent six semesters at Leipzig University, then two at Berlin. Kagan wrote in an autobiographical sketch that already at the end of his fifth semester at Leipzig he realized he was drawn more and more to the Marburg conception of philosophy. Since Cohen had at this time retired from Marburg to teach in Berlin at the *Lehranstalt für Wissenschaft des Judentums*, Kagan decided to go to the German capital rather than directly to Marburg. Moreover, Cassirer was at the time also in Berlin as was Riehl, though the latter left no discernable impression on Kagan. In all, Kagan spent two semesters in Berlin and only one in Marburg itself. Kagan, in one surviving autobiographical sketch dated 3 September 1922 mentioned the title of his thesis as *Zur Geschichte und Systematik des Problems der transzendentalen Apperzeption*, but he curiously omitted any further details concerning, for example, the thesis defense or the doctoral examination, both of which would have been required.[58] He presumably would have submitted it to the philosophy department at Marburg. However, since he acknowledged he spent only one semester at that university, the obvious question becomes how he could have arranged all the procedural matters involved in obtaining a degree in such a short time.[59]

The start of the World War in 1914 prevented Kagan from returning to Russia. As a possible enemy combatant, Kagan was interned for two months but released through the intervention of the Marburg philosophers and allowed to return to Berlin. There, Cohen arranged work for Kagan, but as it was of a "non-neutral character" he rejected the offer, resulting in raising Cohen's ire. Relations improved between them with the Bolshevik Revolution, and Kagan was allowed to return to Russia after the signing of the Brest-Litovsk Treaty. He took up residence in Nevel and there met Bakhtin as well as others who would form what is now commonly referred to as the "Bakhtin Circle." The Orel Proletarian University was created in the fall of 1918 and extended an offer to Kagan to lecture on philosophy there. During its short existence, Kagan taught "Introduction to

---

[58] Kagan (2004b), 23. However in another, later but undated sketch, he simply entitled it *Zur Geschichte der transzendentalen Apperzeption*. See Kagan (2004b), 26.

[59] If he actually submitted such a dissertation, it has been lost. As Dmitrieva writes, "What happened to Kagan's thesis is unknown. It is also unknown whether Kagan managed to pass the doctoral examination before the War and submit the thesis for a defense. No information about this was found in the Marburg archives." Dmitrieva (2007b), 191. All of this leads us to question whether Kagan actually did complete a degree at all. Kagan, nevertheless, leads us – or wants us – to think that he completed his thesis by writing, "I remained under arrest for around two months. At this time I wrote my first work after the thesis...." Kagan (2004b), 26.

Philosophy," the lecture notes for which were preserved and have been published. The transformation of this university into a Higher Pedagogical Institute apparently left Kagan without employment, and he made his way to Moscow, "making occasional trips to Leningrad but, because of professional commitments, unable to see his old friends with any frequency."[60] Although his relations with Gustav Shpet were already cool at this time, he was able to secure a position in the philosophy section of the State Academy for the Study of the Arts (GAKhN).[61]

Kagan's professed attachment to the Marburg approach and to Cohen personally surely lay behind a long essay – in effect an obituary – on Cohen. Although published only in 1922, we can confidently surmise that it was written shortly after Cohen's death in 1918.[62] Kagan, again in one of his autobiographical sketches, tells us that his basic scholarly interest was and remained for some years the philosophy of history. This is borne out not just by later essays, but also in this essay on Cohen in which he remarked that Hegel had pointed out Kant's lack of attention to the history of philosophy. Hegel too, though, is at fault, in Kagan's opinion, for not portraying the systematic progress of philosophy. Neo-Kantianism posed such a portrayal as a task, but it approached the understanding of Kant from two different directions, placing different emphases on Kant's philosophical predecessors. Were they chiefly the British empiricists or the rationalists, such as Leibniz and Descartes going back to Plato? Kagan was dismissive of Riehl's general position, seeing it as a continuation of the psychological approach to philosophical problems stemming from Locke and Hume. However, he also impugned any attempt, such as Cohen's, to ground all other sciences on a single specialized science.[63]

---

60 Kagan (1998), 6. These words are from a biographical sketch by Kagan's daughter.
61 In a letter dated 20 February 1920 and thus while Kagan was still in Orel, Shpet responded to Kagan's submission of an essay for publication in Shpet's relatively new journal *Mysl' i slovo*. Shpet rejected the manuscript, writing "I have a fundamental objection above and beyond all else concerning its *method*. I part company here not so much with you personally as with the school to which you belong. I have the impression that this method was deliberately created in order to cause dispute and doubt even concerning positions that are indisputable in themselves." Kagan (2004b), 629.
62 Kagan (1998), 8; Kagan (2004b), 669.
63 Kagan presented the following specious reasoning for his conclusion. "Otherwise in this respect the particularism and chaos that has dominated and still dominates the interrelationships of separate spheres and domains of culture would remain in force." Kagan (2004a), 198. On the contrary, if the attempt were successful, there would be far less particularism and chaos, since all sciences would be ordered under one.

Kagan endorsed Cohen's approach to philosophy as above all a theory of experience. The problem of the unity of existence, of being, is illuminated through experience understood in terms of an orientation toward the sciences, and understood, in turn, as transcendental synthetic principles. Kagan also voiced support here for Cohen's interpretation of the Kantian thing in itself as an asymptotic limit-concept. The natural sciences, be they mathematical or descriptive, are constantly developing. But, it must be said, Kagan otherwise evinced no particular interest in either further illuminating this early aspect of Cohen's thought or tracing a Marburg-interpretation of "experience" through history in line with his own concern for the philosophy of history.[64]

Kagan had precious little more to say on the philosophy of natural science and nothing at all on the possibility of mathematical physics. He was of a different cast of mind. Kagan's daughter Judith mentions that while still in Germany during the war years 1916 – 1917 her father composed a book in German under the title *Von Gang der Geschichte,* which neither appeared in print nor is a manuscript of it to be found in the family archive.[65] He did, however, publish in 1921 while in Orel a lengthy article in Russian with the distinctly neo-Kantian-sounding title "Kak vozmozhna istorija?" ["How is History Possible?"] originally written at the end of 1919. Kagan's dream of a fourth critique, a "Critique of Historical Reason," differed from Dilthey's. The former surely would have looked at Dilthey's incomplete attempts – just as his own turned out to be – as subjectivistic. Indeed, what Kagan meant by the term "history" is not at all clear, or at least not at the start. He never clearly defined his terms, which likely lay behind Shpet's frustration with Kagan's work. But the text quickly shows that unlike Dilthey Kagan was not interested in the epistemological foundations of the social sciences, but in the ontological foundations of human history, or even more accurately of humanity's being-in-the-world. Kagan's project, thus, like Bakhtin's, borrowed elements from Marburg neo-Kantianism but hardly belonged within it.

Although he did not address at the start what his subject matter was to be, it quickly becomes clear that his interest was only the broad sweep of human history, how humanity could have a history. No philosophy of history can uncover the details and explain in some *a priori* manner, for example, why the French Revolution occurred precisely when it did. The specific dynamics of history are put aside. We also see from this that Kagan's concern is not with history con-

---

[64] Poole correctly notices that Kagan "encountered Cohen when the latter was concerned with ethics and religion, and not with logic and mathematics, which marked the beginning of his career." Poole (1997), 252.

[65] Judith Kagan did not mention the basis of her claim. See Kagan (1998), 15. Her father did not mention it in his autobiographic sketches.

ceived as a social science, as an academic discipline. He wrote, "It is not necessary to prove the reality of history. ... However, if an aspect of reality, as such, is in no need of proof this does not mean that it has no relation or stands outside any relation to an explanation of its fundamental significance."[66] It is such an explanation that Kagan endeavored to elucidate. Kagan gave a possibly misleading analogy in writing that what he seeks for historical reality is comparable to the relationship of mathematics to the being, or reality, of the natural sciences. For mathematics, in the Kantian scheme, consists essentially of *a priori* synthetic propositions that are applied to the *a posteriori* sense manifold, thus making laws, as necessary and universal propositions, possible. Nature is subject to these laws in that we cognize them as nature's formal component. However, Kagan had no intention of uncovering the basis of any *a priori* historical laws that could explain specifically why events happened as they did. Kagan did, though, seek to borrow a page without acknowledgment from Cohen's appropriation of jurisprudence as the "science" upon which to construct ethics.[67] Kagan, not unlike Cohen, sought to ground history in politics and law, both understood broadly. Like the critics of Cohen's approach, we can and must question whether Kagan achieved any firm results. What Kagan sought is some factor or element within human history that makes it possible as history. This element, simply by the fact that it makes history possible, is *a priori* in a new sense and is arguably closer to its usage in Husserl than in Kant. Without it, the contingencies that we associate with history even when taken altogether, would not be history.

The key to the door unlocking Kagan's quandary is through asking what makes human history something real different in contradistinction from material being, which is also real. The latter is simply given. The task of the sciences is to acquire knowledge of it. The scientist is active, whereas the object of the scientist's concern, nature, is passive. The reality of nature is simply given. The other reality, historical reality, however, is or was the product of human volition. "This other possible reality is conceived in creative reason. Here lawfulness and being become and are formed in a creative goal."[68] The human will, generally speaking, operates with an intent or purpose. Our search for a philosophy of history must focus on this purposiveness of human actions. Just as the Marburg neo-Kantians thought the proper philosophical procedure was to inquire from the

---

66 Kagan (2004b), 201.
67 "The logic of pure knowledge was based on mathematics in order for it to be connected with the construction of natural science. Analogously, an attempt is made here to orient ethics on jurisprudence. It is the mathematics of the social sciences [*Geisteswissenschaften*]." Cohen (1904), v. See also Cohen (1904), 63.
68 Kagan (2004b), 210.

fact of natural science into its possibility, so Kagan believed a neo-Kantian turn to history would have to proceed from a fact. The caveat here is that we are not dealing with a passive subject matter.

> What characterizes the fact of history is that it is determined by and becomes reality in a goal that is not immanent in our factual elemental reality in general, an actual fact in general. The goal is a transcendent feature of historical reality, which is impossible without the transcendent reality of a fundamental practical goal. As the pure objective practice of creative action, human history, the history of humanity, cannot be considered objective historical reality without a transcendent principle of purposiveness, free from causality and the subjectiveness of a practical interest in general.[69]

What Kagan in this needlessly convoluted manner of writing hoped to convey is that a philosophy of (human) history, being concerned with purposeful action, is not to be included within "theoretical philosophy" as is the philosophy of physics, but within "practical philosophy." History, in Kagan's sense, is not akin to physics, but is to be seen as an ethical discipline.

Humanity's purposeful action is and historically has been manifested in work, conceived broadly to include not just physical labor but also cultural work. "If we miss the meaning of work in history, history ceases to be for us a reality."[70] Work is never complete, its goal always remains transcendent to our present reality. But Kagan viewed work in religious terms. The need to work is humanity's destiny as a result of the Biblical Fall.[71] It also serves, one might add, as the connecting link between practical and theoretical philosophy in that through work on physical nature we proceed along a road leading from the immanent, i.e., what is passively given to us, toward the transcendent, i.e., the goal of humanity. "Work is an objectification of our value and the dignity of each and every individual person in God."[72] In an earlier but undated essay "O lichnosti v sociologii" ["On the Individual in Sociology"], Kagan wrote that all history is social history and must have its own "logic," its own fundamental science.[73] Just as mathematics serves as the foundation of or the "sci-

---

69 Kagan (2004b), 213.
70 Kagan (2004b), 226.
71 "On the basis of the conclusions in this investigation, history takes on the task of *overcoming the curse of the Fall*. This fact will not escape the attention of the thoughtful, attentive reader." Kagan (2004b), 199. He also wrote in 1923 that work "is a category of the religious purpose of history." Kagan (2004b), 448.
72 Kagan (2004b), 236.
73 Kagan (2004b), 397.

ence" that grounds epistemology and just as jurisprudence provides the same service for ethics, so sociology is the fundamental science for history.[74]

Remaining in the Soviet Union in the early 1920s, we can hardly be surprised that Kagan could not escape a confrontation with Marxism, which also proclaimed, in effect, the sanctity of labor, albeit in less overtly religious terms than did Kagan. He too noticed the similarity. "The Marxist definition of work, on the whole, is essentially no different than that given in the Old Testament."[75] Human history, in Kagan's eyes, begins with work. Thus, he welcomed the prominence given to it in Marxism, even while dissenting from that *ism*'s purely secular understanding of it. Admittedly, Kagan expressed reservations with a literal understanding of the Marxist slogan that being determines consciousness owing to its methodological ambiguity. But to say that the developmental progress of productive forces determines the course of history is, in general, not without foundation. For Kagan, it was important that the Bible formulated the importance of work as a factor in history much earlier and more fundamentally than did Marxism.[76]

Kagan wrote a great deal more than the studies we have just discussed. Many are exceedingly verbose and stray from one topic to another without a clear focus. How he secured a position at GAKhN has not come down to us. There is no reason to think that into the mid-1920s he abandoned his earlier allegiance to Marburg neo-Kantianism, but that allegiance was especially to Cohen, both for his philosophy and his Judaism – perhaps in reverse order. Unfortunately, he found his activities at GAKhN to be neither fruitful nor personally stimulating. Much of his work for a GAKhN-planned dictionary of aesthetic terms was deemed by his associates to be unsuitable. Kagan in a letter to his wife dated 16 August 1924 expressed his efforts as unproductive and that he thought constantly about getting a different job. And then in another letter from 20 May of the following year, he pinned blame on Shpet, "who gives work to his students and those that he needs for diplomatic reasons. ... Frankly, I find the Academy disgusting in many ways, but if it provides a livelihood, then for the time

---

[74] Although *prima facie* Kagan's claim appears plausible, positing sociology as the science of history raises many issues. Since history, for Kagan, belongs under ethics as practical philosophy, sociology as a science must belong under jurisprudence, as the science of ethics. Clearly, Kagan's understanding of these terms must be different from our own routine use of them today.
[75] Kagan (2004b), 447.
[76] Kagan (2004b), 390. "I deeply understand the religious essence of the truth of the atheism that springs from historical materialism." Kagan (2004b), 642.

being it is all the same to me."⁷⁷ Relatively soon, Kagan was able to secure a job dealing with economic statistics in the government's department of energy. With his departure from GAKhN, he left both philosophy and neo-Kantianism behind without ever returning to his earlier interests. Had he had the opportunity, he surely would have said that his new position was totally in keeping with his understanding of neo-Kantian "practical" philosophy. Kagan died in December 1937 after an attack of angina at the age of 47. If his health had permitted, there can be little doubt – as his daughter herself admitted – that he would have been arrested by the "state criminals."⁷⁸ As it was, he passed away at home.

---

77 Kagan (2004b), 657. As an indication of his state of mind at this time, he went on in this letter, writing that he "wanted to live among Jews, to work with them and for them together with you and our Judy."
78 Kagan (1998), 7.

Chapter 10
One-Time Neo-Kantians Who Strayed: Vejdeman, Jakovenko, Sezeman

Not all young scholars in Soviet Russia with some training at Western European universities found themselves against their will on a one-way boat trip out of their homeland in 1922. Some were still too young with too few publications to have attracted the attention of the new regime despite philosophical differences with its official ideology. Some others were broadly sympathetic to what they took to be the moral message lurking behind that ideology and were willing to remain to see how events unfolded. And some, as we saw, adapted to the new order and its way of thinking by changing their own concerns and in this way were able to make a living at least for the time being. Here, we turn to three representatives of the first group, two of whom were for all intents and purposes apolitical. They also had family backgrounds that made it somewhat easier than for others to flee abroad, even if they did not venture far. The individuals in each of the three cases had a distinct background in neo-Kantianism but which in time became diluted with or submerged under contemporary European philosophical trends that arose particularly in the aftermath of the Great War.

10.1 From Pan-Methodism to Pan-Logicism

However startling it may appear to us today, some in St. Petersburg at the time considered a now almost totally forgotten figure to be among Cohen's first disciples in Russia. Aleksandr Vejdeman [Weidemann] (1879–1943) was of Baltic German stock on his father's side. His paternal grandfather Julius in the middle of the nineteenth century left Courland for St. Petersburg. Aleksandr's father Viktor, being raised in German culture and Lutheranism, attended the German Reformed School in the capital but went on to marry an ethnic Russian and Orthodox Christian, Vera. Aleksandr himself was baptized as an infant in an Orthodox ceremony but received a "liberal" religious education in his youth, attended Russian elementary and secondary schools, and was enrolled at St. Petersburg University from 1899–1903, where he was a student of Vvedenskij's. In one short autobiographical sketch composed decades later, Vejdeman, reflecting on his early views, wrote that already in his university days he read a paper to a student philosophical society in March 1901 entitled "Opyt reformy filosofii Kanta" ["An Attempt at Reforming Kant's Philosophy"].

In this report, I already passed from Vvedenskij's Kantianism to Hegelianism. And when in 1903 I founded along with P. A. Kirillov [1874–1910] a philosophical circle, I was already a pure panlogist. My entire philosophical experience clearly testifies to my view that without Kant it is impossible to enter the latest philosophy, but it is also impossible to remain with Kant in it.[1]

Most likely, Vejdeman, after completing his studies in St. Petersburg went to Marburg for the winter semester of 1903/1904 to hear Cohen.[2] Thus, based on the meager facts available to us, we cannot exclude the possibility that Vejdeman passed from Vvedenskij's neo-Kantianism to Hegel's panlogism and then, after his stay in Marburg, passed on yet again for some indefinite period to Cohen's panmethodism.[3]

In many respects, the biographical information that has come down to us leaves us with puzzlement rather than clarification. During the years before the Revolution, Vejdeman appears to have taught at several secondary schools in St. Petersburg. Why he did not pursue further study at the University is not mentioned. He left Soviet Russia in October 1923 with his wife and infant son

---

[1] Vejdeman (1939), 201–202. We must keep in mind that he wrote these lines concerning his youthful philosophical position for posterity decades later. Such autobiographical reflections are notoriously skewed to justify one's later thoughts and earlier actions. The precise year of the publication of Vejdeman's book is not entirely clear. No date is given on the book's cover. Previously in my earlier work *Kant in Imperial Russia* I had assigned it to 1938, since that is the year handwritten on the copy I consulted and which was the copy Vejdeman had sent to Boris Jakovenko, that again based on the unmistakable inscription on the book cover. Moreover, it was not unusual for Vejdeman's books to be postdated, thus leaving the actual publication date of this "1939" book unclear.

[2] Dmitrieva writes, "If we exclude the possibility of a coincidence in family names, Vejdeman stayed in Marburg during the winter semester of 1903/1904 and heard Cohen's lectures on the Kantian system and participated in his seminar on the *Critique of Pure Reason*. We even find the name Vejdeman mentioned in the winter semester 1911/1912." Dmitrieva (2007b), 170. Despite writing several short autobiographical sketches, Vejdeman in none of them mentioned attending classes at Marburg University! However, Gordon in a letter to Fokht dated January 1907 mentioned Vejdeman as a member of a small group in Russia who were concerned about the future of Kantianism and Cohen's philosophy. See Dmitrieva (2007b), 416.

[3] Curiously, in the short extract from his autobiographical sketch quoted above, Vejdeman did not mention any turn to Cohen's neo-Kantianism, the essential feature of which was a "panmethodism," i.e., the employment of Kant's transcendental method throughout all philosophical spheres. Nonetheless, Vejdeman must have been quite impressed with Cohen, for he apparently was influential in Hartmann's decision to go to Marburg. This, at least, is an interpretation of Jakovenko's words that Vejdeman was "the first follower and propagandist of Cohen's philosophy in Russia." Jakovenko also stated that Vejdeman played the role, so to speak, of a "philosophical godfather to Nicolai Hartmann." Jakovenko (2003), 408.

for Latvia, which, after all, was his ancestral home, being able to establish as far as Latvia was concerned that he qualified for citizenship there. Whereas he found employment at private secondary schools during the 1920s, his inability to retain individual teaching positions for long, none of which moreover were in the public sector, was a result of his deficiency in the Latvian language. What does become clear is that Vejdeman remained throughout these years precariously close to financial ruin.[4]

Vejdeman's move to Latvia appears to have stimulated his interest in philosophy, which while not wholly dormant during the previous decade was certainly subdued. For in October 1926, Vejdeman published in Riga his first monograph entitled *Myshlenie i Bytie* [*Thinking and Being*].[5] He tells us in that same autobiography that he had already begun work on this 1926 book during the winter of 1904/1905 and that it developed in subsequent years until in 1916 it was "basically ready."[6] Still, he continued to add to it until finally through a grant he was able to have it published.

Vejdeman conceived his project to be grander even than Kant's, which inquired into the possibility of only mathematical and natural-scientific knowledge (*znanie*). The pressing issue is the possibility of knowledge itself, knowledge in general. Descartes had already seen this and proceeded from a universal doubt. And whereas Kant under the influence of Locke and Hume saw mathematics as grounded in pure intuitions of space and time, Leibniz had earlier already placed the infinitesimal method at the heart of his philosophy, thereby eliminating pure intuition. The German Idealists from Fichte through Hegel recognized the limitations of Kant's quest and turned back to Descartes. In this respect, Fichte's opposition of the I to the non-I and Hegel's being and non-being represent the foundation not just for natural science, but also for all categories of thought. Vejdeman sought, in short, a foundationalist philosophy, whereas Kant and the early German neo-Kantians, in Vejdeman's eyes, did not see science and mathematics as needing a philosophical foundation.[7]

---

4 Apart from the biographical information in the footnotes above, the facts, such as they are, are based on the incredible investigative work of Svetlana Koval'chuk, who over the years has been more than generous in extending her assistance.

5 He acknowledged such, writing "When I arrived in Latvia, I experienced a new upsurge of spiritual strength and philosophical creativity, as a result I plunged headlong into a spiritual freedom that Soviet Russia did not deliver after 1917." Vejdeman (1939), 203. Vejdeman contended in this 1939 work that *Thinking and Being* was actually published in 1926 despite the year given on the book's cover as 1927.

6 Vejdeman (1939), 202.

7 One can easily see Cohen lurking in the background of Vejdeman's thought. Cohen opened his 1904 *Ethik* writing not of "*die Logik der Naturwissenschaften,*" but of "*die Logik der reinen Er-*

An adamant opponent of psychologism, as were the neo-Kantians in general, Vejdeman rejected viewing the laws of logic as natural laws, and thus in his eyes as contingencies. Logic must be an absolute and an autonomous branch of knowledge. Moreover, it has a priority over all other individual sciences. With Kant, the understanding prescribes laws to nature. The understanding conceived as pure thinking constitutes nature and, thus, has a logical priority over nature. In other words, the systematic study of the understanding in cognition is completely distinct from psychology. But Vejdeman again emphasized that Kant asked not how all sciences as a whole, "in general," are possible, but only how mathematics and natural science are possible.[8] Since logic, the theory of cognition in general, must explain all other sciences, it cannot rely on any of them, but must explain its own possibility without help from them. Both Kant and Cohen, by turning to the factual structure of the objects of a specific science, have a theory of experience but, contrary to Cohen's position, not a logic of pure thinking.[9] Vejdeman charged that the philosophies of both Kant and Cohen have a positivistic character, and as they rely on facts given to us they represent peculiar sorts of psychologism.

Vejdeman, like Cohen, rejected what he regarded as the passivity of Kant's position concerning the intuitions of space and time. Although Kant proclaimed the spontaneity of logical thinking, he failed to apply this train of thought to *a priori* intuitions. He did not understand that thinking is not just active, but also creative and that this creativity of thinking entails the rejection of any givenness. "Nothing is and can be given, and the *a priori* not only conditions actively, as it were, but in truth creates all of the content of thought. Thus, thinking bears a creative character."[10] Vejdeman's proposal would appear to lead to an embrace of Cohen's effort to

---

kenntniss." Moreover, whereas Cohen conceived his "logic" as based on mathematics in order thereby to construct science, Vejdeman conceived Cohen's project as insufficiently radical. Vejdeman's proposed "logic" was to construct not just physics, but mathematics as well. For Cohen, see Cohen (1904), v.

8 Vejdeman wanted "logic" to be entirely divorced from all of the sciences. He procedurally parted company with Cohen, who also recognized that all sciences should "flow" from one science. Cohen claimed that what all sciences, properly speaking, have in common is not some content, but method. All of the ambiguity inherent in speaking of methodical thinking is removed "if we start out from a specific science whose methods give a clear presentation of this thinking. Mathematical natural science stands out as this science." Cohen (1902), 17. Contrary to Cohen, though, Vejdeman believed logic can go it alone.

9 Cohen after 1902 would, undoubtedly, have faulted Vejdeman for a poor understanding of what he sought to accomplish as we see from the quotation in the immediately preceding footnote.

10 Vejdeman (1927), 72.

the same end, but he also rejected Cohen's approach. Whereas he earlier had applauded Leibniz's appeal to infinitesimals, Vejdeman faulted Cohen's own use of the infinitesimal method. Indeed, Cohen sought, as did Vejdeman, to eliminate the concept of givenness by transforming it into a never-ending task. Vejdeman, however, saw the source of Cohen's failed procedure as based on the lingering influence of Kant's dichotomy between thought and sensibility. Another approach is necessary, and for this Vejdeman suggested reaching back to Plato, with his account of recollection, and onward to Descartes, whose *Cogito ergo sum* is both analytic while yet a pure intuition. "The initial act of thought is, consequently, its immediate intuition of itself."[11] In short, Vejdeman, as a rationalist, proposed seeing thought as productive and intuitive, an approach that he believed Descartes also had followed. Thought is independent of anything transcendent to it, including sensations. "Thinking must itself create all of its content. ... Thinking, consequently, not only constitutes, but produces or constructs it from its own bowels, from its own elements."[12] In thought, we know not only the world around us not through some subjective glasses, but the world as it exists in itself.

Therefore, we must begin the construction of a philosophical system not with judgment, as did Kant, but with concepts. It is the source of thought, and from them, all else is deduced. The method of deduction is the philosophical method. With such an understanding, Vejdeman rejected the Kantian view of mathematics as based on intuitions of time and space. It is indeed a matter of inferences from simpler to ever more complex ones. However, the intuition of concern in mathematics is not a givenness, i.e., a givenness of time or space, but of a logical idea.[13]

Before proceeding further, we must remark that Vejdeman presented neither examples nor, for that matter, references – none at all – in his work. We cannot even be certain how he understood many of his operative terms and expressions, since he largely provided no definitions of them. We do have, though, a number of instances in which he contrasted his position to that of others, principally Kant and occasionally Cohen. It is through these contrasts that we must look for clarifications of Vejdeman's expressions. For example, we know that Kant held time to be given *a priori*. Vejdeman differed, calling it a category, which he conceived as being a concept, not an intuition or form of intuition as in Kant. Vejdeman, again, shared this in common with both the Marburg neo-Kantians and with Lapshin. Much of the Kantian scheme, then, is unnecessary, since

---

11 Vejdeman (1927), 80.
12 Vejdeman (1927), 77.
13 Vejdeman (1927), 125.

much stems from locating space and time quite apart from the categories of the understanding.

Vejdeman, like the Marburgians, looked to mathematics to clarify his conceptions. He rejected Kant's understanding of it, but also that of the empiricists, thinking that their conception was laced with psychologism. Given that mathematics is *a priori*, Vejdeman thought time arises for us in the category of number conceived as a multitude (*mnozhestvo*), a mathematical series.[14] He also tried to derive other categories through such mental legerdemain, reminiscent more of Hegel than of Kant. He went on, "The categories of unity and multitude substantiate each other: every unity can be only a multitude and every multitude can be only a unity. For otherwise a multitude would not be a category, would not be an *a priori* principle, a principle that constructs being and that substantiates all things."[15] Vejdeman himself saw this derivation of categories as a dialectical process. The category of totality becomes concrete in the category of space. The construction of nature demands time and space be correlative. The external is correlative to the internal. Without space, there would be no time. Only through space does time exist. Vejdeman's efforts, however specious, in the end, come to an affirmation consistent with Marburg neo-Kantianism. Physics receives its true foundation in mathematics. Only through mathematics does nature become precise and rigorous. Physical nature and *a priori* mathematics are not foreign to each other. Nature is mathematics incarnate.

Vejdeman hoped to portray a philosophical system that was more than a logic, a derivation of categories in ever more developed stages. He expressly stated that his sketch presented thought progressing from thesis through antithesis into a synthesis. As the word "progressing" conveys, Vejdeman held this development to be teleological. Having gone this far, Vejdeman then, as it were, changed the subject without explicitly clarifying this teleology. He remarked, instead, that the entire process must have a foundation that comes to an end. This can happen only with a foundation that is its own foundation and is creative. "The logical foundation, as a self-foundation, must have not an intuitive, but a creative character," a contention we have already seen. Vejdeman continued, "Thus, the rejection of logical determinism leads philosophy to establish ethical determinism, and the sphere of logic must be based on the sphere of ethics."[16]

---

14 Vejdeman (1927), 163.
15 Vejdeman (1927), 164.
16 Vejdeman (1927), 241. Vejdeman appears to have differed with Cohen on this point, though Cohen's exact position in this matter is difficult to determine. Cohen wrote, "Ethics presupposes logic, but logic in itself is not ethics. Thus, ethics presupposes natural science of all sorts but

Regardless of the logical superficiality of Vejdeman's "argument" for the primacy of the ethical, of practical reason, we see that he ultimately endorsed, as Kant and Fichte had done more than a century earlier, the primacy of the practical. One cannot help but see the hand of Fichte behind many of Vejdeman's reflections both here and further on. As did Fichte, Vejdeman assigned great importance to the self-constitutive activity of the "I." Thus, the "I," contrary to Kant, is not a category of theoretical reason, but of the practical. The "I" of cognition is not a reality and has no significance apart from its necessary connection to the real "I" of practical reason. "The 'I' of cognition is only a glimpse (*otblesk*) of an object in the world, a glimpse of the 'I' of action. Here is why the entire psychic world, the world of consciousness as such is quite elusive for cognition."[17] Only the "I" of action is a reality.

The world of action is the domain of practical reason, the world of what should be. Although Vejdeman did not directly address the issue of whether there are duties toward oneself, whether in a solipsistic universe there could meaningfully be talk of an ethics, he did write that if there were no being, the question of what should be could not arise. Ethics is, in effect, a rejection of what is. In Vejdeman's eyes, the value and significance of Kantian ethics lies in recognizing this. Nevertheless, there is no absolute contradiction between what is and what should be. In this fashion, Vejdeman distinctly qualified his earlier affirmation of the primacy of practical reason. He refused to depart too far from Cohen.

Just as Cohen asserted at the start of his 1904 *Ethik* that ethics is the positive logic of the *Geisteswissenschaften*, so too Vejdeman wrote, "We understand ethics as a science that lays a foundation, an intrinsic basis for all the human sciences, for all the sciences of human culture."[18] Moreover, just as Cohen saw logic realized in mathematical physics and ethics realized in jurisprudence, so too did Vejdeman invoke without attribution the same analogy. Vejdeman criticized Kant for isolating ethics from law, thereby leaving ethics without any content. On the other hand, law without morality is devoid of a foundation.

Vejdeman's later disclaimers of his debt to Cohen can hardly be taken seriously unless we assume that the many similarities were purely coincidental and that his few reviewers were all mistaken in seeing such a marked influence. He was more candid in a slightly earlier autobiographical sketch entitled "Moj filosofskij put'" ["My Philosophical Path"], writing that in the years before World

---

does not merge into it. Hence, moral ideas can and must be distinguished from theoretical concepts." Cohen (1904), 36.
17 Vejdeman (1927), 238.
18 Cohen (1904), v; Vejdeman (1927), 267.

War I he gave reports that were "an exposition and reworking of Cohenianism. I stood to Cohen's system in approximately the same way as Spinoza stood to Descartes. But I was never a follower of Cohen, a 'Marburgian.' In the eyes of others, I remained and perhaps still remain now a 'Marburgian'."[19] We will return to the question of Vejdeman's position shortly.

Vejdeman in 1927 successfully defended *Thinking and Being* as a doctoral thesis.[20] It would not be his only publication. It would not even be his only published book as we shall see. But first in 1930/1931, Vejdeman published a short article with an ambitious topic spelled out at the start. "How can reason prescribe a law of nature? How can thought generate all of its content out of and through itself? In other words, how is idealism in general possible?"[21] The reader will easily recognize the first of these questions as one which Kant had hoped he had addressed in his first *Critique*, and the second question is one that Cohen's systematic work addressed. Of course, Vejdeman neither truly answered these questions nor did they form the center of his attention in his brief article. But he did clarify his understanding of philosophy, and it did not coincide with the Marburg understanding of the centrality of the transcendental method in philosophical inquiry. Logic, as the term was used among the Marburg philosophers, could be conceived either as the methodology of scientific thought, i.e., epistemology or as metaphysics. Vejdeman chose the latter. Philosophy, properly understood, has no more to do with scientific methodology than with psychology.

Vejdeman, not long afterward in 1931, published in Riga another large work entitled *Mir kak ponjatie* [*The World as a Concept*]. He then later that decade followed up with two more large works. In the preface to the third book, he remarked that it was intended to be the third part of his philosophical system, thereby echoing the three *Critique*s of Kant and the three parts of Cohen's mature systematic works. He also added that he, unlike Kant and Cohen, had a more complete system. They did not account for all spheres of human culture. This third work, entitled *Tragedy as the Essence of Art, Religion, and History*, repre-

---

19 Vejdeman (1938), 27. This autobiography formed, as it were, the first chapter of his book but with separately numbered pages. Vejdeman also had this book published at his own expense in 1936. Vejdeman's reaction was in part due to a review of *Thinking and Being* by a friend of many years, V. Sezeman, who wrote that Vejdeman's system was "an attempt to fill Kant's Criticism with the ontological dialectic of Hegel and to correct Hegel's metaphysics with Kantian Criticism." Sezeman (1928), 169.
20 Koval'chuk (2016), 35.
21 Weidemann (1930/1931), 16.

sented not just a philosophy of religion, but, so he professed, an entire theological system.[22]

Unlike Hegel, but certainly, like Kant and his epigones, Vejdeman gave his chapters distinctively Kantian titles, such as "How is Art Possible?" and "How is Religion Possible?" Vejdeman would have it that aesthetics, ethics, and logic are not separate independent philosophical disciplines, but equal parts of one integral philosophical system. However, if a practical interest in our system is predominant, we can accord primacy to practical reason, i.e., ethics, and likewise, do the same for logic and aesthetics when the respective interest is predominant.[23] Following Cohen, Vejdeman viewed logic as concerned with not just the form of cognition but also its content, thus with what is, i.e., the real. Ethics, clearly, is concerned with what should be, i.e., the ideal. There would, then, seem to be a contradiction between these two spheres. The task of aesthetics is to provide a higher synthesis. As a philosophy of art, the task of aesthetics is to show how the realization of the ideal and the idealization of the real are possible.[24] "If logic constructs the object and ethics constructs the subject, then aesthetics constructs their higher unity, their higher synthesis. It constructs the world as Absolute. This absolute is not a dead, inert Absolute. For we understand the system of philosophy as an idea that is becoming, a developing idea."[25] We must keep in mind that the possibility of a synthesis of nature and spirit, of what is fact and what ought to be, presupposes the efficacy of the will in nature. Only in such a case can nature be an object of art. Thus, aesthetics again must concern itself with the *a priori* principles making such efficacy possible. Vejdeman did not reveal what these principles are.

Vejdeman also concerned himself with the philosophy of religion in his 1938 tract. It is, he wrote, as much a part of philosophy as any other part of a philosophical system. As such, it too is based on general philosophical principles. In his discussion of this topic, he surprisingly refers to Fichte and Hegel, but not to Kant and Cohen. Vejdeman's conception of religion runs along traditional Judeo-Christian lines. As such, we need not linger long on it. Nonetheless, he attempted

---

22 Vejdeman (1938), 3.
23 Those familiar with the philosophical writings of the Russian philosopher Vladimir Solov'ëv will notice a distinct similarity here with his position. Philosophically, this similarity arises from the fact that both Solov'ëv and Vejdeman held to a belief in an integral system, which can be approached equally from different directions. Since Vejdeman neither mentioned Solov'ëv as an influence on his overall viewpoint nor mentioned him in this particular discussion, we can only point out the similarity. That would change in coming years.
24 Vejdeman (1938), 14.
25 Vejdeman (1938), 19. Here again, we find Cohen's idea of the indefinite task.

to fit it into his overall system. He remarked that religion occupies in aesthetics – note no mention of any relation of religion to ethics – the same position as mathematics does in logic and law in social ethics. In doing so, Vejdeman attempted to fill a lacuna in Cohen's system, in which Cohen could not specify the "factual science" which aesthetics could submit to a transcendental investigation of its possibility.[26]

Vejdeman also devoted a considerable number of pages to the topic "How is History Possible?" Unlike previously, Vejdeman now significantly altered his position, writing that economics serves as the paradigmatic social science in a philosophical investigation of its possibility and history plays that role in the investigation of aesthetics. Without devoting considerable attention either to history as a subject-matter or to the methodology of historical investigation, Vejdeman launched into what he saw as defects of Kant's thought, the chief one of which is the division into constitutive and regulative principles. This unnecessary division led to a breach between epistemology and metaphysics. Vejdeman passed, as he acknowledged, from Kantianism to Hegelianism. The overcoming of the absolute ontological differentiation of idea from being means a resurrection of the traditional ontological argument. "From the viewpoint of Hegelianism, which logically completes Kantianism, the very idea of God contains His being."[27] Vejdeman held that prior to Kant the Absolute was conceived as substance, as something given. Kant did make an advance over Spinoza, with the former's conception of the unknowable thing in itself. However, Hegel realized the error of both of his predecessors and conceived the Absolute as idea. We need not proceed further, for with these pronouncements Vejdeman left neo-Kantianism behind despite its marked influence on his thinking, an influence that would remain over all of his constructions.

Vejdeman published one more large work *Opravdanie zla* [*Justification of Evil*] – thereby evoking comparison with Solov'ëv's *Justification of the Moral Good* – in 1939. He explicitly developed in it his panentheism and his conception of another idea borrowed arguably from Solov'ëv, viz., that of Divine humanity.[28]

---

[26] Cohen had raised the issue of a transcendental investigation of religion, writing "We proceed from its facticity and ask of it its right (*nach ihrem Rechte*). Mathematics and physics had to submit to such a transcendental inquisition no less than did law, the state, and finally the cultural fact of art. How could religion, as such a fact, evade the question of the legal basis of its existence and continuance?" Cohen (1915), 8. Vejdeman's choice of aesthetics as an answer to Cohen's quandary does, however, seem quite odd, even perplexing.

[27] Vejdeman (1938), 308.

[28] Vejdeman mentioned Solov'ëv's name and ideas much more often in this 1939 book than in any of his previous tomes.

"The idea of Divine humanity receives its full necessity only in the ultimate form of a religious world outlook, namely in panentheism."²⁹ Nonetheless, Vejdeman continued to write under the influence of Cohen. The Absolute is not something given (*dannost'*); it is something posed as a task (*zadannost'*) and as a task it becomes. The Absolute develops through logical categories. It is not only an object, not only in itself but also a subject, a being for itself. Vejdeman evoked not just Solov'ëv, but Hegel as well.

Vejdeman's years in Latvia were financially difficult. Unable to secure steady employment, he made some money writing for émigré Russian-language publications. The political/military events of 1939/1940 in Latvia and the other Baltic states need not be recounted here. How Vejdeman fared – how he could possibly have fared – during this period is unknown. We know only of his ultimate fate. Koval'chuk writes, "It turned out Vejdeman was buried in Lielupe [, Latvia] in early January 1943 in a small old cemetery."³⁰ We end on this note. His fate, both philosophical and physical, went for all intents and purposes unnoticed and remained so for decades.

10.2 Transcendental Idealism or Skepticism?

The last two figures we will discuss were among the most Western-oriented and talented philosophers at the end of the Imperial Russian era. Boris Jakovenko (1884 – 1948) certainly had another interest besides philosophy, namely political activity. Born to parents who were themselves active participants in the revolutionary populist movement, Boris's godfather was Pavel Bakunin (1820 – 1900), the younger brother of the famed anarchist. Boris attended and completed his secondary school studies, quite possibly his last instance of receiving an academic diploma. He spent time at the Sorbonne in 1902 – 1903 before returning to Russia. He studied during the academic year 1903/1904 at Moscow University but remained in close contact with the populist Socialist Revolutionary Party, including participating in street demonstrations in December. As a result of his arrest early the next year, he was expelled from the University. He left for Heidelberg in December 1905. He remained there for three semesters attending Windelband's classes and then went to Freiburg

---

29 Vejdeman (1939), 7. To be sure, Vejdeman announced his panentheism at least as early as in his 1931 *The World as a Concept*. "From the religious viewpoint, our theory is a theory of panentheism. ... Panentheism is a theory of the unity of God and the world, according to which God without the world is just as absurd as the world without God. Each exists only in the other and through the other and through this in oneself." Vejdeman (1931), 223.
30 Koval'chuk (2016), 39.

for two semesters to hear Rickert. Although he remained in Western Europe until 1911, he played an important role in establishing the Russian language edition of the journal *Logos* and went on to become its most prolific contributor. Not a single issue of the journal appeared without some piece by him. With his return to Russia, he became a member of the Moscow Psychological Society, but he retained his political contacts with the Socialist Revolutionary Party. For that reason, he continued to be under police surveillance, and he was soon enough forced again to leave Russia in September 1913. He lived in Italy until 1924, though with trips to Russia until the Revolution made that impossible. He received a personal invitation from Tomáš Masaryk, the president of Czechoslovakia, that year to go to his country, which in light of political developments in Italy Jakovenko gladly accepted. His daily life and that of his family must have been difficult in the extreme. Although he managed to contribute often to Czechoslovakian philosophical publications, they and his other writings could hardly have brought in much money. Jakovenko received a government allowance, but this was barely enough particularly given that he had a family. As the years passed, his financial situation became increasingly dire.

Jakovenko attempted with Hessen and Stepun to revive the Russian edition of *Logos* in 1925, but it ceased after just one issue. The surviving record fails to clarify whether this cessation was a result of philosophical differences or financial difficulties. At this time he joined a new group in Prague, the Society of Russian Philosophy, to which in December 1924 he presented a report "Sushchnost' pljuralizma" ["The Essence of Pluralism"]. This was the only instance of his addressing the group.[31] Ever the philosophical entrepreneur, Jakovenko spearheaded a German-language journal, *Der russische Gedanke*, intended to help familiarize a Western European audience with Russian, primarily expatriate-Russian, thought. It lasted until 1940. Jakovenko also continued to publish in Italian. How he managed during the turbulent years of the 1940s is unknown. Also unknown is his reaction to the Communist coup in 1948. At the time of his death in

---

31 Ermishin (2009), 72. The Society had just started that month and had at first twenty members that included in addition to Jakovenko Alekseev, Hessen, Lapshin, and Losskij. We can only conjecture why he did not participate more actively in the Society. One possibility is that since he resided a distance from Prague, the travel was difficult; another possibility is that the interests of the Society had diverged from his own. While the latter is certainly a possibility, the record of many of the subsequent addresses to the Society were along traditional philosophical lines. See Chizhevskij (2007). Whatever the case, Jakovenko's recent life-experience, sitting out the Revolution and Civil War in Italy, may have been a factor in his alienation from the Russian émigré community. We also cannot rule out the nationalistic and religious mood among the émigrés, which was foreign to Jakovenko at this time.

January 1949, he was heavily in debt, and to discharge it the most valuable items from his library had to be sold.[32]

As with many of the figures we have discussed, the appropriateness of characterizing Jakovenko as a neo-Kantian is not obvious and as a result, has not gone uncontested.[33] However, his thought over the years remained remarkably consistent. Jakovenko already in 1908 presented his ideas in the form of two papers to the Third International Congress of Philosophy, which convened in Heidelberg. One paper concerned the topic of logicism, but the young Jakovenko, immersed as he was in Cohen's *Logik der reinen Erkenntnis*, understood "logic" not in the sense we have today of formal logic, but in Cohen's sense. Jakovenko rejected the possibility of a reduction of mathematics to logic, but he did not engage in the technicalities found in, for example, Frege and Russell. Transcendental philosophy seeks neither to eliminate mathematics nor to be merely the logic of mathematics as a particular discipline. Rather, it is the logic of science in general.[34]

Of more interest to us here, though, is Jakovenko's second paper, "Was ist die Transzendentale Methode?" in which he again expressed his high opinion of Cohen's *Logik*. Calling it "the most brilliant work of our time" and seeing it as finally demonstrating the "logical impossibility" of psychologism, it demonstrated that just as the individual sciences have a distinctive method so too does philosophy, viz., the transcendental method. Jakovenko, two years later, would remark that Cohen had, with his own system, freed Kant's philosophy from the residues of his dogmatic era. But it must be remembered that Kant rep-

---

32 Shitov (2012), 510.
33 See Shijan (2019), 444. Belov also remarks that the appropriateness of characterizing Jakovenko as a neo-Kantian is "not so simple and unambiguous." Belov (2012a), 289. Belov, nevertheless, attaches Jakovenko to Cohen's Marburg School. Belov (2012a), 294. There was, however, at least one occasion when Jakovenko declared himself to be a neo-Kantian. When Andrej Belyj, the symbolist literary figure, was quickly drifting away from neo-Kantianism he penned a piece "Circular Movement. Forty-two Arabesques" for the symbolist journal *Trudy i dni*, which at the time was also closely associated with the *Logos* philosophers. Belyj's piece fell into Jakovenko's hands prior to its publication and aroused his indignation. Jakovenko wrote to Metner, "Can I, as a representative of neo-Kantianism, be a (close) co-worker of such an organ, one of the articles of which calls the entire neo-Kantian movement an 'idiocy' and in general mocks it?" Quoted in Sapov (2013), 66. Unfortunately, Sapov does not supply a date for the letter. In any case, Belyj's piece did appear in *Trudy i dni* – together with the statements that surely angered Jakovenko. See Belyj (1912). However, it was Fedor Stepun who replied to Belyj in the same issue of the journal immediately following Belyj's piece. Stepun briefly sketched Kant's work and the contributions of, importantly, the Baden School, not Marburg. See Steppun (1912).
34 Jakowenko (1909a), 880.

resented the final link in the chain of thinkers who sought to free philosophy from its historical enslavement to religious faith as well as from any pretense to serve as the guide for scientific development. Only Kant was finally able to determine philosophy's proper subject-matter and how to go about it. Kant saw philosophy as transcendentalism, which determines its transcendental objects by the transcendental method.[35] The unique concern of transcendentalism, or transcendental philosophy, as against other sciences is its investigation and systematization of the objective foundations of knowledge.

Jakovenko in 1910 saw Cohen as a continuer of Kant's philosophical strategy for establishing the independence of philosophy. Cohen's efforts were aimed at establishing the fundamental theses of transcendental philosophy, while removing the extraneous accoutrements in which Kant clothed them, and to systematize these theses into a single integral philosophical system. Cohen broke with a philosophical tradition that looked on consciousness through an interest in metaphysics. Instead, Cohen advanced the transcendental method and in doing so freed philosophy and positivism, in particular, of its naïve psychological and metaphysical presuppositions. Cohen's transcendental idealism, thus, was not only a positivism *par excellence* but the most consistent idealism and "the completion of the entire age-old tradition of philosophy."[36]

We see from the above that Jakovenko viewed Cohen and Marburg neo-Kantianism in general, through an historical lens. Cohen's thought itself, in Jakovenko's estimation, represented a transitional stage. Even though his systematic works appeared in the early twentieth century, they belonged entirely to the nineteenth as the philosophical self-consciousness of that century's second half. Whereas Kant had indicated the proper concern or subject matter of philosophy, he affirmed a functional psychology. Hegel sought to tear philosophy completely away from science and additionally viewed it as a conceptual game. Nonetheless, Hegel served to clarify the logical independence of the proper philosophical method and to establish its consistency. To be sure, Cohen disavowed Hegel's influence on his own thought, but Hegel really was a secret guide, playing an extremely important role in Cohen's philosophical development. For Jakovenko, Cohen's accomplishment was in uniting Kant and Hegel and advancing this unity not merely as the latest achievement of philosophy, but as the basis of philosophy as such.

Cohen's system, nevertheless, is not the end. Jakovenko, even at the very start of his philosophical career found it to be unsatisfactory owing to deficien-

---

[35] Jakovenko (1910b), 201; Jakovenko (2000), 427.
[36] Jakovenko (1910b), 222; Jakovenko (2000), 452.

cies he believed he saw in it, deficiencies that stemmed from its very origins in German Idealism. Already in his 1908 congressional address in Heidelberg, Jakovenko charged much of recent German philosophy with psychologism, albeit of a "transcendental" variety. This included not just Cohen, but also Rickert and even Husserl with his theory of intentionality. All had inherited Kant's doctrine of the Copernican turn in treating objects as projections of cognition.[37] Doing so, all three philosophers looked on cognized objects as if they were mine, albeit mine understood "transcendentally."

> In order to avoid any possible psychologism, we must not merely conceal the empirical illusion of intentionality as transcendental idealism does. We must completely expel it from the sphere of the transcendental. It is not enough to give a transcendental translation of empirically impossible relationships (*Verhältnisse*) in the belief of having thereby freed ourselves of any measure of psychologism. ... We must take a standpoint, where the I-consciousness (*Ichbewusstsein*) is disengaged, where the apparent relationships of life lie outside the actual scientific relationships, where they have, so to speak, been thought away forever.[38]

We must treat objects outside their relationship to consciousness, as if they were not ours, in order to avoid the slightest hint of psychologism. There is, in Jakovenko's broad understanding, a desire, whether conscious or not, to ascribe philosophical significance to psychic factors, to believe that these factors play at least an immediate or even indirect role in the concerns of philosophy. An appeal in a philosophical explanation to a psychological investigation, one that is concerned with mental experiences, is tainted, since this would interfere with the independence of philosophy. Philosophy is a reflection on the possibility of science; thus it expresses a transcendental approach. As such, it cannot actively use what it is to examine. No, philosophy's interest in psychology, its approach and experiences, must be strictly logical, a second-order reflection.

Having accused the most recognizable figures in German philosophy at the time of implicitly employing the very fallacy that they all had emphatically denounced and sought to eradicate, we can hardly be surprised that Jakovenko's

---

[37] "Cohen's philosophical system is psychologistic above all, because it arose on the basis of the philosophies of Kant and Hegel." Jakovenko (2000), 455.
[38] Jakowenko (1909b), 795. Any interpretation of Jakovenko here is surely perilous in light of the paucity of text. Nonetheless, Jakovenko's terminology here is often strikingly similar to that of Husserl in 1910 in a section entitled "Ausschaltung des eigenen Ich" ("Disengagement of one's own I"). The similarity ends there, since Husserl's disengagement is in pursuit of "pure consciousness and, for the time being, my own consciousness." Husserl (2006), 95; cf. Husserl (2006), 48.

thesis was and has been historically ignored.³⁹ Were we to take up Jakovenko's proposal to eliminate from philosophy every vestige of psychology, of reflection on cognition and the performance of the cognitive act, what would we have? Jakovenko himself only hinted at the result, as he saw it at the time. At the end of his 1908 presentation, he called for a turn to pre-objectivized being, "pre-objectivized" meaning pre-reflective. He recognized that such a turn presented a quandary. How could one conceive existence and life without reference to cognition? The answer is a mystery. He termed his standpoint at the time a "transcendental skepticism."⁴⁰

Jakovenko reiterated his portrayal of modern philosophy in another article in the first issue of *Logos* in 1910. Jakovenko welcomed the recent recognition of psychology as an independent natural science that had no special connection to philosophy. This recognition should speed the development of philosophy itself. Psychology will be seen as investigating the process of cognition, and philosophy will devote itself to the sense of that process, its objective significance. With the development of psychology as a science, philosophy would become ever more aware that it must divorce itself from actively employing psychological reflections. With greater recognition of their respective tasks, there would be no fear of psychology intruding into the concerns of philosophy.

Jakovenko discerned, however, that along with the development of psychology as a science there had emerged in the second half of the nineteenth century new concerns within philosophy of a psychological nature even among the neo-Kantians. Even Husserl was not immune to the contagion. In the first volume of his *Logische Untersuchungen,* he expressed the problem with striking clarity, but with his second volume Husserl returned "to the problem, obscuring it with the same zeal that he had showed in the first volume."⁴¹ But Jakovenko remained optimistic. Philosophy faced two open paths, one offered by the Marburg School and the other by the Baden School. The former proposed developing logic independently of psychology as a system of being; the latter proposed the same but as a system of values. Which one offered greater hope of success? Which was

---

39 Jakovenko gets no mention at all in Kusch's otherwise thorough examination of psychologism. See Kusch (1995).
40 Jakowenko (1909b), 799.
41 Jakowenko (1910a), 255. Jakovenko was not the only one to charge Husserl with a relapse into psychologism. Wilhelm Wundt at approximately the same time did so as well on not entirely different grounds than that of Jakovenko. Wundt wrote, "However, since every immediate intuition is a psychological process, this appeal to intuition means a relapse into psychologism, and the difference with the theory of ordinary psychologism consists merely in the fact that it shifts the source of the evidence to the region of feeling." Wundt (1910), 624.

more consistent with the philosophical tradition? Which carried fewer psychological prejudices? If we previously concluded based on his 1908 paper that Jakovenko was absolutely not a neo-Kantian, let alone a Marburg-style neo-Kantianism, we should now pause.

> We must declare at the very start that our hopes lie in the direction of the Marburgers' philosophy, Cohen's philosophy. If this philosophy feeds on psychology, if psychological schemes lie within its foundations (and studying Cohen's youthful and interpretive works, it is not hard to see that they do), then, on the other hand, its orientation toward science and its complete freedom from so-called immediate reality, which it borrows from Hegel, presents us with a most powerful weapon against psychologism.[42]

Baden teleological neo-Kantianism, on the other hand, is devoid of this possibility. It is not particularly concerned with science at all, but with what Kant called "judgments of perception." Baden neo-Kantianism is devoid of any anti-psychologistic guarantees. On the contrary, it is neither in principle nor in fact free from voluntaristic psychologism. Jakovenko, thus, concluded that the future of transcendental idealism lies with Cohen's, not Rickert's, system.

Jakovenko also published a short piece entitled "O zadachakh filosofii v Rossii" ["The Tasks of Philosophy in Russia"] in a Moscow daily newspaper on 30 April 1910. He expressed there his agreement with Kant that philosophy is a line of inquiry independent of other disciplines, a reflection on their possibility. It is the science of science, of morality, of beauty, and of holiness. Such investigations form the sphere of the transcendental. A consequence of this conception, in his eyes, is that it makes no sense to speak of a national philosophical predilection. Already in his own day, there was talk stemming from the Slavophiles of a Russian philosophy with features and concerns that distinguished it from other European philosophies. Jakovenko would have none of it, for such an idea contradicts the very idea of truth. "Absolute truth lies outside any connection with nations, outside any connection with national life."[43] Many nations, to be sure, exhibit different "moods" at different times. However, to assert that a philosophy could be characteristic of a nation would amount to an assertion of relativism in direct contradiction with the definition of philosophy. Jakovenko held – quite correctly – that an acquiescence to talk of "Russian philosophy" or "English philosophy" as more than a mere geographical designation of a certain group of a philosophy's proponents, as a distinctive line of thought unique to a socio-political nation, would be an evasion from the scientific quest. Just as we cannot

---

42 Jakovenko (1910a), 265.
43 Jakovenko (2000), 656.

speak of French trigonometry, English algebra, or German physics, we cannot speak of Russian philosophy. "Just as there is no, and never has been a, national mathematics, so there is no, and has never been a, national philosophy."[44]

Jakovenko returned yet again, and understandably, to the issue of nationalism in philosophy with the enormous anti-German uproar in Russia evoked with the outbreak of hostilities in 1914. Whereas some in Russia denounced Kant, neo-Kantianism, and German culture in general for bringing forth German militarism, Jakovenko rejected any such linkage. If one were simply to look at the state of global philosophy, one could not help but notice distinct ties in all cases to German Idealism.[45] Jakovenko in the January 1915 issue of *Severnye zapiski* (*Northern Notes*) upheld that it is of no importance to philosophy itself in which particular nation some manifestation or movement of it appears. What is of importance, i.e., the main criteria of its value, is its deep traditionalism, i.e., commitment to the age-old idea of philosophy, and bold originality. Jakovenko found that the religiously-inspired metaphysics popular among Russian philosophers veered away from the traditional idea of philosophy, and that the Hegelianism of Boris Chicherin and the Kantianism of Vvedenskij were lacking in originality.[46]

Jakovenko, in effect, continued this train of thought in an essay entitled "On Logos" in the first issue from 1911 of the journal *Logos*. Jakovenko understood philosophy's concern with the time-less, the rational, to mean that its object is the absolute. It is knowledge of the absolute. Philosophy can believe in principles when it is convinced of their complete rationality. "Philosophical belief is a belief based on reason."[47] From Plato through Spinoza and Leibniz, philosophy has historically exhibited a unity, a rational unity. Each new philosophical movement emphasized a new moment in and of this unity along the way to establishing philosophy as a science.[48] But the systems proposed by the names just mentioned were accompanied

---

[44] Jakovenko (2000), 657.
[45] Jakovenko (2000), 738.
[46] Jakovenko (2000), 715.
[47] Jakovenko (1911), 78.
[48] Whereas this certainly may sound as though Jakovenko is proposing the Husserlian ideal, we should recall he was and would continue to be harshly critical of Husserl. Moreover, Jakovenko's "On Logos" was immediately preceded in the same issue of *Logos* by a Russian translation of Husserl's "Philosophy as Rigorous Science"! As late as 1930, Jakovenko wrote, "Phenomenology is based on a two-fold fundamental delusion. First, it believes its considerations and statements are purely descriptive, therefore receptive and without a point of view. ... There is no non-theoretical epistemological opinion and attitude, and there cannot be one. ... Every cognitive act is already marked as such by a specific cognitive direction (intention). Every cognitive content contains, accordingly, specific categorial forms." Jakowenko (1930b), 27. Jakovenko's line of attack,

in philosophy's long history by others that were alien to genuine philosophy.⁴⁹ Philosophy did not succumb or completely halt when confronted but instead moved forward gaining a new element in its unity. Jakovenko saw in his own day neo-Kantianism as arising from the vacillation of its immediate predecessors and competitors. The unprecedented growth of scientific knowledge required a return to the problem of cognition, a return to Kant albeit in a new form. The emphasis on epistemology in Marburg philosophy played only a preliminary, propaedeutic role, clearing the way for a systemization of objective principles of cognition and for a metaphysics of absolute knowledge. "Epistemological analysis serves as the vestibule to ontology."⁵⁰ In this way, Jakovenko revealed his ultimate goal. Not unlike so many at the time and shortly thereafter, he saw epistemology as merely a preliminary inquiry, as a necessary stage along the road to ontology.

Given the level at which he was able to engage with Kant, the German neo-Kantians, and with Husserl, it surely comes as a surprise that Jakovenko held striking metaphysical views and at such an early date.⁵¹ He turned his attention, for example, to Emil Lask, the last Baden School philosopher, in another essay published in 1912. What is distinctive about the essay "What is Philosophy?" is Jakovenko's persistent search for dualisms, a search which we find ever-present in Jakovenko's various critiques of other philosophies. "A philosophical dualism is an intrinsically inadequate conception. Moreover, it is deeply self-contradictory."⁵² Jakovenko saw the presence of a duality, for example in the description of cognition, as making cognition impossible. "Let cognizing and the cognized, reason and Existence (*Sushchee*), the I and the non-I, thought and object, be two irreducible, independent poles. The effort of cognizing, of reason, of the I, to seize the object of cognition, viz., Existence or the non-I, would then be in vain. The unbridged abyss would divide them fundamentally forever and completely in all respects."⁵³

Seeking such a duality was the strategy Jakovenko employed in critiquing the proposals of the Baden philosopher Emil Lask. The latter, in Jakovenko's

---

clearly then, remained a constant over the years, despite Shpet's criticism of it in his *Appearance and Sense*, which originally appeared in 1914. See Shpet (1991), 94.
49 Jakovenko here portrayed the history of philosophy along lines quite similar to those followed later by Gustav Shpet in his 1917 essay "Wisdom or Reason?" See Shpet (2019), 212–265.
50 Jakovenko (1911), 83.
51 Zenkovsky was not totally in error when writing that Jakovenko "was more concerned with criticizing other men's views than with developing his own." But he did hold views despite Zenkovsky's failure to explore them. See Zenkovsky (1953), 704.
52 Jakovenko (2000), 97.
53 Jakovenko (2000), 99.

eyes, "dogmatically" asserted two dualisms: the first between form and matter, and a second dualism between cognition and sense. Jakovenko contended that the elements in each of the dualisms cannot be reconciled, but the two dualisms also cannot be reconciled with each other. They cannot exchange each other's roles. "They must be prompted by some third thing, which is not purely formal and not purely material, which is quite inconceivable."[54] In short, Jakovenko thought that were we to accept Lask's formulations, neither cognition nor sense would be possible. We see from this example, as well as from Jakovenko's critiques of many other figures, the affirmation of his proclamation that "dualism must be transcended at all costs."[55] However, if we confront dualisms repeatedly in the history of philosophy, is there no escape, and what did Jakovenko offer as a replacement?

The fault of all pre-Kantian philosophers was that they conceived cognition as the product of an aggregate of psychic phenomena, causally and temporally determined, within the mind of the individual cognizer. Thus, they lacked the basic epistemological concept of *objective* cognition. Lacking the concepts of "consciousness in general" or "pure consciousness," they all related cognition to the individual. Kant not only introduced these concepts but also with them turned to the objective structure of knowledge. The concern is not with psychic experiences of some individual I, but with cognitive states in general, "the general objective-logical principle of any cognition."[56] Anything, regardless of whether it be physical or psychic, can be meaningfully spoken of only insofar as it is given to a consciousness in general or has direct connections with other already given things. Something apart from any cognitive connection in this way, i.e., something that has no possible relation to a "consciousness in general" is complete nonsense.[57] Jakovenko understood this claim as the essence of Kant's "Copernican method." It is not the object that makes representation possible, but representation that makes the object possible. Thus, the very notion of a "thing in itself," a thing that exists but is *in principle* uncognizable, is nonsensical.

Jakovenko proposed taking Kant's distinction between "judgments of experience" and "judgments of perception" to heart, believing that true scientificity can be achieved only by consciously expelling from our statements any and all reference to the individual cognizing subject. When we acknowledge the veracity of any simple arithmetical proposition, say, seven plus five is twelve, we

---

54 Jakovenko (2000), 103.
55 Jakovenko (2000), 155.
56 Jakovenko (2000), 170.
57 Cf. Husserl (2014), 80–82 (§45).

make no reference to the subject doing the addition. It is "as if" no individual cognizer were ever present. Philosophy, however, is not satisfied with the relative nature implied by this "as if." It, according to Jakovenko, wishes to be certain that there is no delusion involved, that there really is no reference to a cognizer, that no relativism is present. Such is the role of philosophy, namely to ask whether or not a cognizing subject participates, even if only implicitly, in determining whether an arithmetical proposition is true. Jakovenko saw philosophy as critically probing all alleged scientificities.[58] Such critical scientificity is the first of two basic properties of philosophical cognition.

The second basic property of philosophical cognition is immediacy, and the two properties complement each other. Jakovenko held that true scientificity could not be had without the immediate givenness of the object of concern, of the *sushchij*, understanding this term in the broad sense of the existing state of affairs, but so that it includes mathematical propositions and not just physical objects. Not only are arithmetical propositions, once they are understood, obviously true and true for a consciousness in general, but so are other truths. Regrettably, Jakovenko did not probe deeper into this immediacy or directness (*neposredstvennost'*), and even more regrettably, he termed this recognition of the veracity of a state of affairs "mystical," though with a qualification. He wrote, "If philosophical cognition realizes a critical, rational mysticism or, in other words, a critical, rational intuitivism, it is by no means thereby to be identified with what is usually understood as mysticism or intuitivism."[59] In contrast to the usual understanding of mysticism, "rational mysticism" results not in symbols that merely hint at what is, but presents a direct and concrete state of affairs.

Jakovenko, contrasting his proposed "rational mysticism" to the usual understanding of mysticism, added that it is, by its very essence, pluralistic. There are many different types of being and states of affairs, that are irreducible to one another. Thus, true philosophy must also be pluralistic. The spheres dealing with these different states of affairs – the spheres of morality, of beauty, of the holy,

---

58 Jakovenko (2000), 257. He added years later at the Seventh International Congress of Philosophy in 1930 that, following Kant, "it is not a matter of the *quaestio facti*, but rather of the *quaestio juris*." Jakowenko (1930b), 25.

59 Jakovenko (2000), 260. Shijan writes regarding Jakovenko, "We ask ourselves the Kantian question, how is mystical intuition possible? It is difficult to find an explicit answer to this question in Jakovenko's published texts." Shijan (2019), 453. Shijan is correct that Jakovenko does not supply an explicit answer, but at the same time she is looking too hard. The rational, or mystical, intuition Jakovenko had in mind was the obvious insight we have when some judgment appears obvious to us, such as seven plus five equals twelve. When we learn something, we question why we did not see it before. After learning how to do something, successive performances seem obvious. Such is Jakovenko's rational intuition.

etc. – are autonomous and self-sufficient. Psychologism, in attempting to reduce the sphere of logic and mathematics to the sphere of the psyche, is an example of the opposite. The pluralism of which Jakovenko viewed himself to be an advocate advanced a critical and rational intuition, but, importantly, not sense intuition, as the sole means of acquiring truth. In contrast to Russian religious philosophers, such as Solov'ëv, there is no one absolute Truth, but multiple truths. Similarly, in ethics, there are multiple moral truths. Jakovenko failed to elaborate his many points, leaving us with trenchant criticisms of others but with no fully developed account of his own conception. He characterized Cohen's mature system as a "transcendent immanentism" in contrast to, for example, Husserl's "immanent transcendentism" and believed that both contained the same element of transcendency. It is hard to believe, as Jakovenko contended, that Cohen was insufficiently monistic, that he had not attempted to do enough to eliminate sense intuition. Nevertheless, Jakovenko saw Cohen's philosophical constructions as the most refined and differentiated attempt thus far to overcome transcendency. It is in this sense and *to this extent* that Jakovenko was a supporter of Marburg neo-Kantianism.

If pressed to attach a label to Jakovenko, we could call him a neo-Cohenian, a neo-neo-Kantian. He believed he stood, in his own mind, to Cohen's mature system in a similar relation as Cohen stood to Kant. In his later years, he continued to write on philosophy and reiterated the basics of his position without substantial alteration and without illuminating them. He wrote that transcendental idealism was an epistemological metaphysics and ontologism, but what precisely that meant remains quite unclear.[60] It is with these vagaries in mind that we pass to the final figure in our study.

10.3 From Neo-Kantianism to Phenomenology

Vasilij Sezeman (Sesemann, Sezemanas) (1884–1950) is another largely forgotten figure, but who has recently gathered measured attention.[61] Born in Vyborg, Finland to a Swedish father and a German mother, he studied until 1902 at a German secondary school attached to a Lutheran church in St. Petersburg, to which his parents moved shortly after his birth.[62] His first intention was to be a medical doctor,

---

60 Jakowenko (1930a), 327.
61 Botz-Bornstein has pointed out a difficulty in dealing with Sezeman from the start. Sezeman wrote in Russian, German, and Lithuanian and rendered his name in each case according to the traditional manner for that language. See Botz-Bornstein (2002), 511; Botz-Bornstein (2006), 7.
62 Povilajtis (2007), 234. Botz-Bornstein writes that the Sezeman family moved to St. Petersburg already in 1871. See Botz-Bornstein (2006), 9. Why there is a difference in the Sezeman family

like his father, but after merely a year he left the Military-Medical Academy to study philosophy and classical philology at St. Petersburg University.⁶³ Sezeman went to Marburg in 1906 and returned there in the fall of 1909 after graduation from St. Petersburg, having studied with Losskij and Lapshin. The mentioned stays in Marburg were made at his own expense, but in 1910 he was awarded a governmental fellowship to study in Germany. Upon his return, he taught in secondary schools in St. Petersburg and defended a *magister*'s thesis in 1913/1914.⁶⁴ Sezeman volunteered as a medical orderly when hostilities broke out in 1914, but already from 1915, he lectured in philosophy as a *privat-docent* at the University. For several years thereafter, he moved from one teaching assignment to another until finally in 1923 the new University of Kaunas in Lithuania extended an invitation. The University, needing qualified faculty, had to look abroad for professors. The Dean of the Humanities, seeking someone to fill the professorship in philosophy, appealed first to Hartmann, but in rejecting the offer Hartmann recommended Sezeman. The University stipulated that the professorship was contingent on learning the language in three years, but this presented no obstacle for Sezeman, who quickly mastered Lithuanian. The professorship provided him with a satisfactory and stable salary, a situation quite rare among the émigré philosophers.

With Sezeman, we have in the writings of a single individual one of the clearest examples of the depth and scope, on the one hand, and the philosophical fate, on the other, of Russian neo-Kantianism. Although not his first published writing and not particularly original, his contribution to a 1912 *Festschrift* for Hermann Cohen demonstrates the importance the young Russian philosophical

---

residency is itself unclear. Neither Povilajtis nor Botz-Bornstein provides backup for their respective claims.

63 Botz-Bornstein writes that Nicolai Hartmann "persuaded Sesemann to switch from medicine to philosophy." Botz-Bornstein (2006), 10. If correct, this would place Sezeman's friendship with Hartmann while both were in St. Petersburg. However, Povilajtis writes that Sezeman "become familiar with Hartmann" while studying in Marburg. Povilajtis (2007), 234. The evidence is slim but leans toward accepting a friendship between Hartmann and Sezeman already from their early days in St. Petersburg. That they would know each other from their early days is quite understandable. Hartmann also attended the same German-language secondary school. Hartmann's biographer, Wolfgang Harich, writes that Hartmann and Sezeman were best friends engaging in endless conversations over religious and philosophical problems. See Harich (2000), 3.

64 Information about this *magister*'s thesis is surprisingly sparse. Botz-Bornstein writes, "In 1914 Sesemann received his Master['s] Degree for a dissertation on *The Philosophy of Gymnastics*." Botz-Bornstein (2006), 10. I have not found any corroboration for Botz-Bornstein's statement. An editorial statement in Chubarov (2000), 268 reads – "In 1913, Sezeman defended a *magister*'s exam in philosophy (*magister*'s thesis on Platonism)...." Unfortunately, Chubarov provides neither a reference for this claim nor does he provide a title of the thesis/dissertation.

community attached to the study of Plato in the Marburg School. The inclusion of Sezeman's essay, "Die Ethik Platos und das Problem des Bösen," also shows that Natorp, the editor of this collection, viewed Sezeman as one of the Marburg School's most valued students.

Even earlier, however, in his first published essay, Sezeman revealed his attachment to Marburg neo-Kantianism while at the same time displaying a subtle distancing from a central tenet of Cohen's position. This piece, "Das Rationale und das Irrationale im System der Philosophie," appeared in both the German and Russian editions of the journal *Logos* in the respective languages.[65] As we have seen, the Marburg School, particularly as manifested in Cohen's systematic works, emphasized a methodological rationalism at the expense of any alleged irreducible given with its apparent contingency and thus irrationality. Sezeman claimed that one of the hallmarks of pre-Kantian rationalism was its exclusion of irrationality even from the finite. This dogmatic inclusion of the finite within the sphere of reason was challenged by Kantian transcendental idealism. The cognitive object's rationality is not a given and cannot be exhaustively cognized by a finite analysis. The emerging rationality of the world is an ongoing process, and philosophical thought must turn to its own resources, resulting in the rational forms of cognition.[66]

Philosophy in antiquity already foreshadowed the correlative connection between form and matter, real things in the world being based on a synthesis of the two. The correlative connection between these contrasting elements is, thus, a necessary condition for objective cognition, but only modern philosophy has been able to reveal the significance of that correlativity. If there is a necessary correlativity for cognition, then the task of philosophy is to show the intrinsic unity of form and matter, not their opposition, for the purposes of cognition. Such is the path Sezeman hints that Cohen and the Marburg School took, although he did not specifically mention Cohen. If the rationality of thought is conceived merely in terms of an opposition to the irrational, to what lies currently outside knowledge and certainly beyond the bounds of possible knowledge,

---

[65] See Sesemann (1911/1912) and Sezeman (1911/1912). References throughout the discussion of this essay will be to the Russian-language text. Sezeman's reference in the title of his essay to "System der Philosophie" is an allusion to Cohen's three-part work, the 1902 *Logik der reinen Erkenntnis*, technically being only the first part followed in 1904 by his *Ethik der reinen Willens* and then in 1912 the *Ästhetik der reinen Gefühls*. Although Sezeman's essay undoubtedly is an expression of his own position, Dmitrieva, through her own investigations in archival material, has determined that that position "arose out of Marburg conversations with Hartmann and Natorp." Dmitrieva (2007b), 175.

[66] Sezeman (1911/1912), 96.

then what sense can we make of the interrelation of the irreconcilable opposites? Pre-Kantian rationalism, in Sezeman's reading of intellectual history, hoped to avoid the problem by excluding the irrational from concern and thus placed the finite within the sphere of the rational.

Sezeman associated what he had described as a return to the Platonic viewpoint without specifying what that return is, but the "principle of infinity," which asserts that cognition in the strict sense has no limits, is one of its principal merits.[67] An unlimited analysis of an object becomes a synthesis that creates rational forms of cognition from within the heart of the cognitive process itself. "Only synthetic knowledge, i.e., knowledge erected in stages of ideal unconditionality and endlessness, can embrace and grasp objective reality."[68] Sezeman called this emphasis on methodology in terms of a continuous constructive process of the cognitive object methodologism. It is, in his eyes, the final result of the historical development of Platonic idealism. Sezeman contended that empirical knowledge can never, not even asymptotically, attain perfect rationality. It will always contain irrational elements.[69] "These irrational elements are not contingent elements that can be eliminated in one way or another but, on the contrary, are a *necessary correlate* of the infinite nature and problematic character of objective knowledge."[70] The objects of empirical knowledge cannot be thoroughly resolved into rational connections. To be given and to be irrational are inseparably linked. Certainly, Sezeman added that the distinction between the rational and the irrational, the formal and the material, is relative.

The boundary between these components of knowledge is itself conditional, but Sezeman held back from venturing far into delineating what the systematic unity is between the various contrasting elements of cognition. He recognized, as

---

67 Although he did not use the expression frequently, Cohen did write "the principle of infinity" in his *Logik*. See, for example, Cohen (1902), 30.

68 Sezeman (1911/1912), 96.

69 There is no need for us to follow Botz-Bornstein in identifying this feature of Sezeman's thought with an "obvious" influence from Rickert, who, Botz-Bornstein claims, developed around 1902 the thesis of the irrationality of reality. Botz-Bornstein (2006), 30–31. One could also point to Lask's position on the irrationality of the given. Lask in his *Die Logik der Philosophie und die Kategorienlehre* wrote, "One can designate the impenetrability, incomprehensibility, and unclarifiability – this 'givenness' and irreducibility to the logical – as the irrationality of the material." Lask (1923), 76. However, Sezeman's thought could just as well have been influenced in his own formulation by Cohen's treatment of the given and of the thing in itself in his commentaries on Kant's *Critiques*. True, an explanation of the given would always remain a stumbling block for Cohen, and he would go on to formulate a highly idealistic interpretation of the given. It is this that Sezeman would always resist.

70 Sezeman (1911/1912), 99.

did so many others, the untenability of the thing in itself, but like the Marburg philosophers, he saw it as a Kantian idea, not as a transcendent reality. The thing in itself, then, is resolved in terms of the structure of consciousness, wherein the irrational is an ever diminishing element in the steadily increasing rationality that is human knowledge. Although much of this resembles what we find in Cohen, Sezeman withheld from attempting to dissolve away the given. If there were no opposition between the rational and the irrational in knowledge, there would be no knowledge. "The possibility of the logical development [of knowledge] is rooted in the very structure of objective knowledge: in its problematic character, *in its rational-irrational essence.*"[71] The irrational given is necessary for knowledge.[72]

Sezeman also published in 1913 a survey on "The Theoretical Philosophy of the Marburg School" in an occasional series under the general heading *New Ideas in Philosophy*. The essay does little more than reinforce our image of him as a supporter of the general Marburg standpoint. He pointed out in it that unlike the German Idealists in the post-Kantian era the Marburg philosophers had no interest in dictating to the natural sciences what should be and, thus, infringing on their autonomy. The Marburg philosophers sought merely to discover the intrinsic, systematic unity of science that accompanies a transformation and restructuring of Kant's philosophy. Natural science has progressed since his day revealing new categorial forms and syntheses. It has been able to do so through the mathematization of the scientific inquiry of nature, and mathematics in turn has progressed in modernity through its recognition of the infinite, which was foreign to antiquity. This mathematization is intimately connected with characterizing nature not in terms of substances, but in terms of functional or relational concepts. "All principles of knowledge are reduced and should be reducible ultimately into relational *categories*. For only relational categories can provide knowledge with a rigorous systematic character."[73] The logical supremacy of relational concepts gives knowledge a definite idealist coloring. In short, Sezeman

---

[71] Sezeman (1911/1912), 122.
[72] Henry Lanz, already at Stanford University, would pen an essay on "The Irrationality of Reasoning," in which he wrote, "logicians and epistemologists are inclined to regard as rational only those contents which can be logically analysed, i.e., reduced to elements comparatively more simple and more fundamentally familiar than those which are analysed." Could the similarity between the idea expressed here and Sezeman's be purely coincidental? Lanz, then, went on to write that contents that cannot be further resolved are believed to be irrational. Lanz (1926), 340. Lanz, perhaps mindful of the philosophical climate at the time in the U.S., made no reference to the German or Russian neo-Kantians.
[73] Sezeman (1913), 32.

at this time agreed with the Marburg School that the scientific ideal could only be conceived in terms of an endless development of knowledge giving an increasingly idealist worldview.

The sparse secondary literature devoted to Sezeman paid little attention to these early writings. It can certainly be justified in that Sezeman did not reveal any particular originality or insight in them. However, his philosophical stand later did take a turn away from these largely Marburg-inspired early writings. In this sense, we can speak at a minimum of two periods in Sezeman's philosophical career. We cannot determine precisely when the "turn" away from Marburg took place, but it is evident already in a 1923 essay "Survey of the Newest German Philosophical Literature" published in Berlin. Sezeman first charged that Rickert's 1921 *System der Philosophie* presented a disappointment to all including his most faithful followers. Rickert had nothing new to say, and the younger representatives of the Baden direction were distancing themselves from neo-Kantianism while moving toward post-Kantian metaphysics. Then without indicating his own earlier affiliation with Marburg, Sezeman wrote that that School also showed signs of internal decay. Whereas Cassirer with his recent work consistently upheld the Marburg viewpoint, Natorp had, from Sezeman's vantage point, moved in the direction of post-Kantian metaphysics. Thus, we see that Sezeman in 1923 recognized a widening gap between himself and both of the major neo-Kantian Schools. Additionally, he invoked Hartmann in this essay, writing that his friend had *correctly* shown with his 1921 text *Grundzüge einer Metaphysik der Erkenntnis* that Cohen's resolution of the problem of knowledge is a failure.

> In trying to reduce the epistemological problem to a logical one, [Cohen's] scientific idealism not only violates the purity of the sphere of logic, ... but also overlooks the very essence of the problem of epistemology, not noticing that the sense of cognition consists in comprehending the object transcendent to it and essentially distinct from it.[74]

Sezeman, thus, now rejects his previous commitment to the ideal of the asymptotic dissolution of the given into thought. Nevertheless, he is unwilling to break wholeheartedly with the Kantian scheme. Whereas there cannot be a complete identity of being and thought, there can be a partial coincidence of the two, a

---

74 Sezeman (2000b), 491. While conceding that Cohen and Natorp influenced Sezeman, Jonkus likely is correct in saying that Sezeman felt an even greater influence from his long-time friend Hartmann. Both agreed that "the Marburg School neglected the ontological foundations of the theory of knowledge." Jonkus (2017), 84. For further discussion of the Hartmann-Sezeman (Sesemann) relationship, see Tremblay (2017a), 198–199.

coincidence that serves as a condition for the possibility of cognition in general. The given can be known, can pass into the sphere of the cognizable, but – and here Sezeman appeared to be making a concession to some sort of mystical belief – it is impossible to deny the possibility of an absolute irrational being. The important point here, though, is that whereas Cohen rooted being in thought, being as a product of thought, Sezeman in 1923 took an ontological turn, seeing knowledge rooted in being, an objective conditionality that is responsible for knowledge's sense of objectivity.

Sezeman wholeheartedly expressed his appreciation for Hartmann's accomplishment in revealing the ontological foundations of knowledge, but Hartmann had focused on knowledge of externality alone, the world of the natural sciences, not on that of the inner world, which he touched on only briefly. A concern for the latter was the interest in Max Scheler, who Sezeman held in 1923 had established the existence of a world of values, a world as objective and real as that of the sciences.[75] With this caveat to his endorsement of Hartmann, Sezeman also endorsed a phenomenological approach toward ethics.

Sezeman followed up his move away from Marburg neo-Kantianism with a review of Hartmann's *Grundzüge* book in the attempted revival of the journal *Logos* in 1925. Sezeman lauded Hartmann for surmounting Kant's "subjective idealism," thereby further evidencing Sezeman's wholesale abandonment of Cohen's constructive view of human cognition. For Sezeman, Kant's overall theory of experience is obscure. Kant did not fully recognize the independence of moral, aesthetic, and religious experience, for had he done so a new problem would have arisen for the theory of knowledge, viz., "to determine the distinctive traits of each of these forms of experience and their relation to theoretical experience."[76] To each form, a special sort of knowledge corresponds. Sezeman also followed up his earlier neo-Kantian treatment of the irrational in cognition, now holding that what appears to be irrational within the scientific sphere can turn out to be rational in ethical or aesthetic knowledge. He faulted his friend Hartmann for not singling out this observation. Hartmann's understanding of the opposition between the rational and the irrational is one-sidedly oriented toward scientific experience, though this stems from his representative

---

[75] Sezeman toward the end of 1928 would write of Scheler, "his influence on the further development of philosophical thought should prove to be decisive in many respects; he discovered new, hitherto unutilized ways and possibilities for independent philosophical creativity." Sezeman (2000a), 505.
[76] Sezeman (1925), 234.

theory of knowledge. Whether coincidental or not, Hartmann's huge treatise *Ethik* appeared the following year.

In 1927, Sezeman began publishing a series of lengthy articles "Beiträge zum Erkenntnissproblem" on the problem of cognition that unfortunately gained little recognition at the time, despite being written in German.[77] A thorough discussion and examination of them in terms of their originality would lead us far astray from the topic at hand. The essays belong to the history of early phenomenology, though arguably a phenomenology oriented in a direction closer to that of Scheler and Hartmann than to that of Husserl. He criticized the Marburg philosophers for selecting mathematical physics as the paradigm of knowledge, but he also objected to their elimination not just of the cognizing subject in general, but of a *real* cognizing subject. It was phenomenology's merit to have restored the subject in theoretical philosophy. This subject is given in self-consciousness directly. "Consciousness and being here coincide. In a consciousness of myself (self-consciousness), the being of the I is given quite immediately."[78] This already distances Sezeman both from Kant and the Marburg neo-Kantians, but he was also critical of Rickert and the intuitivism of his St. Petersburg professor, Losskij. The Baden neo-Kantians took the secondary moments of cognition, such as the processing of cognition, as primary, whereas Losskij neglected these secondary moments. The latter's intuitivism correctly restored the understanding of cognition as cognition of what is, of being. However, there is a danger here. Intuitivism presupposes an objectivizing attitude without explaining it. The importance and role of the secondary moments in cognition are left in the dark.

Sezeman had broken decisively from his past allegiances. This comes out most strikingly – and most eloquently – in a passage from a 1928 essay that appeared in an obscure Eurasian journal published in Paris. It deserves to be presented at some length.

> The neo-Kantians turned out not to be Kant's successors, but only epigones, living on the legacy of his ideas. They resurrected Kant (but then only in part) but not philosophy itself. Their one-sided orientation toward science or scientificity – due to the variety of the sciences themselves and the methods – inevitably turned out to be oriented toward this or that particular science (exact natural science, biology, history, etc.) so that philosophy unwittingly lost its independence and turned out to be a handmaiden of the positive sciences. Hence, the sectarian spirit of the Kantian schools, their blindness to everything that is beyond the Kantian problematic. Hence, the splintering of the movement into a multitude of

---

77 An exception was Hartmann's review of the compilation of Sezeman's 1927 articles in book form. The review originally appeared in *Kant-Studien* in 1933 See Hartmann (1958).
78 Sesemann (1927), 132.

internecine idealistic interpretations, which in search of the genuine Kant and genuine scientificity, have forgotten the meaning and concerns of philosophy itself.[79]

In Sezeman's depiction of the era he had just witnessed, Bergson attempted to infuse some sense of creativity into philosophy, but Bergson stood apart from the tradition. As such, he could not surmount it. What was needed was someone from within. We find the start of this, according to Sezeman, with Husserl.

> He saw his task as finding and directly approaching the things themselves. Such an approach would provide an undistorted and unobscured grasp of their essence. ... With this positive-scientific understanding of phenomenology, Husserl paid tribute to the tradition against which he had himself fought.[80]

Sezeman, in this telling, saw phenomenology as recognizing that its method must be equal to the challenge of grasping the cognized object, but with it comes a set of new metaphysical problems. Scheler picked up the gauntlet. In brief, Sezeman at this time had fulsome praise for Scheler's philosophical approach that included what he called an emotional-volitional attitude toward the world.[81]

Sezeman penned a number of additional works in the years ahead. Whether he maintained any allegiance to an ontologically realist phenomenology is beyond the scope of the present study. No doubt, he opposed all of the philosophical idealisms that he knew including Husserl's transcendental idealism, which he knew only through the *Ideen* of 1913. He remained and from all indications was quite content in Kaunas and moved in 1940 along with the humanities faculty to Vilnius. When the Nazi occupation closed the University, he taught German at a secondary school but returned after the War. In 1949, he was charged with anti-Soviet activity and participation in Zionist organizations, resulting in his dismissal from the University. The following year, he was arrested and spent six years in a Gulag camp in Siberia. He amazingly survived. His rehabilitation in 1958 allowed him to return to Vilnius and to his job teaching logic. He died in March 1963. During these later years, he prepared a translation into Lithuanian of Aristotle's *De Anima* accompanied with extensive commentary. His treatise *Estetika* appeared posthumously in 1970. Regrettably, he merely men-

---

79 Sezeman (2000a), 500.
80 Sezeman (2000a), 501.
81 Sezeman, also in 1928, reviewed Heidegger's *Sein und Zeit*, finding it to have a level of originality and depth comparable only to the works of Scheler. "Heidegger's book is the first truly systematic work to have emerged from the phenomenological school." Sezeman (2000c), 511.

tions Cohen in it and says nothing significant on neo-Kantianism. It surely was for him a distant memory at best.[82]

---

[82] Sesemann (2007), 250.

Chapter 11
Concluding Remarks

A number of recognized philosophical movements in Europe arose, flourished, and faded in the years covered in this study. The early phenomenological movement spearheaded by Husserl suffered greatly under the strains of World War I. During the succeeding decade or so, that movement became almost unrecognizable from what it had been before the Great War – or at least it did in the eyes of its early participants. Another, though quite different philosophical movement, logical positivism, arose between the World Wars and slowly dissolved through internal criticism. It too faced dissipation when many of its leading proponents emigrated to Britain and the New World with the emergence of Nazism. The fate of yet another philosophical movement, German neo-Kantianism, suffered from the death of Lask in the Great War, Cohen's abandonment of Marburg, and the desertion, in effect, of both Natorp and Rickert in the nineteen-twenties. An even more powerful deathblow to German neo-Kantianism was its close association in the minds of the younger generation with the established pre-War German socio-political order. Ernst Cassirer was left virtually alone as the clouds darkened. As a Jew, he was increasingly marginalized with the rise of an ever-more pronounced anti-Semitism and with the emergence of young voices seemingly more in tune with an era that had seen so much pointless death and destruction. Many would undoubtedly counter that such is arguably the natural fate of any intellectual or cultural trend or fashion – emergence, dissemination, and dissolution.

Unlike the philosophical movements mentioned above, international recognition of Russian neo-Kantianism was and is sparse to non-existent. Even within Russia today, it has received only a modest amount of attention, and then undoubtedly owing more to efforts to reconnect with a past from which the country was forcibly severed for many decades than to the discovery of profound and insightful tracts. The competing movements in the West never needed to fill in such temporal gaps resulting from the decline of one philosophy and the emergence of another. There were an ample number of philosophies to occupy the attention of philosophers and scholars. Early phenomenology transformed, or morphed, seemingly overnight into the hermeneutic and existential phenomenology of Heidegger. And the rise of the latter begged, as it were, for investigations into his intellectual provenance. Although the fate of logical positivism was quite different, it, as a movement, would become intimately linked with the emergence of analytic philosophy. Indeed, one can still read and appreciate many of the positivist writings for their precision and sheer enthusiasm, even if one judges the conclusions to be ultimately wrong-headed.

Russian neo-Kantianism, unlike these other philosophical movements either at the time or afterward, failed to receive international recognition for several reasons.[1] For one thing, unlike Freiburg University, no one from foreign lands ventured to St. Petersburg or Moscow to study with Vvedenskij or Chelpanov, Lappo-Danilevskij or Novgorodcev. No one introduced Russian neo-Kantianism into the English-speaking world in the way that A. J. Ayer promoted logical positivism or Marvin Farber and W. R. Boyce Gibson aided knowledge of phenomenology. Even Heidegger's early rival, Nicolai Hartmann, who knew the philosophical scene in St. Petersburg and who could have transmitted information about Russian neo-Kantianism, was remarkably and curiously silent about his past, as though his realist ontology had roots in Marburg alone. Additionally, in succeeding decades few philosophy students outside Russia would make the necessary effort to learn the Russian language, an effort that offered questionable rewards either intellectually or in terms of career advancement. Even those who, for one reason or another, had knowledge of the language would have found significant obstacles to the acquisition of original texts. After the Bolshevik coup in 1917, the Soviet government made no effort to disseminate or republish texts – or even permit either – that it saw as intimately linked to what it took to be a spent worldview. Another factor in the neglect of Russian neo-Kantianism on the world stage was the absence of an individual spokesperson or representative for and of it. There were no ambassadors of Russian neo-Kantianism who fled Soviet Russia and took up residence at the universities of Cambridge, Paris, or Harvard. An ample number of inheritors of the logical positivist movement found homes in the United States and England. Even Russian existentialism found a voice in Paris in Berdjaev's popular philosophical writings. Cassirer's presence at Yale and Columbia Universities was surely much too short to influence dramatically and unalterably the American scene, but Russian neo-Kantianism did not have even a Cassirer.

Yet another factor in the overlooking of Russian neo-Kantianism was its intimate linkage with German neo-Kantianism particularly during the years just before World War I. With the uncontestable disappearance at least until quite recently of German neo-Kantianism from the pages of the history of modern phi-

---

1 The reflections that follow are intended to be merely a set of observations contrasting the Russian neo-Kantian movement, such as it was, to other philosophical movements of the time. In no respect are they an attempt to construct a complete sociology of knowledge. In this matter, no claim is made regarding some intrinsic truth or falsity of neo-Kantianism as a factor in its failure to achieve lasting popularity or even notoriety.

losophy, the so-to-speak subsidiary movement of Russian neo-Kantianism had little chance of recognition on its own. Russian neo-Kantianism was simply not systematically original and comprehensive enough to draw international attention to itself, to set itself against and apart from the attention already focused on German neo-Kantianism. Above all, we must recall that unlike the respective cases of Husserl and Wittgenstein, there was no textual canon to be mined for original ideas, not even a single sizeable commentary or substantial secondary study devoted to Russian neo-Kantianism – not even in Russian – until relatively recently. Although, as we saw, some of the young philosophers from Imperial Russia wrote theses and articles in German, these were largely ignored. In the rare cases when they were noticed, these writings were taken as expressions within the sphere of *German*, not Russian, neo-Kantianism. Undoubtedly, the attention space of the western readership could not accommodate any additional upstarts, particularly ones without multiple network ties.

Moreover, were we to evaluate Russian neo-Kantianism simply in terms of the number of translations of its works into other languages, we would have to award it a failing grade. The Boyce Gibson translation of Husserl's *Ideen*, however flawed, nevertheless, allowed some access to German phenomenology for an English-only audience, and the same could be said with regard to the 1922 English translation of Wittgenstein's *Tractatus*. In fact, one major factor contributing to the paucity of information in the English-speaking philosophical world of the international neo-Kantian movement has been the lack of translations of key texts.

How, then, are we to judge, or even what are we to make of, Russian neo-Kantianism? Without a doubt, it produced no technical works comparable to Frege's *Grundgesetze der Arithmetik*, no startlingly novel treatises like Wittgenstein's *Philosophische Untersuchungen*, no grand, incisive histories of thought comparable to Cassirer's *Das Erkenntnisproblem*, no thoroughly original perspectives on the human condition comparable to Heidegger's *Sein und Zeit* and no practical utilizations of Russian neo-Kantianism comparable to Weber's adoption of Rickert's philosophy for use in his own methodological treatises. Russian neo-Kantianism never got beyond "a state of preparatory projects, programmatic declarations, and preliminary drafts."[2] Those who were best positioned to issue a ground-breaking treatise were apparently content to translate minor German-language secondary studies by minor figures into Russian. Yet despite all of these shortcomings and deficiencies we surely can speak of *the* Russian neo-Kantian movement that endured for several decades.

---

[2] Belov (2012b), 38.

As a movement, we can trace its origin to Vvedenskij's writings in the late 1880s dealing with philosophical reflection on the physics of the day. Vvedenskij's early explorations were topically comparable to Kant's own work, *Metaphysical Foundations of Natural Science*, published one hundred years earlier. In terms of originality and technical depth, these earliest works by Vvedenskij marked, in one sense, the high point of Russian neo-Kantianism. Vvedenskij's surest possible successor, Ivan Lapshin, had no interest in philosophy of science and demonstrated little interest in technical philosophy after completing his necessary thesis. In other words, he displayed no interest in advancing neo-Kantian philosophy. Chelpanov, in Moscow, kept abreast of western developments, but he was at heart a psychologist and a teacher. He became acquainted with Husserlian phenomenology most likely through the propagandizing efforts of his former student Gustav Shpet, who had no affection for Kant and the neo-Kantians. Whether Chelpanov would have been lured away from the tenuous hold on him of neo-Kantianism is unanswerable. Had a liberal-democratic Russia emerged from World War I and achieved some semblance of socio-economic stability, Chelpanov would have most likely prospered as an experimental psychologist simply leaving philosophical reflection to others such as Shpet. In any case, he, like Lapshin, felt no calling to advance neo-Kantianism as a research program in philosophy.

The Russian neo-Kantian philosophers of law might have persisted for a time in this imaginary liberal-democratic world. Saval'skij, as we saw, died an early natural death, but to the end, he remained a neo-Kantian. Kistjakovskij showed the greatest promise, but his relatively early death, like that of Lask in Germany, leaves us with unanswerable questions. Others simply drifted away. The events that followed make any prediction perilous. Among those in that generation, Stepun was not truly a committed neo-Kantian, but he did value highly the international exchange of ideas. Hessen turned his back on technical philosophy, and there is no reason to think he ever had second thoughts. True, Fokht remained, as far as we know, a neo-Kantian to the end. Perhaps, he would have made a difference with contributions to philosophy. But his thesis on Kant, on which he seemingly worked for years, reveals no spark of genius, insight, or originality. The two Rubinstejns drifted away from neo-Kantianism. Certainly, at least in Moisej's case, his departure was manifest already before it became politically necessary. Sergej, conceivably, would not have abandoned philosophy for psychology, but based on his writings he appeared to be content with his decision. Of the last three figures we discussed, Vejdeman stands as the most vivid example of the general fate of Russian neo-Kantianism, moving from Marburg to a curious blend of metaphysical idealisms along with the spirit of the age. Sezeman too drifted with the spirit of German philosophy at the time, though in his case from neo-Kantianism to the metaphysical realist phenomenologies of Scheler and Hartmann.

The Russian neo-Kantian movement effectively ended with defections and utter silence some forty years after it began. As in the case of German neo-Kantianism, many of its adherents moved in the direction of Hegelian idealism or at least a disavowal of their earlier Critical stands in favor of an explicitly religious and metaphysical attitude. It allowed them to retain an adamant rejection of psychologism and a belief in concrete, objective values. In a number of cases, a parting with earlier views had its roots in a desire to be more active in social causes than these young, activist philosophers thought followed from (neo-)Kantianism. In short, it would be inaccurate to blame the dissipation of the Russian movement entirely on purely external factors. For already in the years immediately preceding the consolidation of the Soviet state, we see an emerging and growing dissatisfaction with the movement. The initial enthusiasm and rebellious spirit so natural in youth led the Russian students to imbibe what they perceived as rigorous philosophy as against the broad conjecturing of their elders. However, once back in Russia the intellectual network linking these neo-Kantians to Germany's own evolving neo-Kantian movement came under severe strain and in most but not all cases was severed, Kistjakovskij being a prominent exception. Our study of Russian neo-Kantianism also shows that there was little of Kant in this movement, particularly after the first writings of Vvedenskij and Lapshin. Few devoted much attention to the details of any of Kant's three *Critiques*, and none grappled with the thorny transcendental deduction. Although we have said that Jakovenko could be termed a neo-neo-Kantian, much of the Russian movement could justifiably be characterized in that fashion. That is, all of the figures discussed in the last several chapters make up, in effect, a Russian neo-neo-Kantian movement.

We saw at the beginning of our study that Vvedenskij, though remaining a neo-Kantian, changed his concern from philosophy of science to more traditional issues that allowed him to discuss and argue with his compatriots who were of a decidedly non-Kantian outlook. Rather than be a lone voice crying in the wilderness, Vvedenskij chose to adapt in order to engage with others. This was the choice he made. In his essay "The Fate of Philosophy in Russia" from 1898, Vvedenskij briefly traced the history of philosophy in Russia along the lines of his country's higher-educational institutions, in which philosophy was taught and promoted. He advanced the thesis that in order for philosophy to have a broad impact on society, it needed a decided measure of organization and consolidation.³ The contending branches of the German neo-Kantian movement, in fact, realized this to a large extent as demonstrated by their efforts – many of which were unsuccessful – to have their students placed in university positions. The Russian *Logos* circle, implicitly, realized that in order for their idea of philosophy to remain alive and vibrant even within academia it needed the publication they proposed. However, unlike Windelband and Cohen, the younger generation of Russian neo-Kantians had no ensconced university positions from which they could lure and train prospective disciples. There were just too few universities, thereby putting the onus squarely on the financially imperiled journalistic enterprise. Without communication between fellow adherents, the movement, such as it was, would wither and ultimately perish. In fact, this is what happened undoubtedly through no fault of their own. Neo-Kantianism, whether on German, French, Italian, or Russian soil, needed interactions between its supporters. In the case of Russian neo-Kantianism, this interaction became increasingly difficult even socially as the years leading up to the Bolshevik Revolution approached.

In any attempt to account for the short lifespan of Russian neo-Kantianism, we should also not dismiss the simple desire of the individual to be a member of a community, to conform. Admittedly, this is a psychological factor, but it surely played a role in Vvedenskij's abandonment of philosophy of science. As Collins has pointed out so extensively in his study of a global history of philosophy, there is only a limited amount of "intellectual attention space" in one era.⁴ But in order for that restricted space to open up, there must be cracks that over time become larger until the underlying structure breaks. This does not happen, as it were, overnight. To determine whether Russian philosophy, on the whole, had such cracks in the first decades of the twentieth century would require another investigation. However, if there were fault lines, Russian neo-Kantianism was not given the needed time to weaken the reigning Russian philosophical paradigm – and in fact, many of its most creative participants showed no particular interest in doing so. Unlike, say Heidegger's challenge to the tottering German neo-Kantian movement, Russian neo-Kantianism offered no paradigm-shifting treatise. Particularly in the years subsequent to the Bolshevik Revolution, it did not radically depart from or challenge the traditional Russian orientation that accorded primacy toward ontological and ethical concerns. Russian neo-Kantianism remains to this very day merely a curious phenomenon, deemed

---

3 Vvedenskij (1898), 36.
4 Collins (2002). The expression "intellectual attention space" is a key phrase throughout Collins' book. Although it has received considerable recognition, few philosophers are willing to admit that what *they* see as the truth is *at all* socially conditioned.

by most of marginal significance in the history of philosophy. In contrast to German neo-Kantianism, it lacked the solidity of even a single distinct institution or locale in the sense that we speak of Marburg or Baden neo-Kantianisms. However, in contrast to Italian and French neo-Kantianism, there were many more neo-Kantians in Imperial Russia than elsewhere except in Germany, and there was far more cohesion between the Russians than elsewhere except among the Germans. Simply given all the forces working against its survival, let alone dissemination, Russian neo-Kantianism never stood a chance.

Bibliography

Abramov, A. I. (1994): "O russkom kantianstve i neokantianstve v zhurnale *Logos*". In *Kant i filosofija v Rossii*. A. I. Abramov (Ed.). Moscow: Nauka, pp. 227–247.
Abul'khanova-Slavskaja, Ksenija A./Brushlinskij, Andrej V. (1989): *Filosofsko-psikhologicheskaja koncepcija S. L. Rubinshtejna*. Moscow: Nauka.
Aksel'rod (Ortodoks), Ljubov (1906): "Opyt kritiki kriticizma". In *Filosofskie ocherki*. St. Petersburg: M. M. Druzhininoj i A. N. Maksimovoj, pp. 186–223.
Alekseev, Nikolaj N. (1909): "Social'naja filosofija Rudol'fa Shtammlera". *Voprosy filosofii i psikhologii*, kn. 96, pp. 1–26.
Alekseev, Nikolaj N. (1912a): *Nauki obshchestvennyja i estestvennyja v istoricheskom vsaimootnoshenii ikh metodov*. Moscow: Tip. Imperatorskago universiteta.
Alekseev, Nikolaj N. (1912b): "Opyt postoenija filosofskoj sistemy na ponjati khozjajstva". *Voprosy filosofii i psikhologii*, kn. 115, pp. 704–735.
Alekseev, Nikolaj N. (1918): *Vvedenie v izuchenie prava*. Moscow: Izd. Moskovskoj Prosvetitel'noj Komissii.
Alekseev, Nikolaj N. (1919a): *Obshchee uchenie o prave*. Simferopol: n.p.
Alekseev, Nikolaj N. (1919b): *Ocherki po obshchej teorii gosudarstva*. Moscow: Moskovskoe Nauchnoe Izd.
Alekseev, Nikolaj N. (1937): "O vysshem ponjatii filosofii". *Put'*, no. 53, pp. 37–52.
Allison, Henry E. (2004): *Kant's Transcendental Idealism*. New Haven: Yale University Press.
Allison, Henry E. (2006): "Kant on freedom of the will". In *The Cambridge Companion to Kant and Modern Philosophy*. Paul Guyer (Ed.). Cambridge: Cambridge University Press, pp. 381–415.
Allison, Henry E. (2015): *Kant's Transcendental Deduction: An Analytical-Historical Commentary*. Oxford: Oxford University Press.
Al'bov, A. P. (2000): "Pavel Ivanovich Novgorodcev". In Novgorodcev, Pavel I. (2000): *Kant i Gegel' v ikh uchenijakh o prave i gosudarstve*. St. Petersburg: Aletejja, pp. 5–29.
Ameriks, Karl (2000): *Kant and the Fate of Autonomy*. Cambridge: Cambridge University Press.
Askol'dov, Sergej (1904): "Teorii novejshago kriticizma. (Po povodu dvukh knig)". *Voprosy filosofii i psikhologii*, kn. 74, pp. 520–549.
Bakhtin, Mikhail M. (1990): "Author and Hero in Aesthetic Activity". In *Art and Answerability. Early Philosophical Essays*. Michael Holquist/Vadim Liapunov (Eds.). Austin, Texas: University of Texas Press, pp. 4–256.
Bakhtin, Mikhail M. (1993): *Toward a Philosophy of the Act*. Vadim Liapunov/Michael Holquist (Eds.). Austin, Texas: University of Texas Press.
Bakhtin, Mikhail M. (1996): *Besedy V. D. Duvakina s M. M. Bakhtinym*. Moscow: Izd. Progress.
Barnes, Christopher (1989): *Boris Pasternak. A Literary Biography*. Vol. 1, *1890–1928*. Cambridge: Cambridge University Press.
Beiser, Frederick C. (2002): *German Idealism: The Struggle against Subjectivism, 1781–1801*. Cambridge, MA: Harvard University Press.
Beiser, Frederick C. (2011): *The German Historicist Tradition*. Oxford: Oxford University Press.
Beiser, Frederick C. (2014): *The Genesis of Neo-Kantianism, 1796–1880*. Oxford: Oxford University Press.

Belov, Vladimir N. (2004): "Filosofija Germana Kogena i russkoe neokantianstvo". In *Istoriko-filosofskij ezhegodnik 2003*. Moscow: Nauka, pp. 333–353.
Belov, Vladimir N. (2012a): "B. V. Jakovenko i neokantianstvo". In *Boris Valentinovich Jakovenko*. Aleksandr A. Ermichev (Ed.). Moscow: ROSSPEN, pp. 288–305.
Belov, Vladimir N. (2012b): "Neokantianstvo: Istorija i osobennosti razvitija". *Kantovskij sbornik*, 1(39), pp. 27–39.
Belov, Vladimir N. (2016a): "Hermann Cohen in the History of Russian Neo-Kantianism". *Russian Studies in Philosophy*. Vol. 54, no. 5, pp. 395–407.
Belov, Vladimir N. (2016b): "Hermann Cohens Lehre in Russland: Besonderheit der Rezeption". *Folia Philosophica*. Vol. 35, no. 1, pp. 7–27.
Belov, Vladimir N. (2019): "Neokantianstvo P. I. Novgorodceva: k polemike Novogorodceva i Saval'skogo". *Lex Russica*, no. 2, pp. 151–162.
Belyj, Andrej (1912): "Krugovoe dvizhenie. Sorok dve arabeski". *Trudy i Dni*, no. 4–5, pp. 51–73.
Belyj, Andrej (1990a): *Mezhdu dvukh revoljucij*. Moscow: Khudozhestvennaja literatura.
Belyj, Andrej (1990b): *Nachalo veka*. Moscow: Khudozhestvennaja literatura.
Berdiaev, N. A. (2003): "The Ethical Problem in the Light of Philosophical Idealism". In *Problems of Idealism: Essays in Russian Social Philosophy*. Randall A. Poole (Ed.). New Haven, CT: Yale University Press, pp. 161–197.
Berdiajew, Nikoli (1900): "Friedrich Albert Lange und die kritische Philosophie in ihren Beziehungen zum Sozialismus". *Die Neue Zeit*. Vol. 18, band 2, pp. 132–140.
Berdjaev, Nikolaj A. (1900): "F. A. Lange i kriticheskaja filosofija". *Mir Bozhij*, no. 7, otd. 1, pp. 224–254.
Berdjaev, Nikolaj A. (1901): "Bor'ba za idealism". In *Sub specie aeternitatis. Opyty filosofkie, social'nye i literaturnye (1900–1906gg.)* (1907). St. Petersburg: Izd. Pirozhkova, pp. 5–34.
Berdjaev, Nikolaj A. (1904): "O novom russkom idealizme". In *Sub specie aeternitatis. Opyty filosofkie, social'nye i literaturnye (1900–1906gg.)* (1907). St. Petersburg: Izd. Pirozhkova, pp. 152–190.
Berdjaev, Nikolaj A. (1910): "Gnoseologicheskaja problema". *Voprosy filosofii i psikhologii*, kn. 105, pp. 281–308.
Berdjaev, Nikolaj A. (1935): "Russkij dukhovnyj renessans nachala XXv. i zhurnal «Put'»". *Put'*, nr. 49, oktjabr-dekabr, pp. 3–22.
Berdjaev, Nikolaj A. (1981): "Pis'ma molodogo Berdjaeva". *Pamjat'. Istoricheskij sbornik*, vyp. 4, pp. 211–219.
Berdjaev, Nikolaj A. (1991): *Samopoznanie. Opyt filosofskoj avtobiografii*. Moscow: Kniga.
Berdjaev, Nikolaj A. (1993a): "N. A. Berdjaev v nachale pyti (Pis'ma k P. B. i N. A. Struve)". In *Lica. Biograficheskij al'manakh 3*. A. V. Lavrov (Ed.). Moscow-St. Petersburg: Feneke, Atheneum, pp. 119–154.
Berdjaev, Nikolaj A. (1993b): "Pis'ma k K. Kautskomu". *Vestnik SPbGU. Filosofija i konflikgologija* Tom 34, vyp. 2, pp. 180–183.
Berdjaev, Nikolaj A. (1999): *Sub"ektizm i individualism v obshchestvennoj filosofii*. Moscow: OI Reabilitacija.
Berdyaev, Nicholas (1951): *Dream and Reality: An Essay in Autobiography*. Katherine Lampert (Trans.). New York: Macmillan Co.

Berest, Julia (2011): *The Emergence of Russian Liberalism: Alexander Kunitsyn in Context, 1783–1840*. New York: Palgrave Macmillan.
Bezrodnyj, Michail V. (1992): "Zur Geschichte des russischen Neukantianismus. Die Zeitschrift 'Logos' und ihre Redakteure". *Zeitschrift für Slawistik*. Vol. 37, pp. 489–511.
Bocharova, E. B./Borob'ev, V. S. (2012): "Maloizvestnye stranicy zhizni G. I. Chelpanova: k 150-letiju so dnja rozhdenija". *Teoreticheskaja i eksperimental'naja psikhologija* Tom 5, no. 3, pp. 81–102.
Boobbyer, Philip (1995): *S. L. Frank: The Life of a Russian Philosopher*. Athens, OH: Ohio University Press.
Botz-Bornstein, Thorsten (2002): "Vasily Sesemann: Neo-Kantianism, Formalism and the Question of Being". *Slavic and East European Journal*. Vol. 46, no. 3, pp. 511–549.
Botz-Bornstein, Thorsten (2006): *Vasily Sesemann. Experience, Formalism, and the Question of Being*. Amsterdam-New York: Rodopi.
Brown, Stuart (1962): "Has Kant a Philosophy of Law?". *Philosophical Review*. Vol. 71, pp. 33–48.
Bulgakov, Sergej (1902): "Ivan Karamazov (v romane Dostoevskogo «Brat'ja Karamazovy» kak filosofskij tip". *Voprosy filosofii i psikhologii*, kn. 61, pp. 826–863.
Bulgakov, Sergej (1903): *Ot marksizma k idealizmu. Sbornik statej (1896–1903)*. St. Petersburg: Tovarishchestvo «Obshchestvennaja Pol'za».
Bulgakov, Sergej (1914): "Transcendental'nja problema religii". *Voprosy filosofii i psikhologii*, kn. 124, pp. 580–652.
Bulgakov, Sergej (2000): *Philosophy of Economy: the world as household*. Catherine Evtuhov (Trans.). New Haven, CT: Yale University Press.
Bulgakov, Sergej (2003): "Basic Problems of the Theory of Progress". In *Problems of Idealism: Essays in Russian Social Philosophy*. Randall A. Poole (Ed.). New Haven, CT: Yale University Press, pp. 85–123.
Cassirer, Ernst (1923): *Substance and Function and Einstein's Theory of Relativity*. William Curtis Swabey/Marie Collins Swabey (Trans.). Chicago: Open Court Publishing Company.
Cassirer, Ernst (1981): *Kant's Life and Thought*. James Haden (Trans.). New Haven, CT: Yale University Press.
Chamberlain, Lesley (2007): *Lenin's Private War: the Voyage of the Philosophy Steamer and the Exile of the Intelligentsia*. New York: St. Martin's Press.
Chelpanov, E. I. (1891): "Gel'mgol'c kak filosof i psikholog". *Voprosy filosofii i psikhologii*, kn. 10, pp. 41–51.
Chelpanov, E. I. (1893): "O prirode vremeni". *Voprosy filosofii i psikhologii*, kn. 19, pp. 36–54.
Chelpanov, G. I. (1896): *Problema vosprijatija prostranstva v svjazi s ucheniem ob apriornosti i vrozhdennosti. Chast I: Predstavlenie prostranstva s tochki zrenija psikhologii*. Kiev: Tip. Kushnerev.
Chelpanov, G. I. (1897a): "K voprosu o vosprijatii prostranstva". *Voprosy filosofii i psikhologii*, kn. 37, pp. 276–287.
Chelpanov, G. I. (1897b): "O svobode voli". *Mir Bozhij*, no. 11, pp. 2–24; no. 12, pp. 42–65.
Chelpanov, G. I. (1901a): "Filosofija Kanta. Stat'ja 1-ja (Teoreticheskaja filosofija)". *Mir Bozhii*, no. 3, pp. 1–23; no. 4, pp. 165–187; no. 5, pp. 55–80.
Chelpanov, G. I. (1901b): *Kurs lekcij po logike*. Kiev: Tip. I. I. Chokolova.

Chelpanov, G. I. (1901c): "Ob apriornykh elementakh poznanija (Ponjatie chisla)". *Voprosy filosofii i psikhologii*, kn. 59, pp. 529–559.

Chelpanov, G. I. (1901d): "Ob apriornykh elementakh poznanija (Ponjatie vremeni, prichinosti, prostranstva)". *Voprosy filosofii i psikhologii*, kn. 60, pp. 699–747.

Chelpanov, G. I. (1902a): "Evoljucionnyj i kriticheskij metod v teorii poznanija". *Mir Bozhij*, no. 8, pp. 94–117.

Chelpanov, G. I. (1902b): "Neogeometrija i ee znachenie dlja teorii poznanija". *Voprosy filosofii i psikhologii*, kn. 65, pp. 1379–1408.

Chelpanov, G. I. (1902c): *O sovremennykh filosofskikh napravlenijakh*. Gustav Shpet (Ed.). Kiev: Tip. Petr Barskij.

Chelpanov, G. I. (1903): "Psikhologija i teorija poznanii". *Voprosy filosofii i psikhologii*, kn. 66, pp. 97–124; kn. 67, pp. 167–189.

Chelpanov, G. I. (1904): *Problema vosprijatija prostranstva v svjazi s ucheniem ob apriornosti i vrozhdennosti. Chast II: Predstavlenie prostranstva s tochki zrenija gnoseologii*. Kiev: Tip. Kushnerev.

Chelpanov, G. I. (1906): *Mozg i dusha. Kritika materializme i ocherk sovremennykh uchenij o dushe*. Kiev: V. A. Prosjanichenko.

Chelpanov, G. I. (1911): "Pamjati prof. N. Ja. Grot". In *Nikolaj Jakovlevich Grot v ocherkakh, vospominanijakh i pis'makh*. St. Petersburg: Tip. Kushnerev, pp. 185–196.

Chelpanov, G. I. (1912): *Mozg i dusha. Kritika materializme i ocherk sovremennykh uchenij o dushe*. Moscow: Kushnerev.

Chelpanov, G. I. (1913): *Kratkij povtoritel'nyj kurs psikhologii*. Moscow: "Sotr. shk." A. Zalesskoj.

Chelpanov, G. I. (1916): *Vvedenie v filosofiju*. Moscow: Tip. V. V. Dumnov.

Chizhevskij, Dmitrij I. (2007): "Filosofskoe obshchestvo v Prage [1924–1927]". In *Issledovanija po istorii russkoj mysli. Ezhegodnik 2004/2005* (2007). Modest A. Kolerov/ Nikolaj S. Plotnikov (Eds.).Moscow: Modest Kolerov, pp. 176–180.

Cohen, Hermann (1871): *Kants Theorie der Erfahrung*. Berlin: Ferd. Dümmler.

Cohen, Hermann (1877): *Kants Begründung der Ethik*. Berlin: Ferd. Dümmler.

Cohen, Hermann (1885): *Kants Theorie der Erfahrung*. Berlin: Ferd. Dümmler.

Cohen, Hermann (1902): *Logik der reinen Erkenntnis*. Berlin: Bruno Cassirer.

Cohen, Hermann (1904): *Ethik des reinen Willens*. Berlin: Bruno Cassirer.

Cohen, Hermann (1907): *Kommentar zu Immanuel Kants Kritik der reinen Vernunft*. Leipzig: Verlag der Dürr'schen Buchhandlung.

Cohen, Hermann (1912): *Äesthetik des reinen Gefühls*. Erster Band. Berlin: Bruno Cassirer.

Cohen, Hermann (1915): *Der Begriff der Religion im System der Philosophie*. Giessen: Alfred Töpelmann.

Collins, Randall (2002): *The Sociology of Philosophies: A Global Theory of Intellectual Change*. Cambridge, MA: Belnap Press of Harvard University.

Cvetaev, Lev A. (1816): *Pervyja nachala prava estestvennago, izdannyja dlja rukovodstva uchashchikhsja*. Moscow: Univ. tip.

Davydov, Iosif A. (1905): *Istoricheskij materialism i kriticheskaja filosofija*. St. Petersburg: Tip. Tovarishchestva "Obshchestvennaja Pol'za".

Davydov, Iosif A. (1907): "Predislovie". In Shtammler, Rudol'f: *Khozjajstvo i parvo s tochki zrenija materialisticheskogo ponimanija istorii*. Tom II. St. Petersburg: Nachalo, pp. iii-lxxii.

Descartes, René (1982): *Principles of Philosophy*. Valentine Rodger Miller/Reese P. Miller (Trans.). Dordrecht: Kluwer Academic Publishers.
Dlugach, Tamara B. (1991): "Pasternak v Marburge". *Istoriko-filosofskij ezhegodnik '91*, Moscow: Nauka, pp. 352–355.
Dmitrieva, Nina A. (2003): "Predislovie". In Fokht, Boris A.: *Izbrannoe (iz filosofskogo nasledija)*. Moscow: Progress-Tradicija, pp. 5–48.
Dmitrieva, Nina A. (2006): "Iz istorii neokantianstva v Rossii". In *Eternity's Hostage: Selected Paper from the Stanford International Conference on Boris Pasternak*. Lazar Fleishman (Ed.). Stanford, CA: Dept. of Slavic Languages and Literatures, Stanford University, pp. 467–494.
Dmitrieva, Nina A. (2007a): "Filosofija gumanizma i prosveshchenija: kriterii, specifika, problematika russkogo neokantianstvo". *Filosofskie nauki*, no. 1, pp. 101–119.
Dmitrieva, Nina A. (2007b): *Russkoe neokantianstvo: «Marburg» v Rossii*. Moscow: ROSSPEN.
Dmitrieva, Nina A. (2009): "Obraz russkogo neokantianca v pis'makh (1905–1909)". In *Issledovanija po istorii russkoj mysli. Ezhegodnik 2006–2007*. M. A. Kolerov/ N. S. Plotnikov (Eds.). Moscow: Modest Kolerov, pp. 369–396.
Dmitrieva, Nina A. (2013): "Kriticizm ili misticizm? A. I. Vvedenskij i russkoe neokantianstvo". In *NeoKantianstvo v Rossii. Aleksandr Ivanovich Vvedenskij, Ivan Ivanovich Lapshin*. V. N. Brjushkin/V. S. Popova (Eds.). Moscow: ROSSPEN, pp. 40–73.
Dmitrieva, Nina A. (2016a): "Back to Kant, or Forward to Enlightenment: The Particularities and Issues of Russian Neo-Kantianism". *Russian Studies in Philosophy*. Vol. 54, no. 5, pp. 378–394.
Dmitrieva, Nina A. (2016b): "Novye shtrikhi k portretam filosofov. Dva pis'ma E. Kassirera k S. L. Rubinshtejnu". *Voprosy filosofii*, no. 2, pp. 127–136.
Dmitrieva, Nina A. (2016c): "S. L. Rubinshtejn kak chitatel' «Fenomenologii dukha»". *Problemy sovremennogo obrazovanija*, no. 4, pp. 9–19.
Durkheim, Emile (1995): *The Elementary Forms of Religious Life*, trans. Karen E. Fields. New York: The Free Press.
Durkheim, Emile (1997): *The Division of Labor in Society*. W. D. Halls (Trans.). New York: The Free Press.
Ermichin. Oleg T. (2009): "B. V. Jakovenko i filosofija russkoj emigracii". *Filosofskie nauki*, no. 6, pp. 71–79.
Evtuhov, Catherine (1997): *The Cross and the Sickle: Sergei Bulgakov and the Fate of Russian Religious Philosophy*. Ithaca, NY: Cornell University Press.
Filatov, V. P. (1999): "Spektorskij Evgenij Vasil'evich". In *Filosofy Rossii XIX-XX stoletij. Biografii, idei, trudy*. Pëtr V. Alekseev (Ed.). Moscow: Akademicheskij Proekt, pp. 756–757.
Finkel, Stuart (2007): *On the Ideological Front. The Russian Intelligentsia and the Making of the Soviet Public Sphere*. New Haven: Yale University Press.
Fokht, Boris A. (1909): [Review of] V. Saval'skij. *Osnovy filosofii prava v nauchnom idealizme*. *Kriticheskoe obozrenie*, vyp. II, pp. 66–74.
Fokht, Boris A. (2003): *Izbrannoe (iz filosofskogo nasledija)*. Nina A. Dmitrieva (Ed.). Moscow: Progress-Tradicija.
Frank, Semion L. (1900): *Teorija cennosti Marksa i eja znachenie*. St. Petersburg: Tip. V. A. Tikhanova.

Frank, Semion L. (1903): [Review of] P. Struve. *Na raznye temy (1893–1901gg.). Voprosy filosofii i psikhologii*, kn. 66, pp. 96–102.
Frank, Semion L. (1904): "O kriticheskom idealism". *Mir Bozhij*, no. 12, pp. 224–264.
Frank, Semion L. (1910): "Lichnost' i veshch'". In *Filosofija i zhizn'*. St. Petersburg: D. E. Zhukovskogo, pp. 164–217.
Frank, Semion L. (1913): "Novaja kniga «Logosa»". *Russkaja molva*, no. 28, p. 5.
Frank, Semion L. (1996): "Predsmertnoe. Vospominanija i mysli". In *Russkoe mirovozzrenie*, A. A. Ermichev (Ed.). St. Petersburg: Nauka, pp. 39–58.
Frank, Semion L. (2007): "Predislovie perevodchika". In Vindel'band, Vil'gel'm. *Preljudii*. Moscow: Giperboreja, pp. 5–6.
Glukhikh, Ja. A. (2002): "Problemy prava i nravstvennosti v filosofii P. I. Novgorodceva". *Vestnik MGTU* Tom 5, no. 3, pp. 321–328.
Gordon, Gavriil O. (1995): "Iz vospominanij o G. I. Chelpanove". *Voprosy psikhologii*, no. 1, pp. 84–96.
Grekhova, G. I. (1976): "Epistoljarnoe nasledie A. S. Lappo-Danilevskogo". In *Vspomogatel'nye istoricheskie discipliny* Tom VIII. Leningrad: Izd. Nauka, pp. 262–273.
Grot, N. Ja. (1896): [Review of] Chelpanov. *Problema vosprijatija prostranstva v svjazi s ucheniem ob apriornosti i vrozhdennosti. Voprosy filosofii i psikhologii*, kn. 35, pp. 651–656.
Gunter, Sadi (1906): *Materialisticheskoe ponimanie istorii i prakticheskij idealizm*. Kiev: Izd. E. Gorskoj.
Gurvitch, George (1932): "Bogdan Alexandrovich Kistyakovsky". In *Encyclopedia of the Social Sciences*. Vol. 8. R. A. Seligman & Alvin Johnson (Eds.). New York: The Macmillan Company, pp. 575–576.
Guyer, Paul (1987) *Kant and the Claims of Knowledge*. Cambridge: Cambridge University Press.
Harich, Wolfgang (2000): *Nicolai Hartmann. Leben, Werk, Wirkung*. Würzburg: Königshausen & Neumann.
Hartmann, Nicolai (1958): "Zu Wilhelm Sesemann". In *Kleinere Schriften*. Band III: *Vom Neukantianismus zu Ontologie*. Berlin: de Gruyter, pp. 368–374.
Helmholtz, Hermann (1876): "The Origin and Meaning of Geometrical Axioms". *Mind* 3(1), pp. 301–321.
Hessen, Sergius (1909): *Individuelle Kausalität. Studien zum transzendentalen Empirismus*. Berlin: Reuther & Reichard.
Hessen, Sergej I. (1910a): "Mistika i metafizika". *Logos*, kn. 1, pp. 118–156.
Hessen, Sergej I. (1910b): [Review]. E. Husserl. *Logicheskie issledovanija*. Ch. 1. *Prolegomeny k chistoj logike. Voprosy filosofii i psikhologii*, kn. 102: 185–188.
Hessen, Sergej I. (1993): "Pis'ma redaktora nepartijnogo zhurnala". *Vestnik Rossijskoj akademii nauk*. Tom 63, no. 6, pp. 526–535.
Hessen, Sergej I. (1994): "Moe zhizneopisanie". *Voprosy filosofii*, no. 7–8, pp. 152–187.
Hessen, Sergej I. (1995): *Osnovy pedagogiki. Vvedenie v prikladnuju filosofiju*. Moscow: Shkola-Press.
Hessen, Sergej I./Metner, Emilij K./Stepun, Fedor A. (1910): "Ot redakcii". *Logos*, kn. 1, pp. 1–16.
Heuman, Susan (1998): *Kistiakovsky. The Struggle for National and Constitutional Rights in the Last Years of Tsarism*. Cambridge, MA: Harvard University Press.

Husserl, Edmund (1970): *Logical Investigations*. J. N. Findlay (Trans.). New York: Humanities Press.
Husserl, Edmund (2006): *The Basic Problems of Phenomenology*. Dordrecht: Springer.
Husserl, Edmund (2014): *Ideas for a Pure Phenomenology and Phenomenological Philosophy. First Book: General Introduction to Pure Phenomenology*. Daniel O. Dahlstrom (Trans.). Indianapolis/Cambridge: Hackett Publishing Company.
Jakovenko, Boris (1910a): "O sovremennom sostojanii nemeckoj filosofii". *Logos*, kn. 1, pp. 250–267.
Jakovenko, Boris (1910b): "O teoreticheskoj filosofii Germana Kogena". *Logos*, kn. 1, pp. 199–249.
Jakovenko, Boris (1911): "O Logose". *Logos*, kn. 1, pp. 57–92.
Jakovenko, Boris (2000): *Moshch' filosofii*. St. Petersburg: Nauka.
Jakovenko, Boris (2003): *Istorija russkoj filosofii*. Moscow: Respublika.
Jakowenko, Boris (1909a): "Die Logistik und die transzendentale Begründung der Mathematik". In *Bericht über den III. Internationalen Kongress für Philosophie*. Theodor Elsenhaus (Ed.). Heidelberg: Carl Winter's Universitätsbuchhandlung, pp. 868–880.
Jakowenko, Boris (1909b): "Was ist die transzendentale Methode?". In *Bericht über den III. Internationalen Kongress für Philosophie*. Theodor Elsenhaus (Ed.). Heidelberg: Carl Winter's Universitätsbuchhandlung, pp. 787–799.
Jakowenko, Boris (1930a): "Dreißig Jahre russischer Philosophie (1900–1929)". *Der russische Gedanke*. Jahr 1, no. 3, pp. 325–349.
Jakowenko, Boris (1930b): "Kritische Bemerkungen über die Phänomenologie". *Der Russische Gedanke*. Jahr 2, no. 1, pp. 25–32.
Jonkus, Dalius (2017): "Filosofija Vasilija Sezemana: Neo-Kantianstvo, intuitivism i fenomenologija". *Horizon* 6(1), pp. 79–96.
Kagan, Iudif' (1998): "People Not of Our Time". In *The Contexts of Bakhtin: Philosophy, Authorship, Aesthetics*. David Shepherd (Ed.). Amsterdam: Harwood Academic Publishers, pp. 3–16.
Kagan, Matvej I. (2004a): "Hermann Cohen". In *The Bakhtin Circle in the Master's Absence*. Craig Brandist/David Shepherd/Galin Tihanov (Eds.). Manchester: Manchester University Press, pp. 193–211.
Kagan, Matvej I. (2004b): *O khode istorii*. V. L. Makhlin (Ed.). Moscow: Jazyki slavjanskoj kul'tury.
Kant, Immanuel (1992): *Theoretical Philosophy, 1755–1770*. David Walford/Ralf Meerbote (Trans.). Cambridge: Cambridge University Press.
Kant, Immanuel (1996): *Practical Philosophy*. Cambridge: Cambridge University Press.
Kant, Immanuel (1997): *Critique of Pure Reason*. Paul Guyer/Allen W. Wood (Trans.). Cambridge: Cambridge University Press.
Kant, Immanuel (2000): *Critique of the Power of Judgment*. Paul Guyer/Eric Matthews (Trans.). Cambridge: Cambridge University Press.
Kant, Immanuel (2002): *Theoretical Philosophy after 1781*. Cambridge: Cambridge University Press.
Kant, Immanuel (2007): *Anthropology, History, and Education.*, Günter Zöller/Robert B. Louden (Eds.). Cambridge: Cambridge University Press.
Kareev, Nikolaj (1897): "Ekonomicheskij materialism i zakonomernost' social'nykh javlenij". *Voprosy filosofii i psikhologii*, kn. 36, pp. 107–119.

Kareev, Nikolaj (1920): "Istor.-teoreticheskie trudy A. S. Lappo-Danilevskago". *Russkij istoricheskij zhurnal*, kn. 6, pp. 112–131.
Kejdan, Vladimir I. (Ed.) (1997): *Vzyskujushchie grada. Khronika chastnoj zhizni russkikh religioznykh filosofov v pis'makh i dnevnikakh.* Moscow: Shkola jazyka russkoj kul'tury.
Kejdan, Vladimir I. (Ed.) (2019): *Vzyskujushchie grada. Khronika russkikh literaturnykh, religiozno-filosofskikh i obshchestvenno-politicheskikh dvizhenij v chastnykh pis'makh i dnevnikakh.* Moscow: Modest Kolerov.
Kemp Smith, Norman (1962): *A Commentary to Kant's 'Critique of Pure Reason'.* Atlantic Highlands, NJ: Humanities Press.
Khvostov, Veniamin M. (1909): "K voprosu o svobode voli". *Voprosy filosofii i psikhologii*, kn. 96, pp. 31–54.
Khvostov, Veniamin M. (1910): "Nauki ob obshchem i nauki ob individual'nom". *Voprosy filosofii i psikhologii*, kn. 103, pp. 340–367.
Khvostov, Veniamin M. (1913): *Ocherk istorii eticheskikh uchenij.* Moscow: Tip. Vil'de.
Khvostov, Veniamin M. (1914): *Teorija istoricheskago processa.* Moscow: Tip. Vil'de.
Khvostov, Veniamin M. (1917): "Klassifikacija nauki i mesto sociologii v sisteme nauchnago znanija". *Voprosy filosofii i psikhologii*, kn. 139, pp. 69–126.
Khvostov, Veniamin M. (1918): "Social'naja svjaz'". *Voprosy filosofii i psikhologii*, kn. 141–142, pp. 49–84.
Kiejzik, Lilianna (2018): "Kriticheskij marksizm N. A. Berdjaeva – period formirovanija". *Vestnik SPbGU. Filosofija i konfliktologija.* Tom 34, vyp.2, pp. 177–185.
Kireevskij, Ivan V. (1911): *Polnoe sobranie sochinenij v dvukh tomakh*, ed. Gershenzon. Moscow: Tip. Moskovskogo universiteta. Tom I.
Kistiakowski, Theodor (1899): *Gesellschaft und Einzelwesen. Eine methodologische Studie.* Berlin: Otto Liebman.
Kistjakovskij, Bogdan A. (1907): "V zashchitu nauchno-filosofskago idealizma". *Voprosy filosofii i psikhologii*, kn. 86, pp. 57–109.
Kistjakovskij, Bogdan A. (1916): *Social'nyja nauki i parvo.* Moscow: Izd. Sabashnikovykh.
Kolerov, Modest A. (2002): *Sbornik «Problemy idealizma». [1902]. Istorija i kontekst.* Moscow: Tri kvadrata.
Kolerov, Modest A. (2018): "Russkij Kogen: nekrologi 1918 goda". *Kantovskij sbornik.* Tom 37, no. 2, pp. 58–63.
Koval'shuk, Svetlana N. (2016): "Aleksandr Vejdeman – Filosof iz peterburga. Naparadnaja biografija na osnove arkhivnykh istochnikov". *Vestnik SPbGU*, ser. 2, vyp. 1, pp. 31–41.
Kudrinskij, M. A. (1993): "Arkhivnaja Istorija sbornika 'Problemy idealizma' (1902)". *Voprosy filosofii*, no. 4, pp. 157–165.
Kusch, Martin (1995): *Psychologism: A Case Study in the Sociology of Philosophical Knowledge.* London/New York: Routledge.
Ladyzhenskij, Aleksandr M. (1966): "Pamjati A. L. Sakketti". *Pravovedenie*, no. 2, pp. 154.
Lanc, Genrikh (1909a): "Edmund Gusserl' i psikhologisty nashikh dnej". *Voprosy filosofii i psikhologii*, no. 98, pp. 393–443.
Lanc, Genrikh (1909b): "Vil'ge'm Shuppe i ideja universal'noj immanentnosti". *Voprosy filosofii i psikhologii*, no. 100, pp. 757–799.
Lanc, Genrikh (1914): "Bytie i znanie v filosofii Fikhte". *Voprosy filosofii i psikhologii*, no.122, pp. 227–282.

Lange, Fr. Al'b. (1881–1883): *Istorija materializma i kritika ego znachenija v nastojashchee vremja.* N. N. Strakhov (Trans.). St. Petersburg: L. F. Panteleeva.
Lange, Fridrikh Al'bert (1899–1900): *Istorija materializme i kritika ego znachenija v nastojashchem.* V. S. Solov'ëv (Ed.). Kiev, Kharkov: F. A. Ioganson.
Lanz, Heinrich (1912): *Das Problem der Gegenständlichkeit in der modernen Logik.* Berlin: Reuther & Reichard.
Lanz, Henry (1926): "The Irrationality of Reasoning". *The Philosophical Review.* Vol. 35, no. 4, pp. 340–359.
Lappo-Danilevskij, Aleksandr S. (1890): "Istoricheskie disputy v 1890 g.". *Istoricheskoe obozrenie.* Tom 1, pp. 276–310.
Lappo-Danilevskij, Aleksandr S. (1918): "Metodologija istorii". *Izvestija Rossijskoj akademii nauk,* vi serija, tom xii, no. 5, pp. 239–260.
Lappo-Danilevskij, Aleksandr S. (2006): *Metodologija istorii.* Moscow: Izd. dom «Territorija budushchego».
Lappo-Danilevskij, Aleksandr S. (2013): "Obshchee obozrenie (Summa) osnovnykh principov obshchestvovedenija. Kurs 1902–1903gg.". *Voprosy filosofii,* no. 12, pp. 96–105.
Lapschin, Ivan (1909): "Denkgesetze und Erkenntnisformen". *Kant-Studien.* Vol.14, no. 1–3, pp. 89–90.
Lapshin, Ivan (1896): "Filosofskoe znachenie psikhologicheskikh vozzrenij Dzhemsa". In Dzhems, Uill'jam. *Psikhologija. (Text Book of Psychology).* Ivan I. Lapshin (Trans.). St. Petersburg: Tip. Sojkina, pp. 1–35.
Lapshin, Ivan (1906a): "O trusnosti v myshlenii". In *Zakony myshlenija i formy poznanija.* St. Petersburg: Tip. Bezobrazova, pp. 273–327.
Lapshin, Ivan (1906b): *Zakony myshlenija i formy poznanija.* St. Petersburg: Tip. Bezobrazova.
Lapshin, Ivan (1910): *Problema "chuzhogo Ja" v novejshej filosofii.* St. Petersburg: Senatskaja tip.
Lapshin, Ivan (1911): *Vselenskoe chuvstvo.* St. Petersburg: M. O. Vol'f.
Lapshin, Ivan (1923): *Oproverzhenie solipsizma.* Prague: Slovo.
Lapshin, Ivan (2006): "Chto est' istina?". In *Niezdannyj Ivan Lapshin.* L. G. Barsova (Ed.). St. Petersburg: SPbGATI, pp. 280–323.
Lask, Emil (1905): *Rechtsphilosophie.* Heidelberg: Carl Winter.
Lask, Emil (1923): *Gesammelte Schriften,* Band II. Tübingen: J. C. B. Mohr.
Levin, Iosif D. (1991): "Shestoj plan". In *Istoriko-filosofskij ezhegodnik 1991.* Moscow: Nauka, pp. 271–306.
Levchenko, Valerij V. (2012): "Sergej Leonidovich Rubinshtejn: grani intellektual'noj biografii Odesskogo period (1910–1920-e gg.)". *Vestnik ONU. Ser: Bibliotekoznavstvo, bibliografoznavstvo, knigoznavstvo.* Tom 17, no. 2(8), pp.109–122.
Longuenesse, Béatrice (1998): *Kant and the Capacity to Judge.* Charles T. Wolfe (Trans.). Princeton, NJ: Princeton University Press.
Lossky, Nicholas O. (1972): *History of Russian Philosophy.* New York: International Universities Press.
Malinov, Aleksej V./Pogodin, Sergej N. (2001): *Aleksandr Lappo-Danilevskij: istorik i filosof.* St. Petersburg: Iskusstvo—SPB.
Marx, Karl (1994): *Selected Writings.* Lawrence H. Simon (Ed.). Indianapolis, IN: Hackett.

Meerson, Michael A. (1995): "Put' against Logos: The Critique of Kant and Neo-Kantians by Russian Religious Philosophers in the Beginning of the Twentieth Century". *Studies in East European Thought*. Vol. 47, pp. 225–243.

Mohr, Friedrich (1869): *Allgemeine Theorie der Bewegung und Kraft, als Grundlage der Physik und Chemie*. Braunschweig: Friedrich Vieweg und Sohn.

Natorp, Paul (1910): *Die logischen Grundlagen der exacten Wissenschaften*. Leipzig and Berlin: Teuber.

Natorp, Paul (1965): *Allgemeine Psychologie nach kritischer Methode*. Amsterdam: E. J. Bonset.

Nemeth, Thomas (1995): "Aleksandr I. Vvedenskij on Other Minds". *Studies in East European Thought*. Vol. 47, no. 3–4, pp. 155–177.

Nemeth, Thomas (2017): *Kant in Imperial Russia*. Cham, Switzerland: Springer.

Nemeth, Thomas (2018): "Positivism in Late Tsarist Russia: Its Introduction, Penetration, and Diffusion". In *The Worlds of Positivism*. Johannes Feichtinger/Jan Surman/Franz L. Fillafer (Eds.). Cham, Switzerland: Palgrave Macmillan, pp. 273–291.

Nethercott, Frances (1995): *Une recontre philosophique. Bergson en Russie (1907–1917)*. Paris: Éditions L'Harmattan.

Novgorodcev, Pavel I. (1896): *Istoricheskaja shkola juristov, eja proiskhozhdenie i sud'ba*. Moscow: Universitetskaja tipografija.

Novgorodcev, Pavel I. (1902a): "Moral' i poznanie". *Voprosy filosofii i psikhologii*, kn. 64, pp. 824–838.

Novgorodcev, Pavel I. (1902b): "Nravstvennyj idealizm v filosofii prava". In *Problemy Idealizma*. Pavel I. Novgorodcev (Ed.). Moscow: Izd. Moskovskago psikhol. O-va, pp. 236–296.

Novgorodcev, Pavel I. (1907): "Pamjati Kuno Fishera". *Voprosy filosofii i psikhologii*, kn. 89, pp. v-viii.

Novgorodcev, Pavel I. (1909): "Russkij posledovatel' Germana Kogena". *Voprosy filosofii i psikhologii*, kn. 99, pp. 636–661.

Novgorodcev, Pavel I. (1910): "G. Saval'skij o samom sebe i o drugikh". *Voprosy filosofii i psikhologii*, kn. 105, pp. 382–389.

Novgorodcev, Pavel I. (1913): "Nauki obshchestvennyja i estestvennyja. (Neskol'ko zamechanij po povodu knigi N. N. Alekseeva)". *Voprosy filosofii i psikhologii*, kn. 120, pp. 716–722.

Novgorodcev, Pavel I. (1991): *Ob obshchestvennom ideale*. Moscow: Izd. Pressa.

Novgorodcev, Pavel I. (2000): *Kant i Gegel' v ikh uchenijakh o prave i gosudarstve*. St. Petersburg: Aletejja.

Novgorodtsev, Pavel I. (2003): "Ethical Idealism in the Philosophy of Law". In *Problems of Idealism: Essays in Russian Social Philosophy*. Randall A. Poole (Ed.). New Haven, CT: Yale University Press, pp. 274–324.

Pasternak, Boris (1959): *Safe Conduct. An Autobiography and Other Writings*. New York: Signet Books.

Pasternak, Evgenij B. (1997): *Boris Pasternak. Biografija*. Moscow: Citadel'.

Pasternak, Evgenij B./Polivanov, K. M. (1990): "Pis'ma Borisa Pasternaka iz Marburga". In *Pamjatniki kul'tury: novye otkrutija. Ezhegodnik 1989*. Moscow: Akademija nauk SSSR.

Paulsen, Friedrich (1902): *Immanuel Kant: His Life and Doctrine*, trans. J. E Creighton and Albert Lefevre. New York: Charles Scribner's Sons.

Pipes, Richard (1970): *Struve: Liberal on the Left, 1870–1905*. Cambridge, MA: Harvard University Press.
Plotnikov, Nikolaj S. (2006): "*Logos* v istorii evropejskoj filosofii: proekt i pamjatnik". In *«Logos» v istorii evropejskoj filosofii: Proekt i pamjatnik*. N. S. Plotnikov (Ed.). Moscow: Isd. Dom Territorija budushchego, pp. 7–12.
Poole, Brian (1997): "Matvei Kagan". *Experiment/эксперимент*, no. 3, pp. 248–266.
Poole, Randall A. (1999): "The Neo-Idealist Reception of Kant in the Moscow Psychological Society". *Journal of the History of Ideas*. Vol. 60, no.2, pp. 319–343.
Popova, Ol'ga A. (2006): "Predislovie k publikacii". *Logos* 6 (57), pp. 185–187.
Popova, Ol'ga A. (2009): "Vlijanie neokantianstva na filosofskie vozzrenija Genrikha Lanca". *Nauchnye vedomosti BelGU*, No. 10(65), pp. 210–217.
Povilajtis, Vladis. (2007): "O filosofii Vasilija Sezemana". *Issledovanija po istorii russkoj mysli. Ezhegodnik 2004–2005*. M. A. Kolerov/N. S. Plotnikov (Eds.). Moscow: Modest Kolerov, pp. 234–248.
"Predpolagaemoe soderzhanie blizhajshikh knig" (1911): *Logos*, no. 1, pp. 239–240.
Preobrazhenskij, V. (1889): [Review of] H. Cohen. Kants Begründung der Aesthetik. *Voprosy filosofii i psikhologii*, kn. 1, pp. 95–96.
Presnjakov, Aleksandr E. (1920): "A. S. Lappo-Danilevskij kak uchenyj i myslitel'". *Russkij istoricheskij zhurnal*, kn. 6, pp. 82–96.
Presnjakov, Aleksandr E. (1922): *Aleksandr Sergeevich Lappo-Danilevskij*. Petersburg: Kolos.
Pustarnakov, Vladimir F. (1999): "I. I. Lapshin kak filosof, issledovatel' na uchnogo i khudozhestvennogo tvorchestva". In Lapshin, Ivan: *Filosofija izobretenija i izobretenie v filosofii: Vvedenie v istoriju filosofii*. Moscow: Respublika, pp. 340–355.
Pustarnakov, Vladimir F. (2003): *Universitetskaja filosofija v Rossii*. St. Petersburg: Izd. Russkogo khristianskogo gumanitarnogo instituta.
Putnam George F. (1977): *Russian Alternatives to Marxism*. Knoxville, TN: The University of Tennessee Press.
Rezvykh, Tat'jana (2018): "O cennosti: podkhod Semena Franka (1898–1908)". *Issledovanija po istorii russkoj mysli. Ezhegodnik 2018*. Moscow: Modest Kolerov, pp. 47–83.
Rickert, Heinrich (1902): *Die Grenzen der naturwissenschaftlichen Begriffsbildung*. Tübingen und Leipzig: J.C.B. Mohr.
Rickert, Heinrich (1904): *Der Gegenstand der Erkenntnis*: Einführung in die Transzendentalphilosophie. Tübingen und Leipzig: J.C.B. Mohr.
Rickert, Heinrich (1905): Gutachten zur Dissertation von Moses Rubinshtejn «Die logischen Grundlagen des Hegelschen Systems und das Ende der Geschichte». In Rubinshtejn, Moisej M. (2008): *O smysle zhizni*. N. S. Plotnikov/K. V. Faradzhev (Eds.). Moscow: Izd. Territorija budushchego, pp. 355–356.
Riehl, Alois (1879): *Der philosophische Kriticismus und seine Bedeutung für die positive Wisenschaft*. Zweiter Band, Erster Theil. Leipzig: Wilhelm Engelmann.
Riehl, Alois (1887): *Der philosophische Kriticismus und seine Bedeutung für die positive Wisenschaft*. Zweiter Band, Zweiter Theil. Leipzig: Wilhelm Engelmann.
Riehl, Alois (1894): *Introduction to the Theory of Science and Metaphysics*. London: Kegan Paul, Trench, Trübner, & Co.
Rikkert, Genrikh (1911): *Nauki o prirode i nauki o kul'ture*. Sergej Hessen (Trans.). Moscow: Obrazovanie.

Royce, Josiah (1909): "The Problem of Truth in the Light of Recent Discussion". In *Bericht über den III. Internationalen Kongres für Philosophie*. Theodor Elsenhans (Ed.). Heidelberg: Carl Winter's Universitätsbuchhandlung, pp. 62–90.

Rubcov, V. V./ Serova, O. E. Serova/Guseva, E. P. (2012): "K 150-letiju so dnja rozhdenija Georgija Ivanovicha Chelpanova". *Kul'turno-istoricheskaja psikhologija*, no. 1, pp. 92–109.

Rubinshtejn, Moisej M. (1905): "Logicheskija osnovy sistemy Gegelja i konec istorii". *Voprosy filosofii i psikhologii*, kn. 80, pp. 695–764.

Rubinshtejn, Moisej M. (1907): "Genrikh Rikkert. Ocherk teoretiko-poznavatel'ago idealizma". *Voprosy filosofii i psikhologii*, kn. 86, pp. 1–61.

Rubinshtejn, Moisej M. (1909): "K voprosu o metodologii i gnoseologii Dekarta". *Voprosy filosofii i psikhologii*, kn. 97, pp. 143–169.

Rubinshtejn, Moisej M. (1911): "K voprosu o transcendentoj real'nosti". *Voprosy filosofii i psikhologii*, kn. 106, pp. 19–54.

Rubinshtejn, Moisej M. (1923): *Problema "Ja" kak iskhodnyj punkt filosofii*. Irkutsk: Tip. Okruzhinogo voenno-redakcionnogo soveta.

Rubinshtejn, Moisej M. (2008): *O smysle zhizni*. N. S. Plotnikov/K. V. Faradzhev (Eds.). Moscow: Izd. Territorija budushchego.

Rubinshtejn, Sergej L. (1986): "Princip tvorcheskoj samodejatel'nosti". *Voprosy psikhologii*, kn. 4, pp. 101–108.

Rubinshtejn, Sergej L. (1989): *Ocherki, vospominanija, materialy*. B. F. Lomov (Ed.). Moscow: Nauka.

Rubinshtejn, Sergej L. (2003): "O filosofskoj Sistema g. Kogena". In *Bytie i soznanie. Chelovek i mir*. St. Petersburg: Piter, pp. 428–451.

Rubinstein, Moses (1906): *Die logischen Grundlagen des Hegelschen Systems und das Ende der Geschichte*. Halle: A. Kaemmerer & Co.

Rubinstein, Sergej (1914): *Eine Studie zum Problem der Methode*. Marburg: Alfred Töpelmann.

Rumjanceva, Marina F. (2012): "Russkaja versija neokantianstva: k postanovke problem". *Uchenye zapiski Kazanskogo universiteta*. Tom 154, kn. 1, pp. 130–141.

Rumjanceva, Marina. F. (2014): "Fenomenologija vs neokantianstvo v koncepcii A. S. Lappo-Danilevskogo". In *Dialog so vremenem 46*. Moscow: Institut vseobshchej, pp. 7–16.

Russell, Bertrand (1948): *Human Knowledge: Its Scope and Limits*. London: George Allen & Unwin.

Sakketti, Aleksandr L. (2017): "Uchenie G. Kogena o forme muzyki. Tekst doklada". In *Iskusstvo kak jazyk – jazyki iskusstva. Gosudarstvennaja akademija khudozhestvennykh nauk i esteticeskaja teorija 1920-x goda*. Tom II. N. S. Plotnikov/N. P. Podzemskaja (Eds.). Moscow: Novoe literaturnoe obozrenie, pp. 412–445.

Sandler, Sergeiy (2015): "A Strange Kind of Kantian: Bakhtin's Re-interpretation of Kant and the Marburg School". *Studies in East European Thought*. Volume 67, no. 3–4, pp. 165–182.

Sapov, V. V. (2013): "Zhurnal «Logos» – prervannyj na polouslove dialog". In Stepun, Fedor A.: *Pis'ma*. V. K. Kantor (Ed.). Moscow: ROSSPEN, pp. 60–70.

Saval'skij, Vasilij A. (1905): "Kritika ponjatija solidarnosti v sociologii O. Konta". *Zhurnal Ministerstva narodnago prosveshchenija*, chast' CCCLXI, otd. 2, pp. 94–106.

Saval'skij, Vasilij A. (1907): "Vvedenie v filosofiju prava". *Kriticheskoe obozrenie*, vyp. 5, pp. 5–13.

Saval'skij, Vasilij A. (1909): *Osnovy filosofii prava v nauchom idealizme. Marburgskaja shkola filosofii: Kogen, Natorp, Shtammler i dr.* Moscow: Tip. Imperatorskogo Moskovskogo universiteta.
Saval'skij, Vasilij A. (1910): "Otvet prof. Novgorodcev". *Voprosy filosofii i psikhologii*, kn. 105, pp. 341–381.
Saval'skij, Vasilij A. (1912): *Gosudarstvennoe pravo. Obshchee i russkoe. Konspekt lekcij.* Warsaw: Izdanie P. Saval'skago.
Savenko, Georgij V. (2002): "Aleksandr Liverievich Sakketti". *Izvestija vysshikh uchebnykh zavedenij. Pravovedenie*, no. 5, pp. 232–241.
Sesemann, Vasily (2007): *Aesthetics*. Mykolas Drunga (Trans.). Amsterdam/New York: Rodopi.
Sesemann, Wilhelm (1911/1912): "Das Rationale und das Irrationale im System der Philosophie". *Logos: international Zeitschrift für Philosophie*. Band 2, pp. 208–241.
Sesemann, Wilhelm (1927): *Beiträge zum Erkenntnissproblem*. Kaunas: Lietuvos universiteto Humanitariniu mokslu fakulteto rastai.
Sezeman, Vasilij (1911/1912): "Racional'noe i irracional'noe v sisteme filosofii". *Logos*, kn. 1: 93–122.
Sezeman, Vasilij (1913): "Teoreticheskaja filosofija Marburgskoj shkoly". In *Novye idei v filosofii*, sbornik 5. N.O. Losskij/E. L Radlov (Eds.). St. Petersburg: Izd. Obrazovanie, pp. 1–34.
Sezeman, Vasilij (1925): [Review of] Nicolai Hartmann. *Grundzüge einer Metaphysik der Erkenntnis. Logos*, no. 1, pp. 229–235.
Sezeman, Vasilij (1928): [Review of] Aleksandr Vejdeman. *Myshlenie i bytie. Versty*, no. 3, pp. 163–172.
Sezeman, Vasilij (2000a): "Filosofija. Maks Sceler". In *Antologija fenomenologicheskoj filosofii v Rossii*. Tom 2. Igor M. Chubarov (Ed.). Moscow: Izd. «Logos», izd. «Progress-Tradicija», pp. 500–505.
Sezeman, Vasilij (2000b): "Obozrenie novejshej germanskoj filosofskoj literatury". In *Antologija fenomenologicheskoj filosofii v Rossii*. Tom 2. Igor M. Chubarov (Ed.). Moscow: Izd. «Logos», izd. «Progress-Tradicija», pp. 488–499.
Sezeman, Vasilij (2000c): [Review of] Martin Heidegger. *Sein und Zeit*. In *Antologija fenomenologicheskoj filosofii v Rossii*. Tom 2. Igor M. Chubarov (Ed.). Moscow: Izd. «Logos», izd. «Progress-Tradicija», pp. 505–512.
Shijan, A. A. (2019): "Transcendentalizm Jakovenko v filosofskom kontekste ego vremeni: fenomenologija i/ili neokantianstvo". *Vestnik RUDN* Vol 23, no. 4, pp. 443–460.
Shitov, Anatolij V. (2012): "Arkhiv B. V. Jakovenko v Prage". In *Boris Valentinovich Jakovenko*. Aleksandr A. Ermichev (Ed.). Moscow: ROSSPEN, pp. 509–514.
Shitov, Anatolij V. (2015): "Chitaja I. I. Lapshina". *Vestnik russkoj Khristianskoj gumanitarnoj akademii*. Tom 16, vyp. 3, pp. 268–271.
Shershenevich, Gabriel F. (1911): *Obshchaja teorija prava*. Moscow: Izd. Br. Bashmakovykh.
Shpet, Gustav (1991): *Appearance and Sense. Phenomenology as the Fundamental Science and Its Problems*. Thomas Nemeth (Trans.). Dordrecht: Kluwer Academic Publishers.
Shpet, Gustav (2004): "Opyt populjarizacii filosofii Gegelja". In Shchedrina, T. G. 2004. *"Ja pishu kak echo drugogo...": Ocherki intellektual'noj biografii Gustava Shpeta*. Moscow: Progress-Tradicija, pp. 281–322.
Shpet, Gustav (2010): [Review of] Khr. Zigwart. *Logika*. In *Filosofskaja kritika: otzyvy, redenzii, obzory*, 84–86. Moscow: ROSSPEN.

Shpet, Gustav (2019): *Hermeneutics and Its Problems. With Selected Essays in Phenomenology*. Thomas Nemeth (Trans.). Cham: Springer.
Shul'ca, Ioganna (1910): *Raz"jasnjajushchee izlozhenie "Kritiki chistogo razuma"*. Moscow: Tip. Somovoj.
Sieg, Ulrich (1994): *Aufstieg und Niedergang des Marburg Neukantianismus*. Band 1. Würzburg: Königshausen und Neumann.
Solov'ëv, Vladimir S. (1988): *Sochinenija v 2 t.* Tom 2. A. V. Gulygi/A. F. Loseva (Eds.). Moscow: Mysl'.
Spektorskij, Evgenij V. (1905): "Iz oblasti chistoj etiki". *Voprosy filosofii i psikhologii*, kn. 78, pp. 384–411.
Stange, Karl (1906): *Khod myslej v "Kritike chistago razuma"*. Moscow: Tip. Shushukina.
Steppuhn, Friedrich (1910): "Wladimir Ssolowjew". *Zeitschrift für Philosophie und philosophische Kritik* Band 138, pp. 1–79; 239–291.
Steppun, Fedor A. (1912): "Otkrytoe pis'mo Andreju Belomu po povudu stat'i «Krugovoe dvizhenie»". *Trudy i dni*, no. 4–5, pp. 74–86.
Stepun, F[edor] A. (1913): "Zhizn' i tvorchestvo". *Logos*, kn. 3–4, pp. 71–82.
Stepun, F[edor] A. (1923): *Zhizn' i tvorchestvo*. Berlin: Obelisk.
Stepun, F[edor] A. (1994): *Byvshee i nesbyvsheesja*, tom 1. St. Petersburg: Izd. Aletejja.
Stepun, F[edor] A. (2000): *Sochinenija*. Moscow: ROSSPEN.
Struve, Petr (1894): *Kriticheskija zametki k voprosu ob ekonomicheskom razvitii Rossii*. St. Petersburg: Tip. I. N. Skorokhodova.
Struve, Petr (1897a): "Eshche o svobode i neobkhodimosti". *Novoe slovo*, kn. 8 (Maj), pp. 200–208.
Struve, Petr (1897b): "Svoboda i istoricheskaja neobkhodimost'". *Voprosy filosofii i psikhologii*, kn. 36, pp. 120–139.
Struve, Peter von (1899): "Die Marxsche Theorie der sozialen Entwicklung. Ein kritischer Versuch". *Archiv für soziale Gesetzebung und Statistik*. Band 14, pp. 658–704.
Struve, Petr (1902): *Na raznyja temy. (1893–1901gg.). Sbornik statej*. St. Petersburg: Tip. A. E. Kolpinskogo.
Struve, Petr (1911): "Facies hippocratica". In *Patriotica. Politika, kul'tura, religija,socialism. Sbornik statej za pjat' let' (1905–1910gg.)*. St. Petersburg: Izd. D. E. Zhukovskogo, pp. 575–596.
Struve, Petr (1999): "Predislovie". In Berdjaev, Nikolaj A.: *Sub"ektizm i individualism v obshchestvennoj filosofii*. Moscow: OI Reabilitacija, pp. 5–78.
Struve, Peter (2003): "Toward Characterization of Our Philosophical Development". In *Problems of Idealism: Essays in Russian Social Philosophy*. Randall A. Poole (Ed.). New Haven, CT: Yale University Press, pp. 143–160.
Stumpf, Carl (1873): *Über den psychologischen Ursprung der Raumvorstellung*. Leipzig: Hirzel.
Swoboda, Philip J. (1995): "Windelband's Influence on S. L. Frank". *Studies in East European Thought* Volume 47, nos. 3–4, pp. 259–290.
Swoboda, Philip J. (2010): Semën Frank's Expressivist Humanism. In *A History of Russian Philosophy 1830–1930*. Gary M. Hamburg/Randall A. Poole (Eds.). Cambridge: Cambridge University Press, pp. 205–223.
Tikhonova, Elena V. (2010): "Russkij intelligent i istinnyj dzhentl'men Ivan Ivanovich Lapshin (1870–1952)". *Methodologija i istorija psikhologii*. Tom 5, vyp. 3, pp. 154–171.

Tikhonova, Elena V. (2013): "Kronika zhizni i tvorchestva A. I. Vvedenskogo (1856–1925)". In *NeoKantianstvo v Rossii. Aleksandr Ivanovich Vvedenskij, Ivan Ivanovich Lapshin*. V. N. Brjushkin/V. S. Popova (Eds.). Moscow: ROSSPEN, pp. 201–211.
Toporkov, Aleksej K. (1912a): "Ideja". *Trudy i dni*, no. 1, pp. 73–82.
Toporkov, Aleksej K. (1912b): "Sic et Non". *Trudy i dni*, no. 4–5, pp. 121–131.
Toporkov, Aleksej K. (1928): *Elementy dialekticheskoj logiki*. Moscow: Rabotnik prosveshchenija.
Treiber Hubert (1995): "Fedor Steppuhn in Heidelberg (1903–1955). Über Freundschafts- und Spätbürgertreffen in einer deutschen Kleinstadt". In *Heidelberg im Schnittpunkt intellektueller Kreise*. Hubert Treiber/K. Sauerland (Eds.). Opladen: Westdeutscher Verlag, pp. 70–118.
Tremblay, Frédéric (2017a): "Historical Introduction to Nicolai Hartmann's Concept of Possibility". *Axiomathes*. Vol. 27, no. 2, pp. 193–207.
Tremblay, Frédéric (2017b): "Nikolai Lossky and Henri Bergson". *Studies in East European Thought*. Vol. 69, no. 1, pp. 3–16.
Tremblay, Frédéric (2019): "Ontological Axiology in Nikolai Losky, Max Scheler, and Nicolai Hartmann". In *Nicolai Hartmanns Neue Ontologie und die Philosophische Anthropologie: Menschliches Leben in Natur und Geist*. Moritz Kalckreuth, et al. (Eds.). Berlin/Boston: Walter de Gruyter, pp. 193–232.
Trubeckoj, Evgenij N. (1909): "Panmetodizm v etike. (K kharakteristike uchenija Kogena)". *Voprosy filosofii i psikhologii*, kn. 97, pp. 121–165.
Trubeckoj, Evgenij N. (1917): *Metafizicheskija predpolozhenija poznanija*. Moscow: Russkaja pechatnja.
Vanchugov, Vasilij V. (2016): "N. O. Losskij kak interpretator istorii russkoj filosofii". In *Nikolaj Onufrievich Losskij*. Vladimir P. Filatov (Ed.). Moscow: Politicheskaja enciklopedija, pp. 178–215.
Vashestov, Andrej G. (1991): "Zhizn i trudy B. A. Fokhta". In *Istoriko-filosofskij ezhegodnik 1991*. Moscow: Nauka, pp. 223–231.
Vasil'ev, Jurij A. (2017): "Byl li A. S. Lappo-Danilevskij neokantiancem v istorii?". *Vlast'*. Tom 25, no. 3, pp. 186–191.
Vejdeman, Aleksandr (1927): *Myshlenie i bytie. (Logika dostatochnogo osnovanija.)* Riga: Akc. O-vo Pechatnogo Dela Salamandra.
Vejdeman, Aleksandr (1931): *Mir kak ponjatie (Myshlenie i bytie)*. Riga: Izdanie avtora.
Vejdeman, Aleksandr (1938): *Tragika kak sushchnost' iskusstva, religii i istorii*. Riga: A. V. Vejdeman.
Vejdeman, Aleksandr (1939): *Opravdanie zla*. Riga: Logos.
Vikhnovich, Vsevolod L.(2014): "Peterburgskij epizod Germana Kogena i ne tol'ko". *Vestnik Russkoj khristianskoj gumanitarnoj akademii*. Tom 15, vyp. 1, pp. 199–209.
Vucinich, Alexander (1976): *Social Thought in Tsarist Russia*. Chicago: University of Chicago Press.
Vvedenskij, Aleksandr I. (1886): "Uchenie Lejbnica o materii v svjazi s monadologiej". *Zhurnal Ministerstva narodnago prosveshchenija*. Chast' CCXLIII, otd. 2, pp. 1–49.
Vvedenskij, Aleksandr I. (1888): *Opyt postroenija teorii materii na principakh kriticheskoj filosofii*. St. Petersburg: Tip. Bezobrazova.

Vvedenskij, Aleksandr I. (1889): "Kritiko-filosofskij analiz massy i svjaz vysshikh zakonov materii v zakon proporcional'nosti". *Zhurnal Ministerstva narodnago prosveshchenija.* Chast' CCLXII, otd. 2, pp. 1–44.
Vvedenskij, Aleksandr I. (1890a): "K voprosu o stroenii materii". *Zhurnal Ministerstva narodnago prosveshchenija.* Chast' CCLXX, otd. 2, pp. 18–65.
Vvedenskij, Aleksandr I. (1890b): "K voprosu o stroenii materii". *Zhurnal Ministerstva narodnago prosveshchenija.* Chast' CCLXX, otd. 2, pp. 191–220.
Vvedenskij, Aleksandr I. (1892): "O predelakh i priznakakh odushevlenija". *Zhurnal Ministerstva narodnago prosveshchenija.* Chast' CCLXXXI, pp. 73–112.
Vvedenskij, Aleksandr I. (1898): *Sud'by filosofii v Rossii.* Moscow: Tip. Kushnerev.
Vvedenskij, Aleksandr I. (1908): "Prof. G. Chelpanova uchebnik logiki". *Zhurnal Ministerstva narodnago prosveshchenija.* Chast' 15 novaja serija, pp. 222–240.
Vvedenskij, Aleksandr I. (1909a): "Chto takoe filosofskij kriticizm". *Novoe slovo.* No. 1, pp. 22–27.
Vvedenskij, Aleksandr I. (1909b): "Novoe i legoe dokazatel'stvo filosofskago kriticizma". *Zhurnal Ministerstva narodnago prosveshchenija.* Chast' 20 novaja serija, pp. 122–144.
Vvedenskij, Aleksandr I. (1912):. *Logika, kak chast' teorii poznanija.* St. Petersburg: Tip. M. M. Stasjulevicha.
Vvedenskij, Aleksandr I. (1915): *Psikhologija bez vsjakoj metafiziki.* St. Petersburg: Tip. M. M. Stasjulevicha.
Vvedenskij, Aleksandr I. (1917): *Logika, kak chast' teorii poznanija.* Petrograd: Tip. M. M. Stasjulevicha.
Vvedenskij, Aleksandr I. (1922): *Logika, kak chast' teorii poznanija.* Petersburg: Gosudarstvennoe izd.
Vvedenskij, Aleksandr I. (1924): *Filosofskie ocherki.* Prague: Plamja.
Vysheslavcev, Boris P. (1914): *Etika Fikhte. Osnovy prava i nravstvennosti v sisteme transcendental'noj filosofii.* Moscow: Tip. Imperatorskogo Moskovskogo universiteta.
Walicki, Andrzej (1987): *Legal Philosophies of Russian Liberalism.* Oxford: Clarendon Press.
Weber, Max (1949): *The Methodology of the Social Sciences.* E. A. Shils/H. A. Finch (Eds.). New York: Free Press.
Weber, Max (1977): *Critique of Stammler.* Guy Oakes (Trans.). New York: The Free Press.
Weber, Max (2012): "Roscher and Knies and the Logical Problem of Historical Economics". In *Max Weber: Complete Methodological Writings.* Hans Henrick Bruun/Sam Whimster (Eds.). Abingdon, Oxon: Routledge, pp. 3–94.
Wedenskij, Alexander (1910): "Ein neuer und leichter Beweis für den philosophischen Kritizismus". *Archiv für systematische Philosophie.* Vol. xvi, no. 2, pp. 191–216.
Weidemann, Alexander (1930/1931): "Das Denken und sein Schaffen". *Der russische Gedanke.* Band 2, Heft 1, pp. 16–24.
Windelband, Wilhelm (1870): *Die Lehren vom Zufall.* Berlin: A. W. Schade's Buchdruckerei.
Windelband, Wilhelm (1905): *A History of Philosophy.* New York: The Macmillan Company.
Windelband, Wilhelm (1911a): *Präludien. Aufsätze und Reden zur Einführung in die Philosophie.* Erster Band. Tübingen: J. C. B. Mohr.
Windelband, Wilhelm (1911b): *Präludien. Aufsätze und Reden zur Einführung in die Philosophie.* Zweiter Band. Tübingen: J. C. B. Mohr.
Windelband, Wilhelm (2015a): "History and Natural Science". In *The Neo-Kantian Reader.* Sebastian Luft (Ed.). Routledge: New York, pp. 287–298.

Windelband, Wilhelm (2015b): Philosophy of Culture and Transcendental Idealism. In *The Neo-Kantian Reader*. Sebastian Luft (Ed.). Routledge: New York, pp. 317–324.

Wischeslavzeff, Boris (1912): "Recht und Moral". In *Philosophische Abhandlungen. Hermann Cohen zum 70sten Geburtstag (4 Juli 1912) dargebracht*. Berlin: Bruno Cassirer, pp. 190–202.

Wittgenstein, Ludwig (1953): *Philosophical Investigations*. G. E. M. Anscombe (Trans.). New York: The Macmillan Company.

Wortman, Richard S (1976): *The Development of a Russian Legal Consciousness*. Chicago: University of Chicago Press.

Wundt, Wilhelm (1910): "Psychologismus und Logizismus". *Kleine Schriften*. Erster Band. Leipzig: Wilhelm Engelmann, pp. 511–634.

Zenkovsky, V. V. (1953): *A History of Russian Philosophy*. George L. Kline (Trans.). London: Routledge & Kegan Paul Ltd.

Zolotnickij, Vladimir T. (1764): *Sokrashchenie estestvennogo prava*. St. Petersburg: n.p.

Index

aesthetic feeling 241
aesthetics, task of 49, 72, 106, 165 f., 194, 206, 223, 238, 240 f., 255, 257, 293 f.
Aksel'rod, Ljubov 221
Alekseev, Nikolaj N. 99, 191–198, 296
all-unity 211, 215, 227
analogical argument 35, 55, 59 f.
anthropologism 226
apodicticity of geometric axioms 76
a priori elements in experience 237
atomism, untenability of 14, 24
autonomy, human 58, 137 f., 163 f., 169, 240, 257, 262, 310

Baden neo-Kantianism 3, 142, 155, 158, 170 f., 186, 208, 210, 219, 266, 268, 272, 301, 322
Bakhtin Circle 278
Bakhtin, Mikhail 10, 275–278, 280
Belyj, Andrej 57 f., 212–214, 217, 231 f., 246, 297
Berdjaev, Nikolaj 105, 109–119, 126, 163, 189, 201, 216, 227 f., 267, 317
Bor'ba za Logos 201
Bulgakov, Sergej 92–100, 102–104, 106, 108, 111, 114, 118, 126, 194, 201, 228

Cassirer, Ernst 2, 6, 8, 20, 220, 223, 238, 252, 257 f., 261, 278, 311, 316–318
categorical imperative 113, 124, 137, 144, 147, 151 f., 170 f., 179, 240, 274
categories of cognition 44, 54, 79
causality, category of 20, 22, 44–45, 46 f., 75, 78, 81–82, 86, 88, 94, 112–113, 121, 124, 145, 156, 160
causality, law of 15, 19–20, 29, 42, 46, 70–71, 75, 87, 93–94, 97, 106, 108, 124–126, 136, 147, 151, 159 f., 162, 174–175, 179–182, 203, 205, 282
causality, noumenal 87, 102, 179–180, 182
Chelpanov, Georgij I. 9, 41, 64–91, 129, 188 f., 212, 233, 243, 317, 319
Chernyshevskij 103

cognition, three elements necessary for 2 f., 15 f., 20 f., 25, 30, 32, 40–42, 44–46, 48, 50 f., 54, 68, 72, 74–76, 78–82, 87–90, 94, 96 f., 101–107, 110–114, 119 f., 124, 136 f., 143, 145, 152, 160, 165–171, 173 f., 176, 182 f., 185, 189, 194, 196, 208, 214–216, 218, 221, 225–227, 233–235, 237–241, 248–253, 256 f., 259, 261–264, 266, 268–271, 273, 288, 291, 293, 299 f., 303–305, 308 f., 311–313
Cohen, Hermann 2 f., 5 f., 9 f., 28 f., 37, 44, 49 f., 57, 82, 89, 101, 117, 140, 143–145, 147–151, 154 f., 170, 180, 191, 195, 201, 204 f., 207, 214, 220–242, 244, 246–249, 252 f., 255–258, 260–265, 273, 275 f., 278–281, 283, 285–295, 297–299, 301, 306–312, 315 f., 321
Collins, Randall 321
Comte, Auguste 6, 141, 163, 172
consciousness 15–16, 18, 20–21, 28, 40, 49–50, 54–55, 59–62, 66, 73–79, 81–86, 101, 103, 107, 116, 120–124, 185–186, 241, 249–250, 267, 269, 304
consciousness, moral 140, 225
consciousness, scientific 236–237
consciousness, transcendental 95–97, 100–104, 111–114
consciousness, unity of 46–47, 96, 147, 236
constructivism of science 264
conviction in the other's mind, my 56
Copernican Revolution 238, 263, 277
cosmopolitanism 258
Critical Marxism 92, 97 f., 114 f.
critical realism, Chelpanov's 87, 90, 108, 184, 186
Cvetaev, Lev A. 132

Dalton's atomic theory 24
Davydov, Iosif Aleksandrovich 122–127, 191
definition of neo-Kantianism 1, 7

determinism   25, 31, 71, 93, 97–99, 102–104, 114, 124, 180, 290
differences between history and natural science   160
Dmitrieva, Nina   13, 36, 128f., 141, 148, 191, 198, 216, 231–233, 237, 243f., 247, 259, 261, 278, 286, 308
double affection   81, 87
Durkheim, Émile   5f., 44, 58, 61, 63
dynamism proposed by Critical Philosophy   18

empirical idealism   80f.
Enlightenment values, defense of   12, 133
epistemological subject   185f., 269–271
Ern, Vladimir F.   201
ethics in jurisprudence, role of   148, 152
ethics, normative   28, 72, 95, 98, 102, 104, 106f., 112, 119, 121, 123, 128, 133, 137, 140, 143–145, 147, 150, 152, 158, 162f., 166, 169f., 179f., 184, 186, 194–196, 205–207, 209, 216, 221–225, 229, 238–241, 245, 257, 262, 276f., 280f., 283, 290f., 293f., 306, 312
Eurasianism   198
existentialism   109, 277, 317

Fichte's ethical philosophy   245
Fischer, Kuno   119, 129, 208, 230
Fokht (Vogt), Boris A.   10, 147, 230–243, 286, 319
forms of intuition   30, 41–43, 46, 203
freedom of conscience   161
free will/determinism, issue of   70
French neo-Kantianism   5–7, 63, 322
French vitalistic philosophy   60
From Marxism to Idealism   99, 115

Gordon, Gavriil O.   10, 230, 242–244, 247, 286
Grot, Nikolaj Ja.   64, 66, 69

Hartmann, Nicolai   207, 229f., 243, 247, 286, 307f., 311–313, 317, 320
Hegel, Georg   28, 42f., 48, 57, 105, 109f., 135, 137–140, 144, 158, 169, 179, 220, 223, 245, 257, 259f., 265f., 273, 279, 286f., 290, 292–295, 298f., 301
Hegelian idealism   260, 320
Hessen, Sergej I.   58, 60, 198–208, 296, 319
historical materialism   92, 100, 122f., 126, 283
historical methodology   53, 58f., 172f., 176, 178
Historical School of Law   130, 135
Historicism   130, 146, 168
Husserl, Edmund   38, 40, 57, 59, 65, 85, 196, 199, 204, 206f., 212–214, 248–253, 276, 281, 299f., 302–304, 306, 313f., 316, 318

idiographic point of view   175f., 178
Immanentism   249f., 253, 306
impenetrability   13, 17f., 21, 309
Individuelle Kausalität   203
inductive metaphysics of Wundt   90
Infinitesimal method   287, 289
intentionality, theory of   50, 299
intersubjectivity   103
introspection as a psychological technique   84
intuitivism   37, 305, 313
irrational in cognition, the   312
Italian neo-Kantianism   5f.

Jakovenko, Boris   7, 52, 91, 104, 148, 199, 201–203, 228, 285f., 295–306, 320
James, William   40f., 80, 186
judgments of experience   34, 47, 166, 250, 304
jurisprudence   129–131, 133–135, 143f., 146, 148, 150, 158, 196f., 223–225, 239, 241, 245, 281, 283, 291
justice, category of   113f., 131, 133, 159, 161, 163, 257

Kagan, Matvej I.   10, 237, 255, 277–284
Kantian *a priori*   6, 41, 90
Kantian understanding of matter   23
Kant's „Refutation of Idealism"   16
Kareev, Nikolaj   95f., 171

Khvostov, Venjamin M.   38, 57–63, 179–187, 212
Kistjakovskij, Bogdan   142, 155–170, 187, 191f., 202f., 218, 319f.
Kries, Johannes von Kries   162
Kunicyn, Aleksandr P.   132f.

Lange, Friedrich A.   15f., 49, 89, 105, 258
Lanz, Henry   248–254, 310
Lappo-Danilevskij, Aleksandr S.   38, 53–59, 62, 141, 171–179, 187, 202, 255, 317
Lapshin as a „neo-Kantian"   38
Lapshin, Ivan I.   9, 38–53, 58, 80, 274, 289, 296, 307, 319f.
Lask, Emil   146, 170, 199, 203, 205, 303f., 309, 316, 319
Lassalle, Ferdinand   105, 112, 115f., 123
Legal Marxists   92
Leibniz's theory of matter   13
Lilienfeld, Pavel von   157
Lipps, Theodor   38, 53, 55, 154
logical laws of thought   29, 32
logical positivism   316f.
logic as normative   167–168, 189, 267
logicism, Vvedenskij's   33f., 285, 297
Logos   10, 198–202, 206, 212–214, 216, 218, 233, 244, 252f., 272, 296f., 300, 302, 308, 312, 321
Lopatin, Leo   7, 57f., 66, 182, 201, 212, 217, 220, 231f., 243, 246, 270, 272
Lossky, Nicholas   38–40, 208, 229

Marburg School   2, 10, 101, 119, 125, 128, 140–144, 151, 153, 215, 218–220, 222f., 229, 243, 247, 253, 276, 297, 300, 308, 310f.
Marxism, great deficiency of   9, 92–95, 97–105, 108–111, 113–118, 122–126, 176, 190, 203, 258, 283
materialism, historical   92, 100, 122, 126, 283f.
materialism, metaphysical   102–103, 105, 115
materialism, social   97
materialist philosophy of history   93
mathematical physics   17, 20, 49, 99, 143–145, 193, 204, 214, 222f., 226f., 234–236, 239, 241, 247, 261, 269, 280, 291, 313
mathematics, universal applicability of   6, 8, 25f., 28–32, 49, 52, 71f., 77, 106, 129, 144f., 148, 152, 193, 195, 197, 204f., 207, 220, 224, 234f., 262, 280–282, 287–290, 294, 297, 302, 306, 310
mathematization of nature   9, 234
mechanism, must be eliminated *a priori*   14, 18–20, 24, 48, 126, 168, 252
mechanistic materialism   23
metaphysical deduction   45, 120, 144f., 157, 213, 237
metaphysical sense as connected with morality   36
metaphysics as knowledge, impossibility of   26f., 33
methodologism   309
methodology, historical   53f., 58, 172–173, 176–179
methodology, scientific   173, 262, 292
Metner, Emilij   199–201, 297
minimalist epistemology   25
Mohr, Friedrich   23
moral law   28, 36, 72, 107, 110, 112f., 132, 136–138, 179, 241, 262, 267, 277
morally grounded faith   30, 32
moral sense, a   28, 36, 54, 61

naïve realism   15, 264, 274
nativist theories of space   66–67
Natorp, Paul   2f., 21, 89, 125f., 191, 207, 220f., 223, 230–232, 242–244, 247, 256, 258, 260f., 263, 308, 311, 316
naturalistic interpretation of Kant   79
natural law theory   128, 131
natural rights   131, 134f.
necessity, category of   28, 68, 72, 80, 97, 99, 102–104, 106, 108, 114, 122, 133, 159–161, 163, 165–167, 169, 181f., 186, 203, 214, 222, 225, 238, 245, 267, 272, 295
neo-Fichtean   116, 143f., 147, 253
Nevel School   275
Newtonian mechanics   1, 222f.
Newton's laws   21, 147
nomothetic point of view   173–177

non-Euclidean geometry 76, 78
normative character of logic 168, 267
norms, formal and transcendental 101f., 112, 114, 136, 147, 152, 157, 163–168, 171, 177, 186, 189, 196, 206, 267, 271
norms of ethics 106, 165
noumenal affection 81, 90f.
Novgorodcev, Pavel I. 116, 118, 128–131, 134–142, 144–153, 155, 161, 172, 188f., 191, 195, 218, 244, 317
number, concept of 7–11, 13f., 22, 24, 39, 49, 60, 62, 65, 72–74, 77, 88, 90, 110, 120, 122f., 128, 140, 144, 147, 149f., 153, 155, 165, 174, 177, 184, 188, 190, 192, 195, 198, 207, 212, 217, 220, 222, 243, 246, 257, 289f., 294, 314, 316–318, 320

objectification, Marburg understanding of 16, 21, 282
objective ethics 137–139, 144
objectivism 112, 136, 262–264
organic theory of society 157
Orthodox Church, fear of Kantianism by 12
other I 49–51, 53–55, 57, 202
other minds, problem of 9, 34–38, 48–50, 52f., 55, 57f., 60–62, 170, 178, 274f.
ought, ethical 107f., 121, 138, 147, 164f., 169, 189, 205, 276f., 293

panentheism 294f.
panmethodism 286
Pasternak, Boris 246–248, 252
personal identity 50, 274
phenomenalism 71, 111, 116, 211f.
philosopher's steamer 246
Philosophy of Economy 99, 194
philosophy of history 58, 114, 171, 209, 265, 279–281
philosophy of law 129, 131, 137, 139–143, 148, 193, 221, 228, 255
physical concept of mass 20f.
positive law 130f., 134f., 137, 146
positive unity, doctrine of 208
positivism, Comtian 6, 7, 171
positivism, legal 134, 157
positivism, logical 316, 317

positivism, reaction to 4, 107, 108, 115, 146, 163, 172, 298
post-Kantian German Idealists 105, 143
postulates of practical reason 33, 98, 111
practical reason 36, 61, 96f., 107f., 110, 113, 124, 138f., 181, 205, 239, 241, 291, 293
primacy of practical reason 111, 268f., 277, 291
problematic idealism of Descartes 82
Problems of Idealism 98, 108, 116, 119, 135, 161, 172
psychological time 65f.
psychologism, opposition to 3, 15, 30, 102, 108, 117, 137, 142, 144f., 152, 157, 177, 183f., 186, 195, 205, 211, 215, 218, 222, 226, 245, 248f., 252, 259, 266, 273, 288, 290, 297, 299–301, 306, 320
psychology, transcendental 9f., 32, 36, 40, 44, 64–66, 69f., 74–76, 78, 80, 84f., 87f., 94–96, 104f., 110, 129, 143, 152, 158, 167f., 172, 177, 181, 183, 206, 214, 222f., 248f., 251f., 258, 262, 264f., 288, 292, 298–301, 319
psychophysical parallelism 56

Radishchev, Aleksandr 131f.
rational mysticism 305
relativized *a priori* 226
religion, philosophy of 52, 72, 99, 143, 169, 183, 228, 280, 292–294
responsibility, sense of 70, 88, 241
Rickert, Heinrich 3, 5, 57, 142f., 147, 152, 154, 159, 167, 169–171, 177, 183–186, 189f., 194, 198, 201–205, 207f., 221, 230f., 246, 252f., 266–273, 296, 299, 301, 309, 311, 313, 316, 318
Riehl, Alois 3–5, 9, 36, 82f., 87, 89–91, 100f., 108, 117, 184, 191, 278f.
Rubinshtejn, Moisej 228, 265–274, 319
Rubinshtejn, Sergej 228, 255–265, 319

Sakketti, Aleksandr L. 237, 255–258
Saval'skij, Vasilij A. 140–154, 159, 170, 188f., 191, 195, 220, 223–225, 228, 239, 244, 255, 319
Savigny, Friedrich Karl von 130, 134f.

Scheler, Max   276, 312–314, 320
scientific consciousness   78, 236f.
Self-consciousness   34, 42, 46–47, 56, 62, 88, 172, 225, 298, 313
Sezeman (Sesemann, Sezemanas), Vasilij   285, 292, 306–314, 319
Shershenevich   134
Shpet, Gustav   50, 57f., 61, 122, 212, 214, 224, 237f., 246, 279f., 283, 303, 319
Simmel, Georg   58, 63, 100, 108, 156, 191
Slavophilism   200, 209, 211f.
social consciousness   61f., 102, 114
social materialism   94, 97
social philosophy   137, 139f., 191–193
solipsism   34, 50, 58f., 186, 212, 274
Solov'ëv, Vladimir   4f., 7, 16, 28, 36, 39, 52, 99, 105, 109, 113, 117, 142, 144, 148, 162, 200f., 209–211, 213, 215f., 228, 293–295, 306
space and time, conceptions of   14–16, 20, 29, 31f., 34, 41–48, 71f., 75f., 79f., 83, 85f., 90, 96f., 160, 175, 184f., 235f., 287f., 290
space, psychological conception of   13, 17, 31f., 41–45, 50, 58, 65–69, 74–79, 81–86, 91, 94, 110, 112, 145, 160, 170, 184, 236, 289f., 318, 321
Spektorskij, Evgenij V.   221–224
Spencer, Herbert   78, 88, 228
Spiritualism   39, 116
Stammler, Rudolf   93–95, 97, 102f., 110, 122f., 125f., 139, 142, 146, 149, 191–193
State Academy for the Study of the Arts (GAKhN)   89, 237, 244, 256, 279
Stepun, Fedor A.   198–203, 208–216, 218, 296f., 319
Struve, Peter   4, 92, 96, 100–112, 114–119, 123, 126, 161, 172, 188–190, 195, 202, 267f.
Stumpf, Carl   41, 64, 66–68
subjective idealism   71, 99, 245, 259, 264, 312
Subjectivism and Individualism in Social Philosophy   105, 112
substances as a necessary presupposition of cognition   88, 310

theocentric point of view   185
thing in itself   5, 30, 33, 35, 41, 45f., 48, 50f., 53, 81–83, 111, 121, 136, 142, 144, 146, 185, 197, 213, 251, 261, 269f., 280, 294, 304, 309f.
three types of proof in support of free will   70
time, concept of   3f., 6, 8–10, 14–16, 20, 30f., 35, 37, 39f., 42–45, 47, 50, 54, 56–59, 61f., 64–66, 68, 70f., 73–76, 78, 81, 84–89, 92–94, 96–102, 104–112, 115–119, 121–123, 129f., 132–136, 141, 144f., 153, 155f., 161–163, 168, 171, 179, 186–189, 195–199, 201–203, 206–212, 214, 216–218, 221, 228f., 231f., 236, 238, 240, 243–247, 250, 253–259, 261–263, 265f., 269, 272f., 275–279, 283–285, 289f., 295–297, 299–303, 305, 308, 310f., 313f., 317, 319, 321
time in itself   31
time, mathematical and psychological   3f., 6, 8–10, 14–16, 20, 30f., 35, 37, 39f., 42–45, 47, 50, 54, 56–59, 61f., 64–66, 68, 70f., 73–76, 78, 81, 84–89, 92–94, 96–102, 104–112, 115–119, 121–123, 129f., 132–136, 141, 144f., 153, 155f., 161–163, 168, 171, 179, 186–189, 195–199, 201–203, 206–212, 214, 216–218, 221, 228f., 231f., 236, 238, 240, 243–247, 250, 253–259, 261–263, 265f., 269, 272f., 275–279, 283–285, 289f., 295–297, 299–303, 305, 308, 310f., 313f., 317, 319, 321
Toporkov, Aleksej K.   216–218
transcendental deduction   18, 34, 48, 96, 120, 144f., 235, 251, 320
transcendental ground of ethics   145
transcendental idealism   4, 15, 31, 44–46, 71, 80, 83, 123, 136, 172, 182, 185, 203, 214, 257, 263f., 295, 298f., 301, 306, 308, 314
transcendental ideality of space   69
transcendentalism   117, 159, 213, 251, 256, 259, 298

transcendental method 28, 143, 151f., 169, 194, 204, 214, 222, 224f., 227, 233, 238–240, 261, 286, 292, 297f.
transcendental realism 9, 79, 81, 83, 91
transcendental reduction 65
transcendental skepticism 300
triune God, idea of 30
Trubeckoj, Evgenij N. 148, 224–227
Trubeckoj, Sergej N. 84, 220
two-worlds interpretation 45, 82, 137

unification of the sense manifold 47
unity of consciousness 46, 47, 96, 147, 236
universal invariant theory of experience 20
universality of physical laws 159

Vaihinger, Hans 4
Vejdeman [Weidemann], Aleksandr 285
Volkonskaja, Zinaida A. 255
Vvedenskij, Aleksandr I. 9, 12–40, 49–54, 58, 60, 62, 129, 171, 178, 188, 202, 206, 274, 285f., 302, 317, 319–321
Vysheslavcev, Boris P. 208, 244f.

Weber, Max 5, 11, 21, 93, 162, 171, 202, 318
Western jurisprudence 133
Wildhagen, Kurt 229
work as objectification of our value 282

Zolotnickij, Vladimir T. 131

www.ingramcontent.com/pod-product-compliance
Lightning Source LLC
Chambersburg PA
CBHW020219170426
43201CB00007B/264